Student Solutions

Elementary Linear Algebra
A Matrix Approach

Second Edition

Spence   Insel   Friedberg

Upper Saddle River, NJ 07458

Editorial Director, Computer Science, Engineering, and Advanced Mathematics: *Marcia J. Horton*
Senior Editor: *Holly Stark*
Editorial Assistant: *Jennifer Lonschein*
Senior Managing Editor: *Scott Disanno*
Production Editor: *Craig Little*
Supplement Cover Designer: *Daniel Sandin*
Manufacturing Buyer: *Lisa McDowell*

© 2008 by Pearson Education, Inc.
Pearson Prentice Hall
Pearson Education, Inc.
Upper Saddle River, NJ 07458

All rights reserved. No part of this book may be reproduced in any form or by any means, without permission in writing from the publisher.

The author and publisher of this book have used their best efforts in preparing this book. These efforts include the development, research, and testing of the theories and programs to determine their effectiveness. The author and publisher make no warranty of any kind, expressed or implied, with regard to these programs or the documentation contained in this book. The author and publisher shall not be liable in any event for incidental or consequential damages in connection with, or arising out of, the furnishing, performance, or use of these programs.

Pearson Prentice Hall™ is a trademark of Pearson Education, Inc.

This work is protected by United States copyright laws and is provided solely for the use of instructors in teaching their courses and assessing student learning. Dissemination or sale of any part of this work (including on the World Wide Web) will destroy the integrity of the work and is not permitted. The work and materials from it should never be made available to students except by instructors using the accompanying text in their classes. All recipients of this work are expected to abide by these restrictions and to honor the intended pedagogical purposes and the needs of other instructors who rely on these materials.

Printed in the United States of America

10 9 8 7 6 5 4 3 2 1

ISBN 0-13-239734-X
       978-0-13-239734-6

Pearson Education Ltd., *London*
Pearson Education Australia Pty. Ltd., *Sydney*
Pearson Education Singapore, Pte. Ltd.
Pearson Education North Asia Ltd., *Hong Kong*
Pearson Education Canada, Inc., *Toronto*
Pearson Educación de Mexico, S.A. de C.V.
Pearson Education—Japan, *Tokyo*
Pearson Education Malaysia, Pte. Ltd.
Pearson Education, Inc., *Upper Saddle River, New Jersey*

Contents

1 Matrices, Vectors, and Systems of Linear Equations — 1

- 1.1 Matrices and Vectors ... 1
- 1.2 Linear Combinations, Matrix-Vector Products, and Special Matrices ... 3
- 1.3 Systems of Linear Equations ... 7
- 1.4 Gaussian Elimination ... 9
- 1.5 Applications of Systems of Linear Equations ... 15
- 1.6 The Span of a Set of Vectors ... 18
- 1.7 Linear Dependence and Linear Independence ... 21
- Chapter 1 Review Exercises ... 25
- Chapter 1 MATLAB Exercises ... 28

2 Matrices and Linear Transformations — 29

- 2.1 Matrix Multiplication ... 29
- 2.2 Applications of Matrix Multiplication ... 31
- 2.3 Invertibility and Elementary Matrices ... 34
- 2.4 The Inverse of a Matrix ... 38
- 2.5 Partitioned Matrices and Block Multiplication ... 40
- 2.6 The LU Decomposition of a Matrix ... 42
- 2.7 Linear Transformations and Matrices ... 48
- 2.8 Composition and Invertibility of Linear Transformations ... 53
- Chapter 2 Review Exercises ... 56
- Chapter 2 MATLAB Exercises ... 57

3 Determinants   59

- 3.1 Cofactor Expansion 59
- 3.2 Properties of Determinants 61
  - Chapter 3 Review Exercises 65
  - Chapter 3 MATLAB Exercises 67

4 Subspaces and Their Properties   68

- 4.1 Subspaces 68
- 4.2 Basis and Dimension 72
- 4.3 The Dimension of Subspaces Associated with a Matrix 78
- 4.4 Coordinate Systems 83
- 4.5 Matrix Representations of Linear Operators 87
  - Chapter 4 Review Exercises 93
  - Chapter 4 MATLAB Exercises 96

5 Eigenvalues, Eigenvectors, and Diagonalization   98

- 5.1 Eigenvalues and Eigenvectors 98
- 5.2 The Characteristic Polynomial 100
- 5.3 Diagonalization of Matrices 105
- 5.4 Diagonalization of Linear Operators 109
- 5.5 Applications of Eigenvalues 113
  - Chapter 5 Review Exercises 121
  - Chapter 5 MATLAB Exercises 123

6 Orthogonality   125

- 6.1 The Geometry of Vectors 125
- 6.2 Orthogonal Vectors 130
- 6.3 Orthogonal Projections 134
- 6.4 Least-Squares Approximations and Orthogonal Projection Matrices 139
- 6.5 Orthogonal Matrices and Operators 142
- 6.6 Symmetric Matrices 146
- 6.7 Singular Value Decomposition 150
- 6.8 Principal Component Analysis 157
- 6.9 Rotations of $\mathcal{R}^3$ and Computer Graphics 159
  - Chapter 6 Review Exercises 164
  - Chapter 6 MATLAB Exercises 166

7 Vector Spaces   170

- 7.1 Vector Spaces and Their Subspaces 170
- 7.2 Linear Transformations 174
- 7.3 Basis and Dimension 177
- 7.4 Matrix Representations of Linear Operators 181
- 7.5 Inner Product Spaces 185
- Chapter 7 Review Exercises 189
- Chapter 7 MATLAB Exercises 193

Chapter 1

Matrices, Vectors, and Systems of Linear Equations

1.1 MATRICES AND VECTORS

1. Each entry of $4A$ is 4 times the corresponding entry of A; so

$$4A = \begin{bmatrix} 4(2) & 4(-1) & 4(5) \\ 4(3) & 4(4) & 4(1) \end{bmatrix}$$

$$= \begin{bmatrix} 8 & -4 & 20 \\ 12 & 16 & 4 \end{bmatrix}.$$

3. We have

$$4A - 2B$$

$$= 4\begin{bmatrix} 2 & -1 & 5 \\ 3 & 4 & 1 \end{bmatrix} - 2\begin{bmatrix} 1 & 0 & -2 \\ 2 & 3 & 4 \end{bmatrix}$$

$$= \begin{bmatrix} 8 & -4 & 20 \\ 12 & 16 & 4 \end{bmatrix} + \begin{bmatrix} -2 & 0 & 4 \\ -4 & -6 & -8 \end{bmatrix}$$

$$= \begin{bmatrix} 6 & -4 & 24 \\ 8 & 10 & -4 \end{bmatrix}.$$

5. We have

$$(2B)^T = \begin{bmatrix} 2(1) & 2(0) & 2(-2) \\ 2(2) & 2(3) & 2(4) \end{bmatrix}^T$$

$$= \begin{bmatrix} 2 & 0 & -4 \\ 4 & 6 & 8 \end{bmatrix}^T = \begin{bmatrix} 2 & 4 \\ 0 & 6 \\ -4 & 8 \end{bmatrix}.$$

9. Matrix A^T can be obtained by interchanging the rows of A with the corresponding columns; so

$$A^T = \begin{bmatrix} 2 & 3 \\ -1 & 4 \\ 5 & 1 \end{bmatrix}.$$

13. Matrix $-A$ can be obtained by multiplying each entry of A by -1; hence

$$-A = (-1)A$$

$$= \begin{bmatrix} -1(3) & -1(-1) & -1(2) & -1(4) \\ -1(1) & -1(5) & -1(-6) & -1(-2) \end{bmatrix}$$

$$= \begin{bmatrix} -3 & 1 & -2 & -4 \\ -1 & -5 & 6 & 2 \end{bmatrix}.$$

17. Because A and B have different sizes, $A - B$ is not defined.

21. Because A and B have different sizes, $A + B$ is not defined.

25. The row 1, column 2 entry is -2.

2   Chapter 1   Matrices, Vectors, and Systems of Linear Equations

29. The first column of C is $\begin{bmatrix} 2 \\ 2e \end{bmatrix}$.

33. Let **v** be the vector given by the arrow in Figure 1.7. Because the arrow has length 300, we have

$$v_1 = 300 \sin 30° = 150$$
$$v_2 = 300 \cos 30° = 150\sqrt{3}.$$

For v_3, we use the fact that the speed in the z-direction is 10 mph. So the velocity vector of the plane in $\mathcal{R}^3$ is

$$\mathbf{v} = \begin{bmatrix} 150 \\ 150\sqrt{3} \\ 10 \end{bmatrix} \text{ mph.}$$

37. True   38. True   39. True

40. False, a scalar multiple of the zero matrix is the zero matrix.

41. False, the transpose of an $m \times n$ matrix is an $n \times m$ matrix.

42. True

43. False, the rows of B are 1×4 vectors.

44. False, the $(3,4)$–entry of a matrix lies in row 3 and column 4.

45. True

46. False, an $m \times n$ matrix has mn entries.

47. True   48. True   49. True

50. False, matrices must have the same size to be equal.

51. True   52. True   53. True

54. True   55. True   56. True

57. Suppose that A and B are $m \times n$ matrices.

(a) The jth column of $A + B$ and $\mathbf{a}_j + \mathbf{b}_j$ are $m \times 1$ vectors. Now the ith component of the jth column of $A+B$ is the (i,j)-entry of $A+B$, which is $a_{ij}+b_{ij}$. By definition, the ith components of $\mathbf{a}_j$ and $\mathbf{b}_j$ are a_{ij} and b_{ij}, respectively. So the ith component of $\mathbf{a}_j+\mathbf{b}_j$ is also $a_{ij}+b_{ij}$. Thus the jth column of $A+B$ is $\mathbf{a}_j+\mathbf{b}_j$.

(b) The jth column of cA and $c\mathbf{a}_j$ are $m \times 1$ vectors. The ith component of the jth column of cA is the (i,j)-entry of cA, which is ca_{ij}. The ith component of $c\mathbf{a}_j$ is also ca_{ij}. Thus the jth column of cA is $c\mathbf{a}_j$.

61. If O is the $m \times n$ zero matrix, then both A and $A+O$ are $m \times n$ matrices; so we need only show they have equal corresponding entries. The (i,j)-entry of $A+O$ is $a_{ij}+0 = a_{ij}$, which is the (i,j)-entry of A.

65. The matrices $(sA)^T$ and sA^T are $n \times m$ matrices; so we need only show they have equal corresponding entries. The (i,j)-entry of $(sA)^T$ is the (j,i)-entry of sA, which is sa_{ji}. The (i,j)-entry of sA^T is the product of s and the (i,j)-entry of A^T, which is also sa_{ji}.

69. If B is a diagonal matrix, then B is square. Since B^T is the same size as B in this case, B^T is square. If $i \neq j$, then the (i,j)-entry of B^T is $b_{ji} = 0$. So B^T is a diagonal matrix.

73. Let O be a square zero matrix. The (i,j)-entry of O is zero, whereas the (i,j)-entry of O^T is the (j,i)-entry of O, which is also zero. So $O = O^T$, and hence O is a symmetric matrix.

77. No. Consider $\begin{bmatrix} 2 & 5 & 6 \\ 5 & 7 & 8 \\ 6 & 8 & 4 \end{bmatrix}$ and $\begin{bmatrix} 2 & 6 \\ 5 & 8 \end{bmatrix}$, which is obtained by deleting row 3 and column 2 of the first matrix.

81. Let
$$A_1 = \frac{1}{2}(A + A^T) \text{ and } A_2 = \frac{1}{2}(A - A^T).$$
It is easy to show that $A = A_1 + A_2$. By Theorem 1.2(b), (a), and (c) and Theorem 1.1(a), we have

$$A_1^T = \frac{1}{2}(A + A^T)^T = \frac{1}{2}[A^T + (A^T)^T]$$
$$= \frac{1}{2}(A^T + A) = \frac{1}{2}(A + A^T)$$
$$= A_1.$$

Thus A_1 is symmetric. By similar reasoning, we have

$$A_2^T = \frac{1}{2}(A - A^T)^T = \frac{1}{2}[A^T - (A^T)^T]$$
$$= \frac{1}{2}(A^T - A) = -\frac{1}{2}(A - A^T)$$
$$= -A_2.$$

So A_2 is skew-symmetric.

1.2 LINEAR COMBINATIONS, MATRIX-VECTOR PRODUCTS, AND SPECIAL MATRICES

1. We have
$$\begin{bmatrix} 3 & -2 & 1 \\ 4 & 0 & 2 \end{bmatrix} \begin{bmatrix} 1 \\ -2 \\ 5 \end{bmatrix}$$
$$= 1 \begin{bmatrix} 3 \\ 4 \end{bmatrix} + (-2) \begin{bmatrix} -2 \\ 0 \end{bmatrix} + 5 \begin{bmatrix} 1 \\ 2 \end{bmatrix}$$
$$= \begin{bmatrix} 12 \\ 14 \end{bmatrix}.$$

3. We have
$$\begin{bmatrix} 2 & -1 & 3 \\ 1 & 0 & -1 \\ 0 & 2 & 4 \end{bmatrix} \begin{bmatrix} 2 \\ 1 \\ 2 \end{bmatrix}$$
$$= 2 \begin{bmatrix} 2 \\ 1 \\ 0 \end{bmatrix} + 1 \begin{bmatrix} -1 \\ 0 \\ 2 \end{bmatrix} + 2 \begin{bmatrix} 3 \\ -1 \\ 4 \end{bmatrix}$$
$$= \begin{bmatrix} 9 \\ 0 \\ 10 \end{bmatrix}.$$

7. We have
$$\begin{bmatrix} 3 & 0 \\ 2 & 1 \end{bmatrix}^T \begin{bmatrix} 4 \\ 5 \end{bmatrix} = \begin{bmatrix} 3 & 2 \\ 0 & 1 \end{bmatrix} \begin{bmatrix} 4 \\ 5 \end{bmatrix}$$
$$= 4 \begin{bmatrix} 3 \\ 0 \end{bmatrix} + 5 \begin{bmatrix} 2 \\ 1 \end{bmatrix}$$
$$= \begin{bmatrix} 22 \\ 5 \end{bmatrix}.$$

11. We have
$$\begin{bmatrix} 2 & -3 \\ -4 & 5 \\ 3 & -1 \end{bmatrix} \begin{bmatrix} 4 \\ 2 \end{bmatrix} = 4 \begin{bmatrix} 2 \\ -4 \\ 3 \end{bmatrix} + 2 \begin{bmatrix} -3 \\ 5 \\ -1 \end{bmatrix}$$
$$= \begin{bmatrix} 2 \\ -6 \\ 10 \end{bmatrix}.$$

15. We have
$$\left(\begin{bmatrix} 3 & 0 \\ -2 & 4 \end{bmatrix}^T + \begin{bmatrix} 1 & 2 \\ 3 & -3 \end{bmatrix}^T \right) \begin{bmatrix} 4 \\ 5 \end{bmatrix}$$
$$= \left(\begin{bmatrix} 3 & -2 \\ 0 & 4 \end{bmatrix} + \begin{bmatrix} 1 & 3 \\ 2 & -3 \end{bmatrix} \right) \begin{bmatrix} 4 \\ 5 \end{bmatrix}$$
$$= \begin{bmatrix} 4 & 1 \\ 2 & 1 \end{bmatrix} \begin{bmatrix} 4 \\ 5 \end{bmatrix} = \begin{bmatrix} (4)(4) + (1)(5) \\ (2)(4) + (1)(5) \end{bmatrix}$$
$$= \begin{bmatrix} 21 \\ 13 \end{bmatrix}.$$

4 Chapter 1 Matrices, Vectors, and Systems of Linear Equations

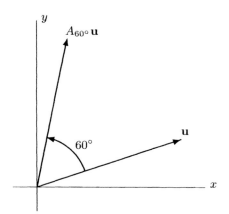

Figure for Exercise 19

19. We have

$$A_{60°} = \begin{bmatrix} \cos 60° & -\sin 60° \\ \sin 60° & \cos 60° \end{bmatrix}$$

$$= \begin{bmatrix} \frac{1}{2} & -\frac{\sqrt{3}}{2} \\ \frac{\sqrt{3}}{2} & \frac{1}{2} \end{bmatrix}$$

$$= \frac{1}{2}\begin{bmatrix} 1 & -\sqrt{3} \\ \sqrt{3} & 1 \end{bmatrix},$$

and hence

$$A_{60°}\mathbf{u} = \frac{1}{2}\begin{bmatrix} 1 & -\sqrt{3} \\ \sqrt{3} & 1 \end{bmatrix}\begin{bmatrix} 3 \\ 1 \end{bmatrix}$$

$$= \frac{1}{2}\begin{bmatrix} 3-\sqrt{3} \\ 3\sqrt{3}+1 \end{bmatrix}.$$

23. We have

$$A_{270°} = \begin{bmatrix} \cos 270° & -\sin 270° \\ \sin 270° & \cos 270° \end{bmatrix}$$

$$= \begin{bmatrix} 0 & 1 \\ -1 & 0 \end{bmatrix},$$

and hence the vector obtained by rotating $\mathbf{u}$ by $270°$ is

$$A_{270°}\mathbf{u} = \begin{bmatrix} 0 & 1 \\ -1 & 0 \end{bmatrix}\begin{bmatrix} -2 \\ 3 \end{bmatrix} = \begin{bmatrix} 3 \\ 2 \end{bmatrix}.$$

27. We have

$$A_{300°} = \begin{bmatrix} \cos 300° & -\sin 300° \\ \sin 300° & \cos 300° \end{bmatrix}$$

$$= \begin{bmatrix} \frac{1}{2} & \frac{\sqrt{3}}{2} \\ -\frac{\sqrt{3}}{2} & \frac{1}{2} \end{bmatrix}$$

$$= \frac{1}{2}\begin{bmatrix} 1 & \sqrt{3} \\ -\sqrt{3} & 1 \end{bmatrix},$$

and hence the vector obtained by rotating $\mathbf{u}$ by $300°$ is

$$A_{300°}\mathbf{u} = \frac{1}{2}\begin{bmatrix} 1 & \sqrt{3} \\ -\sqrt{3} & 1 \end{bmatrix}\begin{bmatrix} 3 \\ 0 \end{bmatrix}$$

$$= \frac{1}{2}\begin{bmatrix} 3 \\ -3\sqrt{3} \end{bmatrix}.$$

31. The vector $\mathbf{u}$ is not a linear combination of the vectors in $\mathcal{S}$. If $\mathbf{u}$ were a linear combination of $\begin{bmatrix} 4 \\ 4 \end{bmatrix}$, then there would be a scalar c such that

$$\begin{bmatrix} -1 \\ 1 \end{bmatrix} = c\begin{bmatrix} 4 \\ 4 \end{bmatrix} = \begin{bmatrix} 4c \\ 4c \end{bmatrix}.$$

But then $1 = 4c$ and $-1 = 4c$. This is impossible.

35. We seek scalars c_1 and c_2 such that

$$\begin{bmatrix} -1 \\ 11 \end{bmatrix} = c_1\begin{bmatrix} 1 \\ 3 \end{bmatrix} + c_2\begin{bmatrix} 2 \\ -1 \end{bmatrix}$$

$$= \begin{bmatrix} c_1 \\ 3c_1 \end{bmatrix} + \begin{bmatrix} 2c_2 \\ -c_2 \end{bmatrix}$$

$$= \begin{bmatrix} c_1 + 2c_2 \\ 3c_1 - c_2 \end{bmatrix}.$$

That is, we seek a solution of the following system of linear equations:

$$c_1 + 2c_2 = -1$$
$$3c_1 - c_2 = 11.$$

Because these equations represent nonparallel lines in the plane, there is exactly one solution, namely, $c_1 = 3$ and $c_2 = -2$. So

$$\begin{bmatrix} -1 \\ 11 \end{bmatrix} = 3 \begin{bmatrix} 1 \\ 3 \end{bmatrix} + (-2) \begin{bmatrix} 2 \\ -1 \end{bmatrix}.$$

39. We seek scalars c_1 and c_2 such that

$$\begin{bmatrix} 3 \\ 5 \\ -5 \end{bmatrix} = c_1 \begin{bmatrix} 2 \\ 0 \\ -1 \end{bmatrix} + c_2 \begin{bmatrix} -1 \\ 1 \\ 0 \end{bmatrix}$$

$$= \begin{bmatrix} 2c_1 - c_2 \\ c_2 \\ -c_1 \end{bmatrix}.$$

Thus we must solve the following system of linear equations:

$$2c_1 - 2c_2 = 3$$
$$c_2 = 5$$
$$-c_1 = -5.$$

From the second and third equations, we see that the only possible solution of this system is $c_1 = 5$ and $c_2 = 5$. Because these values do not satisfy the first equation in the system, the system is inconsistent. Thus **u** is not a linear combination of the vectors in $\mathcal{S}$.

43. We seek scalars c_1, c_2, and c_3 such that

$$\begin{bmatrix} -4 \\ -5 \\ -6 \end{bmatrix} = c_1 \begin{bmatrix} 1 \\ 0 \\ 0 \end{bmatrix} + c_2 \begin{bmatrix} 0 \\ 1 \\ 0 \end{bmatrix} + c_3 \begin{bmatrix} 0 \\ 0 \\ 1 \end{bmatrix}$$

$$= \begin{bmatrix} c_1 \\ c_2 \\ c_3 \end{bmatrix}.$$

Thus we must solve the following system of linear equations:

$$c_1 = -4$$
$$c_2 = -5$$
$$c_3 = -6.$$

Clearly this system is consistent, and so

$$\begin{bmatrix} -4 \\ -5 \\ -6 \end{bmatrix}$$

$$= (-4) \begin{bmatrix} 1 \\ 0 \\ 0 \end{bmatrix} + (-5) \begin{bmatrix} 0 \\ 1 \\ 0 \end{bmatrix} + (-6) \begin{bmatrix} 0 \\ 0 \\ 1 \end{bmatrix}.$$

45. True

46. False. Consider the linear combination

$$3 \begin{bmatrix} 2 \\ 2 \end{bmatrix} + (-6) \begin{bmatrix} 1 \\ 1 \end{bmatrix} = \begin{bmatrix} 0 \\ 0 \end{bmatrix}.$$

If the coefficients were positive, the sum could not equal the zero vector.

47. True **48.** True **49.** True

50. False, the matrix-vector product of a 2×3 matrix and a 3×1 vector is a 2×1 vector.

51. False, the matrix-vector product is a linear combination of the *columns* of the matrix.

52. False, the product of a matrix and a standard vector is a column of the matrix.

53. True

6   Chapter 1   Matrices, Vectors, and Systems of Linear Equations

54. False, the matrix-vector product of an $m \times n$ matrix and a vector in $\mathcal{R}^n$ yields a vector in $\mathcal{R}^m$.

55. False, every vector in $\mathcal{R}^2$ is a linear combination of two *nonparallel* vectors.

56. True

57. False, a standard vector is a vector with a single component equal to 1 and the others equal to 0.

58. True

59. False, consider
$$A = \begin{bmatrix} 1 & -1 \\ -1 & 1 \end{bmatrix} \quad \text{and} \quad \mathbf{u} = \begin{bmatrix} 1 \\ 1 \end{bmatrix}.$$

60. True

61. False, $A_\theta \mathbf{u}$ is the vector obtained by rotating $\mathbf{u}$ by a *counterclockwise* rotation of the angle θ.

62. False. Consider $A = \begin{bmatrix} 1 & -1 \\ -1 & 1 \end{bmatrix}$, and the vectors $\mathbf{u} = \begin{bmatrix} 1 \\ 1 \end{bmatrix}$, and $\mathbf{v} = \begin{bmatrix} 2 \\ 2 \end{bmatrix}$.

63. True   64. True

65. We have
$$A_{0°}\mathbf{v} = \begin{bmatrix} \cos 0° & -\sin 0° \\ \sin 0° & \cos 0° \end{bmatrix} \begin{bmatrix} v_1 \\ v_2 \end{bmatrix}$$
$$= \begin{bmatrix} 1 & 0 \\ 0 & 1 \end{bmatrix} \begin{bmatrix} v_1 \\ v_2 \end{bmatrix} = \begin{bmatrix} v_1 \\ v_2 \end{bmatrix} = \mathbf{v}.$$

69. Let $\mathbf{p} = \begin{bmatrix} 400 \\ 300 \end{bmatrix}$ and $A = \begin{bmatrix} .85 & .03 \\ .15 & .97 \end{bmatrix}$.

(a) We compute
$$A\mathbf{p} = \begin{bmatrix} .85 & .03 \\ .15 & .97 \end{bmatrix} \begin{bmatrix} 400 \\ 300 \end{bmatrix}$$

$$= \begin{bmatrix} (.85)(400) + (.03)(300) \\ (.15)(400) + (.97)(300) \end{bmatrix}$$

$$= \begin{bmatrix} 349 \\ 351 \end{bmatrix}.$$

Thus there are 349,000 in the city and 351,000 in the suburbs.

(b) We compute the result using the answer from (a).

$$A(A\mathbf{p}) = \begin{bmatrix} .85 & .03 \\ .15 & .97 \end{bmatrix} \begin{bmatrix} 349 \\ 351 \end{bmatrix}$$

$$= \begin{bmatrix} (.85)(349) + (.03)(351) \\ (.15)(349) + (.97)(351) \end{bmatrix}$$

$$= \begin{bmatrix} 307.18 \\ 392.82 \end{bmatrix}.$$

Thus there are 307,180 in the city and 392,820 in the suburbs.

73. The reflection of $\mathbf{u}$ about the x-axis is the vector $\begin{bmatrix} a \\ -b \end{bmatrix}$. To obtain this vector, let $B = \begin{bmatrix} 1 & 0 \\ 0 & -1 \end{bmatrix}$. Then

$$B\mathbf{u} = \begin{bmatrix} 1 & 0 \\ 0 & -1 \end{bmatrix} \begin{bmatrix} a \\ b \end{bmatrix} = \begin{bmatrix} a \\ -b \end{bmatrix}.$$

77. Let $\mathbf{v} = \begin{bmatrix} a \\ 0 \end{bmatrix}$. Then

$$A\mathbf{v} = \begin{bmatrix} 1 & 0 \\ 0 & 0 \end{bmatrix} \begin{bmatrix} a \\ 0 \end{bmatrix} = \begin{bmatrix} a \\ 0 \end{bmatrix} = \mathbf{v}.$$

81. We can write $\mathbf{v} = a_1\mathbf{u}_1 + a_2\mathbf{u}_2$ and $\mathbf{w} = b_1\mathbf{u}_1 + b_2\mathbf{u}_2$ for some scalars a_1, a_2, b_1, and b_2. Then a typical linear combination of $\mathbf{v}$ and $\mathbf{w}$ has the form

$$c\mathbf{v} + d\mathbf{w}$$
$$= c(a_1\mathbf{u}_1 + a_2\mathbf{u}_2) + d(b_1\mathbf{u}_1 + b_2\mathbf{u}_2)$$

$$= (ca_1 + db_1)\mathbf{u}_1 + (ca_2 + db_2)\mathbf{u}_2,$$

for some scalars c and d. The preceding calculation shows that this is also a linear combination of $\mathbf{u}_1$ and $\mathbf{u}_2$.

85. We have

$$A\mathbf{e}_j = A \begin{bmatrix} 0 \\ \vdots \\ 0 \\ 1 \\ 0 \\ \vdots \\ 0 \end{bmatrix}$$

$$= 0\mathbf{a}_1 + 0\mathbf{a}_2 + \cdots + 0\mathbf{a}_{j-1} + 1\mathbf{a}_j$$
$$\quad + 0\mathbf{a}_{j+1} + \cdots + 0\mathbf{a}_n$$
$$= \mathbf{a}_j.$$

89. The jth column of I_n is $\mathbf{e}_j$. So

$$I_n\mathbf{v} = v_1\mathbf{e}_1 + v_2\mathbf{e}_2 + \cdots + v_n\mathbf{e}_n = \mathbf{v}.$$

1.3 SYSTEMS OF LINEAR EQUATIONS

1. (a) $\begin{bmatrix} 0 & -1 & 2 \\ 1 & 3 & 0 \end{bmatrix}$

(b) $\begin{bmatrix} 0 & -1 & 2 & 0 \\ 1 & 3 & 0 & -1 \end{bmatrix}$

5. (a) $\begin{bmatrix} 0 & 2 & -3 \\ -1 & 1 & 2 \\ 2 & 0 & 1 \end{bmatrix}$

(b) $\begin{bmatrix} 0 & 2 & -3 & 4 \\ -1 & 1 & 2 & -6 \\ 2 & 0 & 1 & 0 \end{bmatrix}$

9. If we denote the given matrix as A, then the $(2,j)$-entry of the desired matrix equals $2a_{1j} + a_{2j}$ for every j. Thus the desired matrix is

$$\begin{bmatrix} 1 & -1 & 0 & 2 & -3 \\ 0 & 4 & 3 & 3 & -5 \\ 0 & 2 & -4 & 4 & 2 \end{bmatrix}.$$

13. If we denote the given matrix as A, then the $(3,j)$-entry of the desired matrix equals $4a_{2j} + a_{3j}$ for every j. Thus the desired matrix is

$$\begin{bmatrix} 1 & -1 & 0 & 2 & -3 \\ -2 & 6 & 3 & -1 & 1 \\ -8 & 26 & 8 & 0 & 6 \end{bmatrix}.$$

17. As in Exercises 9 and 13, we obtain

$$\begin{bmatrix} 1 & -2 & 0 \\ -1 & 1 & -1 \\ 0 & 0 & 6 \\ -3 & 2 & 1 \end{bmatrix}.$$

21. As in Exercises 9 and 13, we obtain

$$\begin{bmatrix} 1 & -2 & 0 \\ -1 & 1 & -1 \\ 2 & -4 & 6 \\ -1 & 0 & 3 \end{bmatrix}.$$

25. No, because the left side of the second equation yields $1(2) - 2(1) = 0 \neq -3$. Alternatively,

$$\begin{bmatrix} 1 & -4 & 0 & 3 \\ 0 & 0 & 1 & -2 \end{bmatrix} \begin{bmatrix} 3 \\ 0 \\ 2 \\ 1 \end{bmatrix} = \begin{bmatrix} 6 \\ 0 \end{bmatrix} \neq \begin{bmatrix} 6 \\ -3 \end{bmatrix}.$$

29. Because

$$9 - 4(0) \quad + 3(-1) = 6$$
$$-5 - 2(-1) = -3,$$

the given vector satisfies every equation in the system, and so is a solution of the system.

33. Because

$$2 - 2(1) + 1 + 0 + 7(0) = 1$$
$$2 - 2(1) + 2(1) + 10(0) = 2$$
$$2(2) - 4(1) + 4(0) + 8(0) = 0,$$

the given vector satisfies every equation in the system, and so is a solution of the system.

37. Since $0 - 2(3) + (-1) + 3 + 7(0) \neq 1$, the given vector does not satisfy the first equation in the system, and hence is not a solution of the system.

41. The system of linear equations is consistent because the augmented matrix contains no row where the only nonzero entry lies in the last column. The corresponding system of linear equations is

$$x_1 - 2x_2 = 6$$
$$0x_1 + 0x_2 = 0.$$

The general solution is

$$x_1 = 6 + 2x_2$$
$$x_2 \quad \text{free}.$$

45. The system of linear equations is consistent because the augmented matrix contains no row where the only nonzero entry lies in the last column. The corresponding system of linear equations is

$$x_1 - 2x_2 \quad\quad = 4$$
$$x_3 = 3$$
$$0x_1 + 0x_2 + 0x_3 = 0.$$

The general solution is

$$x_1 = 4 + 2x_2$$
$$x_2 \quad \text{free}$$
$$x_3 = 3.$$

49. The system of linear equations is consistent because the augmented matrix contains no row where the only nonzero entry lies in the last column. The corresponding system of linear equations is

$$x_2 \quad\quad\quad = -3$$
$$x_3 \quad\quad = -4$$
$$x_4 = 5.$$

The general solution is

$$x_1 \quad \text{free}$$
$$x_2 = -3$$
$$x_3 = -4$$
$$x_4 = 5.$$

The solution in vector form is

$$\begin{bmatrix} x_1 \\ x_2 \\ x_3 \\ x_4 \end{bmatrix} = x_1 \begin{bmatrix} 1 \\ 0 \\ 0 \\ 0 \end{bmatrix} + \begin{bmatrix} 0 \\ -3 \\ -4 \\ 5 \end{bmatrix}.$$

53. The system of linear equations is not consistent because the second row of the augmented matrix has its only nonzero entry in the last column.

57. False, the system $0x_1 + 0x_2 = 1$ has no solutions.

58. False, see the boxed result on page 29.

59. True

60. False, the matrix $\begin{bmatrix} 2 & 0 \\ 0 & 0 \end{bmatrix}$ is in row echelon form.

61. True   **62.** True

63. False, the matrices $\begin{bmatrix} 2 & 0 \\ 0 & 0 \end{bmatrix}$ and $\begin{bmatrix} 1 & 0 \\ 0 & 0 \end{bmatrix}$ are both row echelon forms of $\begin{bmatrix} 2 & 0 \\ 0 & 0 \end{bmatrix}$.

64. True   **65.** True

66. False, the system
$$0x_1 + 0x_2 = 1$$
$$0x_1 + 0x_2 = 0$$
is inconsistent, but its augmented matrix is $\begin{bmatrix} 0 & 0 & 1 \\ 0 & 0 & 0 \end{bmatrix}$.

67. True   68. True

69. False, the coefficient matrix of a system of m linear equations in n variables is an $m \times n$ matrix.

70. True   71. True   72. True

73. False, multiplying every entry of some row of a matrix by a *nonzero* scalar is an elementary row operation.

74. True

75. False, the system may be inconsistent; consider $0x_1 + 0x_2 = 1$.

76. True

77. If $[R \ \mathbf{c}]$ is in reduced row echelon form, then so is R. If we apply the same row operations to A that were applied to $[A \ \mathbf{b}]$ to produce $[R \ \mathbf{c}]$, we obtain the matrix R. So R is the reduced row echelon form of A.

81. The ranks of the possible reduced row echelon forms are 0, 1, and 2. Considering each of these ranks, we see that there are 7 possible reduced row echelon forms: $\begin{bmatrix} 0 & 0 & 0 \\ 0 & 0 & 0 \end{bmatrix}$, $\begin{bmatrix} 1 & * & * \\ 0 & 0 & 0 \end{bmatrix}$, $\begin{bmatrix} 0 & 1 & * \\ 0 & 0 & 0 \end{bmatrix}$, $\begin{bmatrix} 0 & 0 & 1 \\ 0 & 0 & 0 \end{bmatrix}$, $\begin{bmatrix} 1 & 0 & * \\ 0 & 1 & * \end{bmatrix}$, $\begin{bmatrix} 1 & * & 0 \\ 0 & 0 & 1 \end{bmatrix}$, and $\begin{bmatrix} 0 & 1 & 0 \\ 0 & 0 & 1 \end{bmatrix}$.

85. Multiplying the second equation by c produces a system whose augmented matrix is obtained from the augmented matrix of the original system by the elementary row operation of multiplying the second row by c. From the statement on page 33, the two systems are equivalent.

1.4 GAUSSIAN ELIMINATION

1. The reduced row echelon form of the augmented matrix of the given system is $\begin{bmatrix} 1 & 3 & -2 \end{bmatrix}$. So the general solution of the system is
$$x_1 = -2 - 3x_2$$
$$x_2 \ \text{free}.$$

3. The augmented matrix of the given system is
$$\begin{bmatrix} 1 & -2 & -6 \\ -2 & 3 & 7 \end{bmatrix}.$$
Apply the Gaussian elimination algorithm to this augmented matrix to obtain a matrix in reduced row echelon form:
$$\begin{bmatrix} 1 & -2 & -6 \\ -2 & 3 & 7 \end{bmatrix} \xrightarrow{2\mathbf{r}_1 + \mathbf{r}_2 \to \mathbf{r}_2}$$
$$\begin{bmatrix} 1 & -2 & -6 \\ 0 & -1 & -5 \end{bmatrix} \xrightarrow{-\mathbf{r}_2 \to \mathbf{r}_2}$$
$$\begin{bmatrix} 1 & -2 & -6 \\ 0 & 1 & 5 \end{bmatrix} \xrightarrow{2\mathbf{r}_2 + \mathbf{r}_1 \to \mathbf{r}_1}$$
$$\begin{bmatrix} 1 & 0 & 4 \\ 0 & 1 & 5 \end{bmatrix}.$$
This matrix corresponds to the system
$$x_1 = 4$$
$$x_2 = 5,$$
which yields the solution.

7. The augmented matrix of this system is

$$\begin{bmatrix} 1 & -2 & -1 & -3 \\ 2 & -4 & 2 & 2 \end{bmatrix}.$$

Apply the Gaussian elimination algorithm to this augmented matrix to obtain a matrix in reduced row echelon form:

$$\begin{bmatrix} 1 & -2 & -1 & -3 \\ 2 & -4 & 2 & 2 \end{bmatrix} \xrightarrow{-2r_1+r_2 \to r_2}$$

$$\begin{bmatrix} 1 & -2 & -1 & -3 \\ 0 & 0 & 4 & 8 \end{bmatrix} \xrightarrow{\frac{1}{4}r_2 \to r_2}$$

$$\begin{bmatrix} 1 & -2 & -1 & -3 \\ 0 & 0 & 1 & 2 \end{bmatrix} \xrightarrow{r_2+r_1 \to r_1}$$

$$\begin{bmatrix} 1 & -2 & 0 & -1 \\ 0 & 0 & 1 & 2 \end{bmatrix}.$$

This matrix corresponds to the system

$$\begin{aligned} x_1 - 2x_2 &= -1 \\ x_3 &= 2. \end{aligned}$$

Its general solution is

$$\begin{aligned} x_1 &= -1 + 2x_2 \\ x_2 &\text{ free} \\ x_3 &= 2. \end{aligned}$$

11. The augmented matrix of this system is

$$\begin{bmatrix} 1 & 3 & 1 & 1 & -1 \\ -2 & -6 & -1 & 0 & 5 \\ 1 & 3 & 2 & 3 & 2 \end{bmatrix}.$$

Apply the Gaussian elimination algorithm to this augmented matrix to obtain a matrix in reduced row echelon form:

$$\begin{bmatrix} 1 & 3 & 1 & 1 & -1 \\ -2 & -6 & -1 & 0 & 5 \\ 1 & 3 & 2 & 3 & 2 \end{bmatrix} \xrightarrow{\begin{subarray}{l} 2r_1+r_2 \to r_2 \\ -r_1+r_3 \to r_3 \end{subarray}}$$

$$\begin{bmatrix} 1 & 3 & 1 & 1 & -1 \\ 0 & 0 & 1 & 2 & 3 \\ 0 & 0 & 1 & 2 & 3 \end{bmatrix} \xrightarrow{-r_2+r_3 \to r_3}$$

$$\begin{bmatrix} 1 & 3 & 1 & 1 & -1 \\ 0 & 0 & 1 & 2 & 3 \\ 0 & 0 & 0 & 0 & 0 \end{bmatrix} \xrightarrow{-r_2+r_1 \to r_1}$$

$$\begin{bmatrix} 1 & 3 & 0 & -1 & -4 \\ 0 & 0 & 1 & 2 & 3 \\ 0 & 0 & 0 & 0 & 0 \end{bmatrix}.$$

This matrix corresponds to the system

$$\begin{aligned} x_1 + 3x_2 \quad - \quad x_4 &= -4 \\ x_3 + 2x_4 &= 3. \end{aligned}$$

Its general solution is

$$\begin{aligned} x_1 &= -4 - 3x_2 + x_4 \\ x_2 &\text{ free} \\ x_3 &= 3 - 2x_4 \\ x_4 &\text{ free}. \end{aligned}$$

15. The augmented matrix of this system is

$$\begin{bmatrix} 1 & 0 & -1 & -2 & -8 & -3 \\ -2 & 0 & 1 & 2 & 9 & 5 \\ 3 & 0 & -2 & -3 & -15 & -9 \end{bmatrix}.$$

Apply the Gaussian elimination algorithm to this augmented matrix to obtain a matrix in reduced row echelon form:

$$\begin{bmatrix} 1 & 0 & -1 & -2 & -8 & -3 \\ -2 & 0 & 1 & 2 & 9 & 5 \\ 3 & 0 & -2 & -3 & -15 & -9 \end{bmatrix}$$

$$\xrightarrow[-3r_1 + r_3 \rightarrow r_3]{2r_1 + r_2 \rightarrow r_2}$$

$$\begin{bmatrix} 1 & 0 & -1 & -2 & -8 & -3 \\ 0 & 0 & -1 & -2 & -7 & -1 \\ 0 & 0 & 1 & 3 & 9 & 0 \end{bmatrix}$$

$$\xrightarrow{r_2 + r_3 \rightarrow r_3}$$

$$\begin{bmatrix} 1 & 0 & -1 & -2 & -8 & -3 \\ 0 & 0 & -1 & -2 & -7 & -1 \\ 0 & 0 & 0 & 1 & 2 & -1 \end{bmatrix}$$

$$\xrightarrow[2r_3 + r_1 \rightarrow r_1]{2r_3 + r_2 \rightarrow r_2}$$

$$\begin{bmatrix} 1 & 0 & -1 & 0 & -4 & -5 \\ 0 & 0 & -1 & 0 & -3 & -3 \\ 0 & 0 & 0 & 1 & 2 & -1 \end{bmatrix}$$

$$\xrightarrow{-r_2 \leftrightarrow r_2}$$

$$\begin{bmatrix} 1 & 0 & -1 & 0 & -4 & -5 \\ 0 & 0 & 1 & 0 & 3 & 3 \\ 0 & 0 & 0 & 1 & 2 & -1 \end{bmatrix}$$

$$\xrightarrow{r_2 + r_1 \rightarrow r_1}$$

$$\begin{bmatrix} 1 & 0 & 0 & 0 & -1 & -2 \\ 0 & 0 & 1 & 0 & 3 & 3 \\ 0 & 0 & 0 & 1 & 2 & -1 \end{bmatrix}.$$

This matrix corresponds to the system

$$\begin{aligned} x_1 \quad\quad\quad\quad - x_5 &= -2 \\ x_3 \quad\quad + 3x_5 &= 3 \\ x_4 + 2x_5 &= -1. \end{aligned}$$

Its general solution is

$$\begin{aligned} x_1 &= -2 + x_5 \\ x_2 &\text{ free} \\ x_3 &= 3 - 3x_5 \\ x_4 &= -1 - 2x_5 \\ x_5 &\text{ free.} \end{aligned}$$

19. The augmented matrix of the system is

$$\begin{bmatrix} 1 & -2 & 0 \\ 4 & -8 & r \end{bmatrix}.$$

Adding -4 times row 1 to row 2, we obtain

$$\begin{bmatrix} 1 & -2 & 0 \\ 0 & 0 & r \end{bmatrix}.$$

So the system is inconsistent if $r \neq 0$.

23. The augmented matrix of the system is

$$\begin{bmatrix} -1 & r & 2 \\ r & -9 & 6 \end{bmatrix}.$$

Adding r times row 1 to row 2 produces the matrix

$$\begin{bmatrix} -1 & r & 2 \\ 0 & r^2 - 9 & 2r + 6 \end{bmatrix}.$$

For the system corresponding to this augmented matrix to be inconsistent, the second row must have its only nonzero entry in the last column. Thus we must have $r^2 - 9 = 0$ and $2r + 6 \neq 0$. So $r = \pm 3$ and $r \neq -3$, that is, $r = 3$.

27. The augmented matrix of the system is

$$\begin{bmatrix} 1 & r & 5 \\ 3 & 6 & s \end{bmatrix}.$$

Adding -3 times row 1 to row 2 produces the matrix

$$\begin{bmatrix} 1 & r & 5 \\ 0 & 6 - 3r & s - 15 \end{bmatrix}.$$

(a) As in Exercise 23, for the system to be inconsistent, we need $6 - 3r = 0$ and $s - 15 \neq 0$. So $r = 2$ and $s \neq 15$.
(b) From the second row of the preceding matrix, we have

$$(6 - 3r)x_2 = s - 15.$$

For the system to have a unique solution, we must be able to solve this equation for x_2. Thus we need $6 - 3r \neq 0$, that is, $r \neq 2$.

(c) For the system to have infinitely many solutions, there must be a free variable. Thus $6 - 3r = 0$ and $s - 15 = 0$. So $r = 2$ and $s = 15$.

31. The augmented matrix of the system is

$$\begin{bmatrix} 1 & r & -3 \\ 2 & 5 & s \end{bmatrix}.$$

Adding -2 times row 1 to row 2 produces the matrix

$$\begin{bmatrix} 1 & r & -3 \\ 0 & -2r+5 & 6+s \end{bmatrix}.$$

(a) As in Exercise 23, for the system to be inconsistent, we must have $-2r + 5 = 0$ and $6 + s \neq 0$. So $r = \frac{5}{2}$ and $s \neq -6$.
(b) For the system to have a unique solution, we need $-2r + 5 \neq 0$, that is, $r \neq \frac{5}{2}$.
(c) For the system to have infinitely many solutions, there must be a free variable. Thus $-2r+5 = 0$ and $6 + s = 0$. So $r = \frac{5}{2}$ and $s = -6$.

35. To find the rank and nullity of the given matrix, we first find its reduced row echelon form R:

$$\begin{bmatrix} 1 & -1 & -1 & 0 \\ 2 & -1 & -2 & 1 \\ 1 & -2 & -2 & 2 \\ -4 & 2 & 3 & 1 \\ 1 & -1 & -2 & 3 \end{bmatrix} \begin{array}{l} -2r_1 + r_2 \to r_2 \\ -r_1 + r_3 \to r_3 \\ 4r_1 + r_4 \to r_4 \\ -r_1 + r_5 \to r_5 \end{array}$$

$$\begin{bmatrix} 1 & -1 & -1 & 0 \\ 0 & 1 & 0 & 1 \\ 0 & -1 & -1 & 2 \\ 0 & -2 & -1 & 1 \\ 0 & 0 & -1 & 3 \end{bmatrix} \begin{array}{l} r_2 + r_3 \to r_3 \\ 2r_2 + r_4 \to r_4 \end{array}$$

$$\begin{bmatrix} 1 & -1 & -1 & 0 \\ 0 & 1 & 0 & 1 \\ 0 & 0 & -1 & 3 \\ 0 & 0 & -1 & 3 \\ 0 & 0 & -1 & 3 \end{bmatrix} \begin{array}{l} -r_3 + r_4 \to r_4 \\ -r_3 + r_5 \to r_5 \end{array}$$

$$\begin{bmatrix} 1 & -1 & -1 & 0 \\ 0 & 1 & 0 & 1 \\ 0 & 0 & -1 & 3 \\ 0 & 0 & 0 & 0 \\ 0 & 0 & 0 & 0 \end{bmatrix} \quad -r_3 \leftrightarrow r_3$$

$$\begin{bmatrix} 1 & -1 & -1 & 0 \\ 0 & 1 & 0 & 1 \\ 0 & 0 & 1 & -3 \\ 0 & 0 & 0 & 0 \\ 0 & 0 & 0 & 0 \end{bmatrix} \quad r_3 + r_1 \to r_1$$

$$\begin{bmatrix} 1 & -1 & 0 & -3 \\ 0 & 1 & 0 & 1 \\ 0 & 0 & 1 & -3 \\ 0 & 0 & 0 & 0 \\ 0 & 0 & 0 & 0 \end{bmatrix} \quad r_2 + r_1 \to r_1$$

$$\begin{bmatrix} 1 & 0 & 0 & -2 \\ 0 & 1 & 0 & 1 \\ 0 & 0 & 1 & -3 \\ 0 & 0 & 0 & 0 \\ 0 & 0 & 0 & 0 \end{bmatrix} = R.$$

The rank of the given matrix equals the number of nonzero rows in R, which is

3. The nullity of the given matrix equals its number of columns minus its rank, which is $4 - 3 = 1$.

39. Because the reduced row echelon form of the augmented matrix is
$$\begin{bmatrix} 1 & 0 & -2 & 0 \\ 0 & 1 & 3 & 0 \\ 0 & 0 & 0 & 1 \end{bmatrix},$$
its rank is 3 (the number of nonnzero rows in the matrix above), and its nullity is $4 - 3 = 1$ (the number of columns in the matrix minus its rank).

43. Let x_1, x_2, and x_3 be the number of days that mines 1, 2, and 3, respectively, must operate to supply the desired amounts.

   (a) The requirements may be written with the matrix equation
   $$\begin{bmatrix} 1 & 1 & 2 \\ 1 & 2 & 2 \\ 2 & 1 & 0 \end{bmatrix} \begin{bmatrix} x_1 \\ x_2 \\ x_3 \end{bmatrix} = \begin{bmatrix} 80 \\ 100 \\ 40 \end{bmatrix}.$$
   The reduced row echelon form of the augmented matrix is
   $$\begin{bmatrix} 1 & 0 & 0 & 10 \\ 0 & 1 & 0 & 20 \\ 0 & 0 & 1 & 25 \end{bmatrix};$$
   so $x_1 = 10$, $x_2 = 20$, $x_3 = 25$.

   (b) A similar system of equations yields the reduced row echelon form
   $$\begin{bmatrix} 1 & 0 & 0 & 10 \\ 0 & 1 & 0 & 60 \\ 0 & 0 & 1 & -15 \end{bmatrix}.$$
   Because $x_3 = -15$ is impossible for this problem, the answer is no.

47. We need $f(-1) = 14$, $f(1) = 4$, and $f(3) = 10$. These conditions produce the system
$$\begin{aligned} a - b + c &= 14 \\ a + b + c &= 4 \\ 9a + 3b + c &= 10, \end{aligned}$$
which has the solution $a = 2$, $b = -5$, $c = 7$. So $f(x) = 2x^2 - 5x + 7$.

51. Column j is $\mathbf{e}_3$. Each pivot column of the reduced row echelon form of A has exactly one nonzero entry, which is 1, and hence it is a standard vector. Also, because of the definition of the reduced row echelon form, the pivot columns in order are $\mathbf{e}_1, \mathbf{e}_2, \ldots$. Hence, the third pivot column must be $\mathbf{e}_3$.

53. True

54. False. For example, the matrix $\begin{bmatrix} 0 & 1 \\ 2 & 0 \end{bmatrix}$ can be reduced to I_2 by interchanging its rows and then multiplying the first row by $\frac{1}{2}$, or by multiplying the second row by $\frac{1}{2}$ and then interchanging rows.

55. True   56. True   57. True

58. True

59. False, because rank A + nullity A equals the number of columns of A (by definition of the rank and nullity of a matrix), we cannot have a rank of 3 and a nullity of 2 for a matrix with 8 columns.

60. False, we need only repeat one equation to produce an equivalent system with a different number of equations.

61. True   62. True   63. True

64. False, there is a zero row in the aug-

mented matrix $\begin{bmatrix} 1 & 0 & 2 \\ 0 & 1 & 3 \\ 0 & 0 & 0 \end{bmatrix}$, but the corresponding system has the unique solution $x_1 = 2$, $x_3 = 3$.

65. False, there is a zero row in the augmented matrix $\begin{bmatrix} 0 & 0 & 1 \\ 0 & 0 & 0 \end{bmatrix}$, but the system is not consistent.

66. True   67. True

68. False, the sum of the rank and nullity of a matrix equals the number of *columns* in the matrix.

69. True   70. True

71. False, the third pivot position in a matrix may be in any column to the right of column 2.

72. True

75. The largest possible rank is 4. The reduced row echelon form is a 4×7 matrix and hence has at most 4 nonzero rows. So the rank must be less than or equal to 4. On the other hand, the 4×7 matrix whose first four columns are the distinct standard vectors has rank 4.

79. The largest possible rank is the minimum of m and n. If $m \leq n$, the solution is similar to that of Exercise 75. Suppose that A is an $m \times n$ matrix with $n \leq m$. By the first boxed result on page 48, the rank of a matrix equals the number of pivot columns of the matrix. Clearly, the number of pivot columns of an $m \times n$ matrix cannot exceed n, the number of columns; so rank $A \leq n$. In addition, if every column of the reduced row echelon form of A is a distinct standard vector, then rank $A = n$.

83. There are either no solutions or infinitely many solutions. Let the underdetermined system be $A\mathbf{x} = \mathbf{b}$, and let R be the reduced row echelon form of A. Each nonzero row of R corresponds to a basic variable. Since there are fewer equations than variables, there must be free variables. Therefore the system is either inconsistent or has infinitely many solutions.

87. Yes, $A(c\mathbf{u}) = c(A\mathbf{u}) = c \cdot \mathbf{0} = \mathbf{0}$; so $c\mathbf{u}$ is a solution of $A\mathbf{x} = \mathbf{0}$.

91. If $A\mathbf{x} = \mathbf{b}$ is consistent, then there exists a vector $\mathbf{u}$ such that $A\mathbf{u} = \mathbf{b}$. So $A(c\mathbf{u}) = c(A\mathbf{u}) = c\mathbf{b}$. Hence $c\mathbf{u}$ is a solution of $A\mathbf{x} = c\mathbf{b}$, and therefore $A\mathbf{x} = c\mathbf{b}$ is consistent.

95. By proceeding as in Exercise 7, we see that the general solution of the given system is

$$\begin{aligned} x_1 &= 2.32 + 0.32x_5 \\ x_2 &= -6.44 + 0.56x_5 \\ x_3 &= 0.72 - 0.28x_5 \\ x_4 &= 5.92 + 0.92x_5 \\ x_5 &\quad \text{free.} \end{aligned}$$

99. The reduced row echelon form of the given matrix is (approximately)

$$\begin{bmatrix} 1 & 0 & 0 & 0 & 0.0000 \\ 0 & 1 & 0 & 0 & 1.0599 \\ 0 & 0 & 1 & 0 & 0.8441 \\ 0 & 0 & 0 & 1 & 0.4925 \\ 0 & 0 & 0 & 0 & 0.0000 \end{bmatrix}.$$

The rank equals the number of nonzero rows, 4, and the nullity is found by subtracting the rank from the number of columns, and hence equals $5 - 4 = 1$.

1.5 APPLICATIONS OF SYSTEMS OF LINEAR EQUATIONS

1. True 2. True

3. False, $\mathbf{x} - C\mathbf{x}$ is the net production vector. The vector $C\mathbf{x}$ is the total output of the economy that is consumed during the production plrocess.

4. False, see Kirchoff's voltage law.

5. True 6. True

7. Because $c_{34} = .22$, each dollar of output from the entertainment sector requires an input of \$.22 from the services sector. Thus \$50 million of output from the entertainment sector requires an imput of $.22(\$50 \text{ million}) = \$11$ million from the services sector.

9. The third column of C gives the amounts from the various sectors required to produce one unit of services. The smallest entry in this column, .06, corresponds to the input from the service sector, and hence services is least dependent on the service sector.

13. Let
$$\mathbf{x} = \begin{bmatrix} 30 \\ 40 \\ 30 \\ 20 \end{bmatrix}.$$

The total value of the inputs from each sector consumed during the production process are the components of

$$C\mathbf{x} = \begin{bmatrix} .12 & .11 & .15 & .18 \\ .20 & .08 & .24 & .07 \\ .18 & .16 & .06 & .22 \\ .09 & .07 & .12 & .05 \end{bmatrix} \begin{bmatrix} 30 \\ 40 \\ 30 \\ 20 \end{bmatrix}$$

$$= \begin{bmatrix} 16.1 \\ 17.8 \\ 18.0 \\ 10.1 \end{bmatrix}.$$

Therefore the total value of the inputs from each sector consumed during the production process are \$16.1 million of agriculture, \$17.8 million of manufacturing, \$18 million of services, and \$10.1 million of entertainment.

17. (a) The gross production vector is $\mathbf{x} = \begin{bmatrix} 40 \\ 30 \\ 35 \end{bmatrix}$. If C is the input-output matrix, then the net production vector is

$$\mathbf{x} - C\mathbf{x}$$

$$= \begin{bmatrix} 40 \\ 30 \\ 35 \end{bmatrix} - \begin{bmatrix} .2 & .20 & .3 \\ .4 & .30 & .1 \\ .2 & .25 & .3 \end{bmatrix} \begin{bmatrix} 40 \\ 30 \\ 35 \end{bmatrix}$$

$$= \begin{bmatrix} 15.5 \\ 1.5 \\ 9.0 \end{bmatrix}.$$

So the net productions are \$15.5 million of transportation, \$1.5 million of food, and \$9 million of oil.

(b) Denote the net production vector by
$$\mathbf{d} = \begin{bmatrix} 32 \\ 48 \\ 24 \end{bmatrix},$$

and let $\mathbf{x}$ denote the gross production vector. Then $\mathbf{x}$ is a solution of the system of linear equations $(I_3 - C)\mathbf{x} = \mathbf{d}$. Since

$I_3 - C$

$$= \begin{bmatrix} 1 & 0 & 0 \\ 0 & 1 & 0 \\ 0 & 0 & 1 \end{bmatrix} - \begin{bmatrix} .2 & .20 & .3 \\ .4 & .30 & .1 \\ .2 & .25 & .3 \end{bmatrix}$$

$$= \begin{bmatrix} .80 & -.20 & -.30 \\ -.40 & .70 & -.10 \\ -.20 & -.25 & .70 \end{bmatrix},$$

the augmented matrix of this system is

$$\begin{bmatrix} .80 & -.20 & -.30 & 32 \\ -.40 & .70 & -.10 & 48 \\ -.20 & -.25 & .70 & 24 \end{bmatrix}.$$

The reduced row echelon form of the augmented matrix is

$$\begin{bmatrix} 1 & 0 & 0 & 128 \\ 0 & 1 & 0 & 160 \\ 0 & 0 & 1 & 128 \end{bmatrix},$$

and hence the gross productions required are $128 million of transportation, $160 million of food, and $128 million of oil.

21. The input-output matrix for this economy is

$$C = \begin{bmatrix} .10 & .10 & .15 \\ .20 & .40 & .10 \\ .20 & .20 & .30 \end{bmatrix}.$$

(a) Let

$$\mathbf{x} = \begin{bmatrix} 70 \\ 50 \\ 60 \end{bmatrix}.$$

Then the net production vector is given by

$\mathbf{x} - C\mathbf{x}$

$$= \begin{bmatrix} 70 \\ 50 \\ 60 \end{bmatrix} - \begin{bmatrix} .1 & .1 & .15 \\ .2 & .4 & .10 \\ .2 & .2 & .30 \end{bmatrix} \begin{bmatrix} 70 \\ 50 \\ 60 \end{bmatrix}$$

$$= \begin{bmatrix} 49 \\ 10 \\ 18 \end{bmatrix}.$$

Therefore the net productions are $49 million of finance, $10 million of goods, and $18 million of services.

(b) Let

$$\mathbf{d} = \begin{bmatrix} 40 \\ 50 \\ 30 \end{bmatrix},$$

the net production vector, and let $\mathbf{x}$ denote the gross production vector. Then $\mathbf{x}$ is the solution of the matrix equation $(I_3 - C)\mathbf{x} = \mathbf{d}$. Since

$I_3 - C$

$$= \begin{bmatrix} 1 & 0 & 0 \\ 0 & 1 & 0 \\ 0 & 0 & 1 \end{bmatrix} - \begin{bmatrix} .1 & .1 & .15 \\ .2 & .4 & .10 \\ .2 & .2 & .30 \end{bmatrix}$$

$$= \begin{bmatrix} .90 & -.10 & -.15 \\ -.20 & .60 & -.10 \\ -.20 & -.20 & .70 \end{bmatrix},$$

the augmented matrix of this system is

$$\begin{bmatrix} .90 & -.10 & -.15 & 40 \\ -.20 & .60 & -.10 & 50 \\ -.20 & -.20 & .70 & 30 \end{bmatrix}.$$

The reduced row echelon form of the augmented matrix is

$$\begin{bmatrix} 1 & 0 & 0 & 75 \\ 0 & 1 & 0 & 125 \\ 0 & 0 & 1 & 100 \end{bmatrix},$$

and hence the gross productions are $75 million of finance, $125 million of goods, and $100 million of services.

(c) We proceed as in (b), except that in this case

$$\mathbf{d} = \begin{bmatrix} 40 \\ 36 \\ 44 \end{bmatrix}.$$

In this case, the augmented matrix of $(I_3 - C)\mathbf{x} = \mathbf{d}$ is

$$\begin{bmatrix} .90 & -.10 & -.15 & 40 \\ -.20 & .60 & -.10 & 36 \\ -.20 & -.20 & .70 & 44 \end{bmatrix},$$

which has the reduced row echelon form

$$\begin{bmatrix} 1 & 0 & 0 & 75 \\ 0 & 1 & 0 & 104 \\ 0 & 0 & 1 & 114 \end{bmatrix}.$$

Therefore the gross productions are $75 million of finance, $104 million of goods, and $114 million of services.

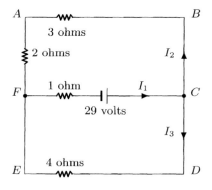

Figure for Exercise 25

25. Applying Kirchoff's voltage law to the closed path $FCBAF$ in the network above, we obtain the equation

$$3I_2 + 2I_2 + 1I_1 = 29.$$

Similarly, from the closed path $FCDEF$, we obtain

$$1I_1 + 4I_3 = 29.$$

At the junction C, Kirchoff's current law yields the equation

$$I_1 = I_2 + I_3.$$

Thus the currents I_1, I_2, and I_3 satisfy the system

$$\begin{aligned} I_1 + 5I_2 & = 29 \\ I_1 + 4I_3 & = 29 \\ I_1 - I_2 - I_3 & = 0. \end{aligned}$$

Since the reduced row echelon form of the augmented matrix of this system is

$$\begin{bmatrix} 1 & 0 & 0 & 9 \\ 0 & 1 & 0 & 4 \\ 0 & 0 & 1 & 5 \end{bmatrix},$$

this system has the unique solution $I_1 = 9$, $I_2 = 4$, $I_3 = 5$.

29. Applying Kirchoff's voltage law to the closed path $ABGHA$ in the network shown on the next page, we obtain the equation

$$2I_3 + 4I_1 = 60.$$

Similarly, from the closed path $BCFGB$, we obtain

$$1I_2 + 1(-I_5) + 1(I_6) + 2(-I_3) = 0,$$

and from the closed path $CDEFC$, we obtain

$$2I_4 + 1I_5 = 30.$$

At the junctions B, C, F, and G, Kirchoff's current law yields the equations

$$\begin{aligned} I_1 &= I_2 + I_3 \\ I_2 + I_5 &= I_4 \\ I_4 &= I_5 + I_6 \end{aligned}$$

18 Chapter 1 Matrices, Vectors, and Systems of Linear Equations

and
$$I_3 + I_6 = I_1.$$

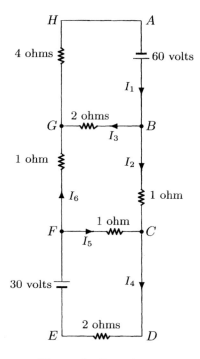

Figure for Exercise 29

Thus the currents I_1, I_2, I_3, I_4, I_5, and I_6 satisfy the system

$$\begin{aligned} 4I_1 \quad\quad + 2I_3 \quad\quad\quad\quad\quad\quad &= 60 \\ I_2 - 2I_3 \quad\quad - I_5 + I_6 &= 0 \\ 2I_4 + I_5 \quad\quad &= 30 \\ I_1 - I_2 - I_3 \quad\quad\quad\quad\quad &= 0 \\ I_2 \quad\quad - I_4 + I_5 \quad\quad &= 0 \\ I_4 - I_5 - I_6 &= 0 \\ I_1 \quad - I_3 \quad\quad\quad - I_6 &= 0. \end{aligned}$$

Solving this system gives $I_1 = 12.5$, $I_2 = 7.5$, $I_3 = 5$, $I_4 = 12.5$, $I_5 = 5$, and $I_6 = 7.5$.

1.6 THE SPAN OF A SET OF VECTORS

1. Let
$$A = \begin{bmatrix} 1 & -1 & 1 \\ 0 & 1 & 1 \\ 1 & 1 & 3 \end{bmatrix} \text{ and } \mathbf{v} = \begin{bmatrix} -1 \\ 4 \\ 7 \end{bmatrix}.$$

Vector $\mathbf{v}$ is in the span if and only if the system $A\mathbf{x} = \mathbf{v}$ is consistent. The reduced row echelon form of the augmented matrix of this system is

$$R = \begin{bmatrix} 1 & 0 & 2 & 3 \\ 0 & 1 & 1 & 4 \\ 0 & 0 & 0 & 0 \end{bmatrix}.$$

Thus the system is consistent, and so $\mathbf{v}$ is in the span of the given set.

3. Let
$$A = \begin{bmatrix} 1 & -1 & 1 \\ 0 & 1 & 1 \\ 1 & 1 & 3 \end{bmatrix} \text{ and } \mathbf{v} = \begin{bmatrix} 0 \\ 5 \\ 2 \end{bmatrix}.$$

Vector $\mathbf{v}$ is in the span if and only if the system $A\mathbf{x} = \mathbf{v}$ is consistent. The reduced row echelon form of the augmented matrix of this system is

$$R = \begin{bmatrix} 1 & 0 & 2 & 0 \\ 0 & 1 & 1 & 0 \\ 0 & 0 & 0 & 1 \end{bmatrix}.$$

Because of the form of the third row of R, the system is inconsistent. Hence $\mathbf{v}$ is not in the span of the given set.

5. Let
$$A = \begin{bmatrix} 1 & -1 & 1 \\ 0 & 1 & 1 \\ 1 & 1 & 3 \end{bmatrix} \text{ and } \mathbf{v} = \begin{bmatrix} -1 \\ 1 \\ 1 \end{bmatrix}.$$

The reduced row echelon form of the augmented matrix of this system is

$$R = \begin{bmatrix} 1 & 0 & 2 & 0 \\ 0 & 1 & 1 & 1 \\ 0 & 0 & 0 & 0 \end{bmatrix}.$$

Thus the system is consistent, and so **v** is in the span of the given set.

9. Let

$$A = \begin{bmatrix} 1 & 2 & -1 \\ 2 & -1 & 2 \\ -1 & 1 & 0 \\ 1 & 0 & 3 \end{bmatrix} \text{ and } \mathbf{v} = \begin{bmatrix} 0 \\ -1 \\ 1 \\ 0 \end{bmatrix}.$$

The reduced row echelon form of the augmented matrix of this system is

$$R = \begin{bmatrix} 1 & 0 & 0 & 0 \\ 0 & 1 & 0 & 0 \\ 0 & 0 & 1 & 0 \\ 0 & 0 & 0 & 1 \end{bmatrix}.$$

Because of the form of the third row of R, the system is inconsistent. Hence **v** is not in the span of the given set.

13. Let

$$A = \begin{bmatrix} 1 & 2 & -1 \\ 2 & -1 & 2 \\ -1 & 1 & 0 \\ 1 & 0 & 3 \end{bmatrix} \text{ and } \mathbf{v} = \begin{bmatrix} 4 \\ 0 \\ 5 \\ 8 \end{bmatrix}.$$

The reduced row echelon form of the augmented matrix of this system is

$$R = \begin{bmatrix} 1 & 0 & 0 & -1 \\ 0 & 1 & 0 & 4 \\ 0 & 0 & 1 & 3 \\ 0 & 0 & 0 & 0 \end{bmatrix}.$$

Thus the system is consistent, and so **v** is in the span of the given set.

17. The vector **v** is in the span of S if and only if the system $A\mathbf{x} = \mathbf{v}$ is consistent, where

$$A = \begin{bmatrix} 1 & -1 \\ 0 & 3 \\ -1 & 2 \end{bmatrix}.$$

The augmented matrix of $A\mathbf{x} = \mathbf{v}$ is

$$\begin{bmatrix} 1 & -1 & 2 \\ 0 & 3 & r \\ -1 & 2 & -1 \end{bmatrix}.$$

Applying elementary row operations to this matrix, we obtain

$$\begin{bmatrix} 1 & -1 & 2 \\ 0 & 3 & r \\ -1 & 2 & -1 \end{bmatrix} \xrightarrow{r_1 + r_3 \to r_3}$$

$$\begin{bmatrix} 1 & -1 & 2 \\ 0 & 3 & r \\ 0 & 1 & 1 \end{bmatrix} \xrightarrow{r_2 \leftrightarrow r_3}$$

$$\begin{bmatrix} 1 & -1 & 2 \\ 0 & 1 & 1 \\ 0 & 3 & r \end{bmatrix} \xrightarrow{-3r_2 + r_3 \to r_3}$$

$$\begin{bmatrix} 1 & -1 & 2 \\ 0 & 1 & 1 \\ 0 & 0 & r-3 \end{bmatrix}.$$

So the system is consistent if and only if $r - 3 = 0$, that is, $r = 3$. Therefore **v** is in the span of S if and only if $r = 3$.

21. No. Let $A = \begin{bmatrix} 1 & -2 \\ -1 & 2 \end{bmatrix}$. The reduced row echelon form of A is $\begin{bmatrix} 1 & -2 \\ 0 & 0 \end{bmatrix}$, which has rank 1. By Theorem 1.6, the set is not a generating set for $\mathcal{R}^2$.

25. Yes. Let

$$A = \begin{bmatrix} 1 & -1 & 1 \\ 0 & 1 & 2 \\ -2 & 4 & -2 \end{bmatrix}.$$

The reduced row echelon form of A is

$$\begin{bmatrix} 1 & 0 & 0 \\ 0 & 1 & 0 \\ 0 & 0 & 1 \end{bmatrix},$$

which has rank 3. By Theorem 1.6, the set is a generating set for $\mathcal{R}^3$.

29. Yes. The reduced row echelon form of A is $\begin{bmatrix} 1 & 0 \\ 0 & 1 \end{bmatrix}$, which has rank 2. By Theorem 1.6, the system $A\mathbf{x} = \mathbf{b}$ is consistent for every $\mathbf{b}$ in $\mathcal{R}^2$.

33. The reduced row echelon form of A is

$$\begin{bmatrix} 1 & 0 \\ 0 & 1 \\ 0 & 0 \end{bmatrix},$$

which has rank 2. By Theorem 1.6, $A\mathbf{x} = \mathbf{b}$ is inconsistent for at least one $\mathbf{b}$ in $\mathcal{R}^3$.

37. The desired set is $\left\{ \begin{bmatrix} 1 \\ 3 \end{bmatrix}, \begin{bmatrix} 0 \\ 1 \end{bmatrix} \right\}$. If we delete either vector, then the span of $\mathcal{S}$ consists of all multiples of the remaining vector. Because neither vector in $\mathcal{S}$ is a multiple of the other, neither can be deleted.

41. One possible set is $\left\{ \begin{bmatrix} 1 \\ -2 \\ 1 \end{bmatrix} \right\}$. The last two vectors in $\mathcal{S}$ are multiples of the first, and so can be deleted without changing the span of $\mathcal{S}$.

45. True   46. True   47. True

48. False, by Theorem 1.6(c), we need rank $A = m$ for $A\mathbf{x} = \mathbf{b}$ to be consistent for every vector $\mathbf{b}$.

49. True   50. True   51. True

52. False, the sets $\mathcal{S}_1 = \{\mathbf{e}_1\}$ and $\mathcal{S}_2 = \{2\mathbf{e}_1\}$ have the same spans, but are not equal.

53. False, the sets $\mathcal{S}_1 = \{\mathbf{e}_1\}$ and $\mathcal{S}_2 = \{\mathbf{e}_1, 2\mathbf{e}_1\}$ have equal spans, but do not contain the same number of elements.

54. False, $\mathcal{S} = \{\mathbf{e}_1\}$ and $\mathcal{S} \cup \{2\mathbf{e}_1\}$ have equal spans, but $2\mathbf{e}_1$ is not in $\mathcal{S}$.

55. True   56. True   57. True
58. True   59. True   60. True
61. True   62. True   63. True
64. True

65. (a) 2
    (b) There are infinitely many vectors because every choice of the scalars a and b yields a different vector $a\mathbf{u}_1 + b\mathbf{u}_2$ in the span.

69. For $r \geq k$, let $\mathcal{S}_1 = \{\mathbf{u}_1, \mathbf{u}_2, \ldots, \mathbf{u}_k\}$ and $\mathcal{S}_2 = \{\mathbf{u}_1, \mathbf{u}_2, \ldots, \mathbf{u}_r\}$, and suppose that $\mathcal{S}_1$ is a generating set for $\mathcal{R}^n$. Let $\mathbf{v}$ be in $\mathcal{R}^n$. Then for some scalars $a_1, a_2, \ldots, a_k$ we can write

$$\mathbf{v} = a_1\mathbf{u}_1 + a_2\mathbf{u}_2 + \cdots + a_k\mathbf{u}_k$$
$$= a_1\mathbf{u}_1 + a_2\mathbf{u}_2 + \cdots + a_k\mathbf{u}_k$$
$$+ 0\mathbf{u}_{k+1} + \cdots + 0\mathbf{u}_r.$$

So $\mathcal{S}_2$ is also a generating set for $\mathcal{R}^n$.

73. No, let $A = \begin{bmatrix} 1 & 0 \\ 1 & 0 \end{bmatrix}$. Then $R = \begin{bmatrix} 1 & 0 \\ 0 & 0 \end{bmatrix}$. The span of the columns of A equals all multiples of $\begin{bmatrix} 1 \\ 1 \end{bmatrix}$, whereas the span of the columns of R equals all multiples of $\mathbf{e}_1$.

77. Let $\mathbf{u}_1, \mathbf{u}_2, \ldots, \mathbf{u}_m$ be the rows of A. We must prove that $\text{Span}\{\mathbf{u}_1, \mathbf{u}_2, \ldots, \mathbf{u}_m\}$ is unchanged if we perform any of the three

types of elementary row operations. For ease of notation, we will consider operations that affect only the first two rows of A.

Case 1 If we interchange rows 1 and 2 of A, then the rows of A are the same as the rows of B (although in a different order), and hence the span of the rows of A equals the span of the rows of B.

Case 2 Suppose we multiply the first row of A by $k \neq 0$. Then the rows of B are $k\mathbf{u}_1, \mathbf{u}_2, \ldots, \mathbf{u}_m$. Any vector in the span of the rows of A can be written

$$c_1\mathbf{u}_1 + c_2\mathbf{u}_2 + \cdots + c_m\mathbf{u}_m$$
$$= \left(\tfrac{c_1}{k}\right)k\mathbf{u}_1 + c_2\mathbf{u}_2 + \cdots + c_m\mathbf{u}_m,$$

which is in the span of the rows of B. Likewise, any vector in the span of the rows of B can be written

$$c_1(k\mathbf{u}_1) + c_2\mathbf{u}_2 + \cdots + c_m\mathbf{u}_m$$
$$= (c_1 k)\mathbf{u}_1 + c_2\mathbf{u}_2 + \cdots + c_m\mathbf{u}_m,$$

which is in the span of the rows of A.

Case 3 Suppose we add k times the second row of A to the first row of A. Then the rows of B are $\mathbf{u}_1 + k\mathbf{u}_2, \mathbf{u}_2, \ldots, \mathbf{u}_m$. Any vector in the span of the rows of A can be written

$$c_1\mathbf{u}_1 + c_2\mathbf{u}_2 + \cdots + c_m\mathbf{u}_m$$
$$= c_1(\mathbf{u}_1 + k\mathbf{u}_2) + (c_2 - kc_1)\mathbf{u}_2$$
$$+ c_3\mathbf{u}_3 + \cdots + c_m\mathbf{u}_m,$$

which is in the span of the rows of B. Likewise, any vector in the span of the rows of B can be written

$$c_1(\mathbf{u}_1 + k\mathbf{u}_2)$$
$$+ c_2\mathbf{u}_2 + \cdots + c_m\mathbf{u}_m$$
$$= c_1\mathbf{u}_1 + (kc_1 + c_2)\mathbf{u}_2 + \cdots$$
$$+ c_m\mathbf{u}_m,$$

which is in the span of the rows of A.

81. By proceeding as in Exercise 1, we see that the given vector is not in the span of the given set.

1.7 LINEAR DEPENDENCE AND LINEAR INDEPENDENCE

1. Because the second vector in the set is a multiple of the first, the given set is linearly dependent.

5. No, the first two vectors are linearly independent because neither is a multiple of the other. The third vector is not a linear combination of the first two because its first component is not zero. So, by Theorem 1.9, the set of 3 vectors is linearly independent.

9. A set consisting of a single nonzero vector is linearly independent.

13. Because the second vector in $\mathcal{S}$ is a multiple of the first, it can be removed from $\mathcal{S}$ without changing the span of the set.

17. Because the second vector in $\mathcal{S}$ is a multiple of the first, it can be removed from $\mathcal{S}$ without changing the span of the set. Neither of the two remaining vectors is a multiple of the other; so a smallest subset of $\mathcal{S}$ having the same span as $\mathcal{S}$ is

$$\left\{ \begin{bmatrix} 2 \\ -3 \\ 5 \end{bmatrix}, \begin{bmatrix} 1 \\ 0 \\ 2 \end{bmatrix} \right\}.$$

22   Chapter 1   Matrices, Vectors, and Systems of Linear Equations

21. Note that
$$2\begin{bmatrix}-2\\0\\3\end{bmatrix}+\frac{1}{4}\begin{bmatrix}0\\4\\0\end{bmatrix}=\begin{bmatrix}-4\\1\\6\end{bmatrix}.$$

Hence, by Theorem 1.7, the third vector in $\mathcal{S}$ can be removed from $\mathcal{S}$ without changing the span of $\mathcal{S}$. Because neither of the two remaining vectors is a multiple of the other, a smallest subset of $\mathcal{S}$ having the same span as $\mathcal{S}$ is

$$\left\{\begin{bmatrix}-2\\0\\3\end{bmatrix},\begin{bmatrix}0\\4\\0\end{bmatrix}\right\}.$$

25. Yes, let
$$A=\begin{bmatrix}1&1&1\\2&-3&2\\0&1&-2\\-1&-2&3\end{bmatrix},$$

which has reduced row echelon form
$$\begin{bmatrix}1&0&0\\0&1&0\\0&0&1\\0&0&0\end{bmatrix}.$$

So rank $A=3$. By Theorem 1.8, the set is linearly independent.

29. No, let
$$A=\begin{bmatrix}1&-1&-1&0\\-1&0&-4&1\\-1&1&1&-2\\2&-1&3&1\end{bmatrix},$$

which has reduced row echelon form
$$\begin{bmatrix}1&0&4&0\\0&1&5&0\\0&0&0&1\\0&0&0&0\end{bmatrix}.$$

So rank $A=3$. By Theorem 1.8, the set is linearly dependent.

33. Let A be the matrix whose columns are the vectors in $\mathcal{S}$. From the reduced row echelon form of A, we see that the general solution of $A\mathbf{x}=\mathbf{0}$ is
$$x_1=-5x_3$$
$$x_2=-4x_3$$
$$x_3\quad\text{free.}$$

So one solution is $x_1=-5$, $x_2=-4$, and $x_3=1$. Therefore
$$-5\begin{bmatrix}0\\1\\1\end{bmatrix}-4\begin{bmatrix}1\\0\\-1\end{bmatrix}+\begin{bmatrix}4\\5\\1\end{bmatrix}=\mathbf{0}.$$

Hence
$$\begin{bmatrix}4\\5\\1\end{bmatrix}=5\begin{bmatrix}0\\1\\1\end{bmatrix}+4\begin{bmatrix}1\\0\\-1\end{bmatrix}.$$

37. The reduced row echelon form of the matrix whose columns are the vectors in $\mathcal{S}$ is
$$\begin{bmatrix}1&0&0&5\\0&1&0&-3\\0&0&1&3\end{bmatrix}.$$

As in Exercise 33, we see that
$$5\begin{bmatrix}1\\2\\-1\end{bmatrix}-3\begin{bmatrix}0\\1\\-1\end{bmatrix}+3\begin{bmatrix}-1\\-2\\0\end{bmatrix}=\begin{bmatrix}2\\1\\-2\end{bmatrix}.$$

41. Using elementary row operations, we can transform the matrix
$$A=\begin{bmatrix}-2&1&-1\\0&1&1\\1&-3&r\end{bmatrix},$$
into
$$\begin{bmatrix}1&0&1\\0&1&1\\0&0&r+2\end{bmatrix}.$$

Then rank $A=2$ if $r=-2$. By Theorem 1.8, the set is linearly dependent if $r=-2$.

45. The given set is linearly dependent if the third vector is a linear combination of the first two. But the first two vectors are nonparallel vectors in $\mathcal{R}^2$, and so the boxed statement on page 17 implies that the third vector is a linear combination of the first two vectors for every value of r.

49. Neither of the first two vectors in the set is a multiple of the other. Hence Theorem 1.8 implies that the given set is linearly dependent if and only if there are scalars c_1 and c_2 such that

$$c_1 \begin{bmatrix} 1 \\ 2 \\ 3 \\ -1 \end{bmatrix} + c_2 \begin{bmatrix} 3 \\ 1 \\ 6 \\ 1 \end{bmatrix} = \begin{bmatrix} -1 \\ 3 \\ -2 \\ r \end{bmatrix}.$$

Thus c_1 and c_2 must satisfy

$$\begin{aligned} c_1 + 2c_2 &= -1 \\ 2c_1 + c_2 &= 3 \\ 3c_1 + 6c_2 &= -2 \\ -c_1 + c_2 &= r. \end{aligned}$$

Consider the system of three equations in c_1 and c_2 consisting of the first three equations above. The reduced row echelon form of the augmented matrix of this system is I_3, and so there are no values of c_1 and c_2 that satisfy the first three equations above. Hence there is no value of r for which the system of four equations is consistent, and therefore there is no value of r for which the given set is linearly dependent.

53. The general solution of the given system is

$$\begin{aligned} x_1 &= -3x_2 - 2x_4 \\ x_2 &\quad \text{free} \\ x_3 &= 6x_4 \\ x_4 &\quad \text{free}. \end{aligned}$$

So the vector form of the general solution is

$$\begin{bmatrix} x_1 \\ x_2 \\ x_3 \\ x_4 \end{bmatrix} = \begin{bmatrix} -3x_2 - 2x_4 \\ x_2 \\ 6x_4 \\ x_4 \end{bmatrix}$$

$$= x_2 \begin{bmatrix} -3 \\ 1 \\ 0 \\ 0 \end{bmatrix} + x_4 \begin{bmatrix} -2 \\ 0 \\ 6 \\ 1 \end{bmatrix}.$$

57. The general solution of the given system is

$$\begin{aligned} x_1 &= -x_4 - 3x_6 \\ x_2 &\quad \text{free} \\ x_3 &= 2x_4 - x_6 \\ x_4 &= \text{free} \\ x_5 &= 0 \\ x_6 &\quad \text{free}. \end{aligned}$$

So the vector form of the general solution is

$$\begin{bmatrix} x_1 \\ x_2 \\ x_3 \\ x_4 \\ x_5 \\ x_6 \end{bmatrix} = \begin{bmatrix} -x_4 - 3x_6 \\ x_2 \\ 2x_4 - x_6 \\ x_4 \\ 0 \\ x_6 \end{bmatrix}$$

$$= x_2 \begin{bmatrix} 0 \\ 1 \\ 0 \\ 0 \\ 0 \\ 0 \end{bmatrix} + x_4 \begin{bmatrix} -1 \\ 0 \\ 2 \\ 1 \\ 0 \\ 0 \end{bmatrix} + x_6 \begin{bmatrix} -3 \\ 0 \\ -1 \\ 0 \\ 0 \\ 1 \end{bmatrix}.$$

61. The general solution of the given system is

$$\begin{aligned} x_1 &= -2x_2 + x_3 - 2x_5 + x_6 \\ x_2 &\quad \text{free} \\ x_3 &\quad \text{free} \\ x_4 &= - 4x_5 - 3x_6 \\ x_5 &\quad \text{free} \\ x_6 &\quad \text{free}. \end{aligned}$$

So the vector form of the general solution is

$$\begin{bmatrix} x_1 \\ x_2 \\ x_3 \\ x_4 \\ x_5 \\ x_6 \end{bmatrix} = \begin{bmatrix} -2x_2 + x_3 - 2x_5 + x_6 \\ x_2 \\ x_3 \\ -4x_5 - 3x_6 \\ x_5 \\ x_6 \end{bmatrix}$$

$$= x_2 \begin{bmatrix} -2 \\ 1 \\ 0 \\ 0 \\ 0 \\ 0 \end{bmatrix} + x_3 \begin{bmatrix} 1 \\ 0 \\ 1 \\ 0 \\ 0 \\ 0 \end{bmatrix}$$

$$+ x_5 \begin{bmatrix} -2 \\ 0 \\ 0 \\ -4 \\ 1 \\ 0 \end{bmatrix} + x_6 \begin{bmatrix} 1 \\ 0 \\ 0 \\ -3 \\ 0 \\ 1 \end{bmatrix}.$$

63. True

64. False, the columns are linearly independent (see Theorem 1.8). Consider the matrix $\begin{bmatrix} 1 & 0 \\ 0 & 1 \\ 0 & 0 \end{bmatrix}$.

65. False, the columns are linearly independent (see Theorem 1.8). See the matrix in the solution to Exercise 64.

66. True **67.** True **68.** True

69. False, consider the equation $I_2 \mathbf{x} = \mathbf{0}$.

70. True

71. False, if $\mathbf{v} \neq \mathbf{0}$.

72. False, consider the set

$$\left\{ \begin{bmatrix} 1 \\ 0 \end{bmatrix}, \begin{bmatrix} 0 \\ 1 \end{bmatrix}, \begin{bmatrix} 1 \\ 1 \end{bmatrix} \right\}.$$

73. False, consider $n = 3$ and the set

$$\left\{ \begin{bmatrix} 1 \\ 0 \\ 0 \end{bmatrix}, \begin{bmatrix} 2 \\ 0 \\ 0 \end{bmatrix} \right\}.$$

74. True **75.** True **76.** True

77. False, see the matrix in Exercise 64.

78. True

79. False, if $c_1 \mathbf{u}_1 + c_2 \mathbf{u}_2 + \cdots + c_k \mathbf{u}_k = \mathbf{0}$ *only when* $c_1 = c_2 = \cdots = c_k = 0$, then $\{\mathbf{u}_1, \mathbf{u}_2, \ldots, \mathbf{u}_k\}$ is linearly independent.

80. True **81.** True **82.** True

85. In $\mathcal{R}^3$, take $\mathbf{u}_1 = \mathbf{e}_1$, $\mathbf{u}_2 = \mathbf{e}_2$, and $\mathbf{v} = \mathbf{e}_1 + \mathbf{e}_2$. Then $\{\mathbf{u}_1, \mathbf{u}_2\}$, and $\{\mathbf{v}\}$ are both linearly independent, but $\{\mathbf{u}_1, \mathbf{u}_2, \mathbf{v}\}$ is not because $\mathbf{v} = \mathbf{u}_1 + \mathbf{u}_2$.

89. Suppose $a_1, a_2, \ldots, a_k$ are scalars such that

$$a_1(c_1 \mathbf{u}_1) + a_2(c_2 \mathbf{u}_2) + \cdots + a_k(c_k \mathbf{u}_k) = \mathbf{0},$$

that is,

$$(a_1 c_1) \mathbf{u}_1 + (a_2 c_2) \mathbf{u}_2 + \cdots + (a_k c_k) \mathbf{u}_k = \mathbf{0}.$$

Because $\{\mathbf{u}_1, \mathbf{u}_2, \ldots, \mathbf{u}_k\}$ is linearly independent, we have

$$a_1 c_1 = a_2 c_2 = \cdots = a_k c_k = 0.$$

Thus, since $c_1, c_2, \ldots, c_k$ are nonzero, it follows that $a_1 = a_2 = \cdots = a_k = 0$.

93. Suppose that $\mathbf{v}$ is in the span of $\mathcal{S}$ and that

$$\mathbf{v} = c_1 \mathbf{u}_1 + c_2 \mathbf{u}_2 + \cdots + c_k \mathbf{u}_k$$

and

$$\mathbf{v} = d_1 \mathbf{u}_1 + d_2 \mathbf{u}_2 + \cdots + d_k \mathbf{u}_k$$

for scalars $c_1, c_2, \ldots, c_k, d_1, d_2, \ldots, d_k$. Subtracting the second equation from the first yields

$$\mathbf{0} = (c_1 - d_1)\mathbf{u}_1 + \cdots + (c_k - d_k)\mathbf{u}_k.$$

Because $\mathcal{S}$ is linearly independent, $c_1 - d_1 = c_2 - d_2 = \cdots = c_k - d_k = 0$. So $c_1 = d_1, c_2 = d_2, \cdots, c_k = d_k$.

97. Suppose

$$c_1 A\mathbf{u}_1 + c_2 A\mathbf{u}_2 + \cdots + c_k A\mathbf{u}_k = \mathbf{0}$$

for some scalars $c_1, c_2, \ldots, c_k$. Then

$$A(c_1\mathbf{u}_1 + c_2\mathbf{u}_2 + \cdots + c_k\mathbf{u}_k) = \mathbf{0}.$$

By Theorem 1.8, it follows that

$$c_1\mathbf{u}_1 + c_2\mathbf{u}_2 + \cdots + c_k\mathbf{u}_k = \mathbf{0}.$$

Because $\mathcal{S}$ is linearly independent, we have $c_1 = c_2 = \cdots = c_k = 0$.

101. Proceeding as in Exercise 37, we see that the set is linearly dependent and $\mathbf{v}_5 = 2\mathbf{v}_1 - \mathbf{v}_3 + \mathbf{v}_4$, where $\mathbf{v}_j$ is the jth vector in the set.

CHAPTER 1 REVIEW

1. False, the columns are 3×1 vectors.

2. True **3.** True **4.** True

5. True **6.** True **7.** True

8. False, the nonzero entry has to be the last entry.

9. False, consider the matrix in reduced row echelon form $\begin{bmatrix} 0 & 0 & 2 \\ 0 & 1 & 3 \\ 0 & 0 & 0 \end{bmatrix}$. The associated system has the unique solution $x_1 = 2, x_2 = 3$.

10. True **11.** True **12.** True

13. False, in $A = \begin{bmatrix} 1 & 2 \end{bmatrix}$, the columns are linearly dependent, but rank $A = 1$, which is the number of rows in A.

14. True **15.** True

16. False, the subset $\left\{ \begin{bmatrix} 1 \\ 2 \\ 3 \end{bmatrix}, \begin{bmatrix} 2 \\ 4 \\ 6 \end{bmatrix} \right\}$ of $\mathcal{R}^3$ is linearly dependent.

17. False, consider the example in Exercise 16.

19. (a) By Theorem 1.8, if A is an $m \times n$ matrix with rank n, then $A\mathbf{x} = \mathbf{b}$ has at most one solution for every $\mathbf{b}$ in $\mathcal{R}^m$.

(b) By Theorem 1.6, if A is an $m \times n$ matrix with rank m, then $A\mathbf{x} = \mathbf{b}$ has at least one solution for every $\mathbf{b}$ in $\mathcal{R}^m$.

21. $A + B = \begin{bmatrix} 3 & 2 \\ -2 & 7 \\ 4 & 3 \end{bmatrix}$

25. We have

$$A^T D^T = \begin{bmatrix} 1 & -2 & 0 \\ 3 & 4 & 2 \end{bmatrix} \begin{bmatrix} 1 \\ -1 \\ 2 \end{bmatrix}$$

$$= \begin{bmatrix} (1)(1) + (-2)(-1) + (0)(2) \\ (3)(1) + (4)(-1) + (2)(2) \end{bmatrix}$$

$$= \begin{bmatrix} 3 \\ 3 \end{bmatrix}.$$

29. The components are the average values of sales at all stores during January of last year for produce, meats, dairy, and processed foods, respectively.

33. We have

$$A_{-30°}\begin{bmatrix} 2 \\ -1 \end{bmatrix}$$

$$= \begin{bmatrix} \cos(-30°) & -\sin(-30°) \\ \sin(-30°) & \cos(-30°) \end{bmatrix} \begin{bmatrix} 2 \\ -1 \end{bmatrix}$$

$$= \begin{bmatrix} \frac{\sqrt{3}}{2} & \frac{1}{2} \\ -\frac{1}{2} & \frac{\sqrt{3}}{2} \end{bmatrix} \begin{bmatrix} 2 \\ -1 \end{bmatrix}$$

$$= \frac{1}{2}\begin{bmatrix} \sqrt{3} & 1 \\ -1 & \sqrt{3} \end{bmatrix} \begin{bmatrix} 2 \\ -1 \end{bmatrix}$$

$$= \frac{1}{2}\begin{bmatrix} (\sqrt{3})(2) + (1)(-1) \\ (-1)(2) + (\sqrt{3})(-1) \end{bmatrix}$$

$$= \frac{1}{2}\begin{bmatrix} 2\sqrt{3} - 1 \\ -2 - \sqrt{3} \end{bmatrix}.$$

37. Let A be the matrix whose columns are the vectors in S. Then $\mathbf{v}$ is a linear combination of the vectors in S if and only if the system $A\mathbf{x} = \mathbf{v}$ is consistent. The reduced row echelon form of the augmented matrix of this system is

$$\begin{bmatrix} 1 & 0 & -\frac{1}{2} & 0 \\ 0 & 1 & \frac{1}{2} & 0 \\ 0 & 0 & 0 & 1 \end{bmatrix}.$$

Because the third row has its only nonzero entry in the last column, this system is not consistent. So $\mathbf{v}$ is not a linear combination of the vectors in S.

41. The reduced row echelon form of the augmented matrix of the system is

$$\begin{bmatrix} 1 & 0 & -\frac{1}{3} & 0 \\ 0 & 1 & \frac{5}{3} & 0 \\ 0 & 0 & 0 & 1 \end{bmatrix}.$$

Because the third row has its only nonzero entry in the last column, the system is not consistent.

45. The reduced row echelon form is $[1 \ 2 \ -3 \ 0 \ 1]$, and so the rank is 1. Thus the nullity is $5 - 1 = 4$.

49. Let x_1, x_2, x_3, respectively, be the appropriate numbers of the three packs. We must solve the system

$$\begin{aligned} 10x_1 + 10x_2 + 5x_3 &= 500 \\ 10x_1 + 15x_2 + 10x_3 &= 750 \\ 10x_2 + 5x_3 &= 300, \end{aligned}$$

where the first equation represents the total number of oranges from the three packs, the second equation represents the total number of grapefruit from the three packs, and the third equation represents the total number of apples from the three packs. We obtain the solution $x_1 = 20$, $x_2 = 10$, $x_3 = 40$.

53. Let A be the matrix whose columns are the vectors in the given set. Then by Theorem 1.6, the set is a generating set for $\mathcal{R}^3$ if and only if rank $A = 3$. The reduced row echelon form of A is

$$\begin{bmatrix} 1 & 0 & \frac{2}{3} & \frac{1}{3} \\ 0 & 1 & \frac{1}{3} & -\frac{1}{3} \\ 0 & 0 & 0 & 0 \end{bmatrix}.$$

Therefore rank $A = 2$, and so the set is not a generating set for $\mathcal{R}^3$.

57. For an $m \times n$ matrix A, the system $A\mathbf{x} = \mathbf{b}$ is consistent for every vector $\mathbf{b}$ in $\mathcal{R}^m$ if and only if the rank of A equals m (Theorem 1.6). Since the reduced row echelon form of the given matrix is I_3, its rank equals 3, and so $A\mathbf{x} = \mathbf{b}$ is consistent for every vector $\mathbf{b}$ in $\mathcal{R}^3$.

61. Let A be the matrix whose columns are the vectors in the given set. By Theorem 1.8, the set is linearly independent if and only if rank $A = 3$. The reduced row echelon form of A is

$$\begin{bmatrix} 1 & 0 & 0 \\ 0 & 1 & 0 \\ 0 & 0 & 1 \\ 0 & 0 & 0 \end{bmatrix},$$

and hence rank $A = 3$. So the set is linearly independent.

65. Let A be the matrix whose columns are the vectors in $\mathcal{S}$. By Theorem 1.8, there exists a nonzero solution of $A\mathbf{x} = \mathbf{0}$. The general solution of this system is

$$\begin{aligned} x_1 &= -2x_3 \\ x_2 &= -x_3 \\ x_3 & \text{ free.} \end{aligned}$$

So one solution is $x_1 = -2$, $x_2 = -1$, $x_3 = 1$. Therefore

$$-2\begin{bmatrix} 1 \\ 2 \\ 3 \end{bmatrix} - \begin{bmatrix} 1 \\ -1 \\ 2 \end{bmatrix} + \begin{bmatrix} 3 \\ 3 \\ 8 \end{bmatrix} = \mathbf{0}.$$

Thus

$$\begin{bmatrix} 3 \\ 3 \\ 8 \end{bmatrix} = 2\begin{bmatrix} 1 \\ 2 \\ 3 \end{bmatrix} + \begin{bmatrix} 1 \\ -1 \\ 2 \end{bmatrix}.$$

69. The general solution of the system is

$$\begin{aligned} x_1 &= -3x_3 \\ x_2 &= 2x_3 \\ x_3 & \text{ free.} \end{aligned}$$

So

$$\begin{bmatrix} x_1 \\ x_2 \\ x_3 \end{bmatrix} = \begin{bmatrix} -3x_3 \\ 2x_3 \\ x_3 \end{bmatrix} = x_3 \begin{bmatrix} -3 \\ 2 \\ 1 \end{bmatrix}.$$

73. We prove the equivalent result: Suppose that $\mathbf{w}_1$ and $\mathbf{w}_2$ are linear combinations of vectors $\mathbf{v}_1$ and $\mathbf{v}_2$. If $\mathbf{v}_1$ and $\mathbf{v}_2$ are linearly dependent, then $\mathbf{w}_1$ and $\mathbf{w}_2$ are linearly dependent.

By assumption, one of $\mathbf{v}_1$ or $\mathbf{v}_2$ is a multiple of the other, say $\mathbf{v}_1 = k\mathbf{v}_2$ for some scalar k. Thus, for some scalars a_1, a_2, b_1, b_2, we have

$$\begin{aligned} \mathbf{w}_1 &= a_1\mathbf{v}_1 + a_2\mathbf{v}_2 \\ &= a_1 k\mathbf{v}_2 + a_2\mathbf{v}_2 = (a_1 k + a_2)\mathbf{v}_2 \end{aligned}$$

and

$$\begin{aligned} \mathbf{w}_2 &= b_1\mathbf{v}_1 + b_2\mathbf{v}_2 = b_1 k\mathbf{v}_2 + b_2\mathbf{v}_2 \\ &= (b_1 k + b_2)\mathbf{v}_2. \end{aligned}$$

Let $c_1 = a_1 k + a_2$ and $c_2 = b_1 k + b_2$. Then $\mathbf{w}_1 = c_1\mathbf{v}_2$ and $\mathbf{w}_2 = c_2\mathbf{v}_2$. If $\mathbf{w}_1 = \mathbf{0}$ or $\mathbf{w}_2 = \mathbf{0}$, then $\mathbf{w}_1$ and $\mathbf{w}_2$ are linearly dependent. Otherwise, $c_2 \neq 0$, and

$$\mathbf{w}_1 = c_1\mathbf{v}_2 = c_1\left(\frac{1}{c_2}\mathbf{w}_2\right) = \frac{c_1}{c_2}\mathbf{w}_2,$$

proving that $\mathbf{w}_1$ and $\mathbf{w}_2$ are linearly dependent.

CHAPTER 1 MATLAB EXERCISES

1. (a) $A \begin{bmatrix} 1.5 \\ -2.2 \\ 2.7 \\ 4 \end{bmatrix} = \begin{bmatrix} 3.38 \\ 8.86 \\ 16.11 \\ 32.32 \\ 15.13 \end{bmatrix}$

   (b) $A \begin{bmatrix} 2 \\ 2.1 \\ 0 \\ -1.1 \end{bmatrix} = \begin{bmatrix} 13.45 \\ -4.30 \\ -1.89 \\ 7.78 \\ 10.69 \end{bmatrix}$

   (c) $A \begin{bmatrix} 0 \\ 3.3 \\ 1.2 \\ -1 \end{bmatrix} = \begin{bmatrix} 20.18 \\ -11.79 \\ 7.71 \\ 8.52 \\ 0.28 \end{bmatrix}$

5. Answers are given correct to 4 places after the decimal point.

   (a) The reduced row echelon form of the augmented matrix $[A \ \mathbf{b}]$ does not contain the row $[0 \ 0 \ 0 \ 0 \ 0 \ 1]$. So the system is consistent, and its general solution is

   $\begin{bmatrix} x_1 \\ x_2 \\ x_3 \\ x_4 \\ x_5 \\ x_6 \end{bmatrix} = \begin{bmatrix} -8.2142 \\ -0.4003 \\ 0.0000 \\ 3.2727 \\ 0.0000 \\ 0.0000 \end{bmatrix} + x_3 \begin{bmatrix} -2 \\ -1 \\ 1 \\ 0 \\ 0 \\ 0 \end{bmatrix}$

   $+ x_5 \begin{bmatrix} -0.1569 \\ -0.8819 \\ 0.0000 \\ 0.2727 \\ 1.0000 \\ 0.0000 \end{bmatrix} + x_6 \begin{bmatrix} -9.2142 \\ 0.5997 \\ 0.0000 \\ 3.2727 \\ 0.0000 \\ 1.0000 \end{bmatrix}.$

   (b) The reduced row echelon form of the augmented matrix $[A \ \mathbf{b}]$ contains the row $[0 \ 0 \ 0 \ 0 \ 0 \ 1]$. So this system is inconsistent.

   (c) The reduced row echelon form of the augmented matrix $[A \ \mathbf{b}]$ does not contain the row $[0 \ 0 \ 0 \ 0 \ 0 \ 1]$. Thus the system is consistent, and its general solution is

   $\begin{bmatrix} x_1 \\ x_2 \\ x_3 \\ x_4 \\ x_5 \\ x_6 \end{bmatrix} = \begin{bmatrix} -9.0573 \\ 1.4815 \\ 0.0000 \\ 4.0000 \\ 0.0000 \\ 0.0000 \end{bmatrix} + x_3 \begin{bmatrix} -2 \\ -1 \\ 1 \\ 0 \\ 0 \\ 0 \end{bmatrix}$

   $+ x_5 \begin{bmatrix} -0.1569 \\ -0.8819 \\ 0.0000 \\ 0.2727 \\ 1.0000 \\ 0.0000 \end{bmatrix}$

   $+ x_6 \begin{bmatrix} -9.2142 \\ 0.5997 \\ 0.0000 \\ 3.2727 \\ 0.0000 \\ 1.0000 \end{bmatrix}.$

   (d) The reduced row echelon form of the augmented matrix $[A \ \mathbf{b}]$ contains the row $[0 \ 0 \ 0 \ 0 \ 0 \ 1]$. So this system is inconsistent.

Chapter 2

Matrices and Linear Transformations

2.1 MATRIX MULTIPLICATION

1. AB is defined and has size 2×2.

5. $C\mathbf{y} = 1\begin{bmatrix} 3 \\ 2 \end{bmatrix} + 3\begin{bmatrix} 8 \\ 0 \end{bmatrix} + (-5)\begin{bmatrix} 1 \\ 4 \end{bmatrix} = \begin{bmatrix} 22 \\ -18 \end{bmatrix}$

9. $AC\mathbf{x}$ is undefined since $AC\mathbf{x} = A(C\mathbf{x})$ and $C\mathbf{x}$ is undefined because C is a 2×3 matrix, $\mathbf{x}$ is a 2×1 matrix, and the number of columns in C does not equal the number of rows in $\mathbf{x}$.

13. The first column of BC is

$$\begin{bmatrix} 7 & 4 \\ 1 & 2 \end{bmatrix}\begin{bmatrix} 3 \\ 2 \end{bmatrix} = 3\begin{bmatrix} 7 \\ 1 \end{bmatrix} + 2\begin{bmatrix} 4 \\ 2 \end{bmatrix} = \begin{bmatrix} 29 \\ 7 \end{bmatrix}.$$

Similarly, the second column of BC is

$$\begin{bmatrix} 7 & 4 \\ 1 & 2 \end{bmatrix}\begin{bmatrix} 8 \\ 0 \end{bmatrix} = 8\begin{bmatrix} 7 \\ 1 \end{bmatrix} + 0\begin{bmatrix} 4 \\ 2 \end{bmatrix} = \begin{bmatrix} 56 \\ 8 \end{bmatrix},$$

and the third column of BC is

$$\begin{bmatrix} 7 & 4 \\ 1 & 2 \end{bmatrix}\begin{bmatrix} 1 \\ 4 \end{bmatrix} = 1\begin{bmatrix} 7 \\ 1 \end{bmatrix} + 4\begin{bmatrix} 4 \\ 2 \end{bmatrix} = \begin{bmatrix} 23 \\ 9 \end{bmatrix}.$$

So $BC = \begin{bmatrix} 29 & 56 & 23 \\ 7 & 8 & 9 \end{bmatrix}$.

17. $A^3 = A(AA)$

$$= A\left(\begin{bmatrix} 1 & -2 \\ 3 & 4 \end{bmatrix}\begin{bmatrix} 1 & -2 \\ 3 & 4 \end{bmatrix}\right)$$

$$= \begin{bmatrix} 1 & -2 \\ 3 & 4 \end{bmatrix}\begin{bmatrix} -5 & -10 \\ 15 & 10 \end{bmatrix}$$

$$= \begin{bmatrix} -35 & -30 \\ 45 & 10 \end{bmatrix}$$

19. $C^2 = CC$ is undefined because C is a 2×3 matrix and two 2×3 matrices cannot be multiplied because the number of columns of C does not equal the number of rows of C.

23. Since

$$AB = \begin{bmatrix} 1 & -2 \\ 3 & 4 \end{bmatrix}\begin{bmatrix} 7 & 4 \\ 1 & 2 \end{bmatrix} = \begin{bmatrix} 5 & 0 \\ 25 & 20 \end{bmatrix},$$

we have $(AB)^T = \begin{bmatrix} 5 & 25 \\ 0 & 20 \end{bmatrix}$. Also

$$B^T A^T = \begin{bmatrix} 7 & 1 \\ 4 & 2 \end{bmatrix}\begin{bmatrix} 1 & 3 \\ -2 & 4 \end{bmatrix} = \begin{bmatrix} 5 & 25 \\ 0 & 20 \end{bmatrix}.$$

27. By the row-column rule, the (2, 3)-entry of CA equals the sum of the products of the corresponding entries from row 2 of C and column 3 of A, which is

$$4(3) + 3(4) + (-2)(0) = 24.$$

31. Column 1 of CA equals the matrix-vector product of C and the first column of A, which is

$$\begin{bmatrix} 2 & 1 & -1 \\ 4 & 3 & -2 \end{bmatrix} \begin{bmatrix} 1 \\ 2 \\ -3 \end{bmatrix}$$

$$= 1 \begin{bmatrix} 2 \\ 4 \end{bmatrix} + 2 \begin{bmatrix} 1 \\ 3 \end{bmatrix} + (-3) \begin{bmatrix} -1 \\ -2 \end{bmatrix}$$

$$= \begin{bmatrix} 7 \\ 16 \end{bmatrix}.$$

33. False, the product is not defined unless $n = m$.

34. False, if A is a 2×3 matrix and B is a 3×4 matrix.

35. False, see Example 5.

36. True

37. False, if A is a 2×3 matrix and B is a 3×2 matrix.

38. False, $(AB)^T = B^T A^T$.

39. True   40. True

41. False, see the box titled "Row-Column Rule for the (i,j)-Entry of a Matrix Product."

42. False, it is the sum of the products of corresponding entries from the ith row of A and the jth column of B.

43. True

44. False, $(A+B)C = AC + BC$.

45. False. If $A = B = \begin{bmatrix} 0 & 1 \\ 1 & 0 \end{bmatrix}$, then $AB = \begin{bmatrix} 1 & 0 \\ 0 & 1 \end{bmatrix}$.

46. True

47. False, let $A = \begin{bmatrix} 1 & 0 \\ 0 & 0 \end{bmatrix}$ and $B = \begin{bmatrix} 0 & 0 \\ 1 & 0 \end{bmatrix}$.

48. True   49. True   50. True

51. (a) The number of people living in single-unit houses is $.70v_1 + .95v_2$. Similarly, the number of people living in multiple-unit housing is $.30v_1 + .05v_2$. These results may be expressed as the matrix equation

$$\begin{bmatrix} .70 & .95 \\ .30 & .05 \end{bmatrix} \begin{bmatrix} v_1 \\ v_2 \end{bmatrix} = \begin{bmatrix} u_1 \\ u_2 \end{bmatrix}.$$

So take

$$B = \begin{bmatrix} .70 & .95 \\ .30 & .05 \end{bmatrix}.$$

(b) Because $A \begin{bmatrix} v_1 \\ v_2 \end{bmatrix}$ represents the number of people living in the city and suburbs after one year, it follows from (a) that $BA \begin{bmatrix} v_1 \\ v_2 \end{bmatrix}$ gives the number of people living in single- and multiple-unit housing after one year.

55. We prove that $C(P+Q) = CP + CQ$. Note that $P+Q$ is an $n \times p$ matrix, and so $C(P+Q)$ is an $m \times p$ matrix. Also CP and CQ are both $m \times p$ matrices; so $CP + CQ$ is an $m \times p$ matrix. Hence the matrices on both sides of the equation have the same size. The jth column of $P+Q$ is $\mathbf{p}_j + \mathbf{q}_j$; so the jth column of $C(P+Q)$ is $C(\mathbf{p}_j + \mathbf{q}_j)$, which equals $C\mathbf{p}_j + C\mathbf{q}_j$ by Theorem 1.3(c). On the other hand, the jth columns of CP and CQ are $C\mathbf{p}_j$ and $C\mathbf{q}_j$, respectively. So the jth column of $CP + CQ$ equals $C\mathbf{p}_j + C\mathbf{q}_j$. Therefore $C(P+Q)$ and $CP + CQ$ have the same corresponding columns, and hence are equal.

59. By the row-column rule, the (i,j)-entry of AB is

$$a_{i1}b_{1j} + a_{i2}b_{2j} + \cdots + a_{in}b_{nj}.$$

63. Let $A = \begin{bmatrix} 1 & 0 \\ 0 & 0 \end{bmatrix}$ and $B = \begin{bmatrix} 0 & 0 \\ 1 & 0 \end{bmatrix}$. Then $AB = O$, but
$$BA = \begin{bmatrix} 0 & 0 \\ 1 & 0 \end{bmatrix} \neq O.$$

67. Using (b) and (g) of Theorem 2.1, we have
$$(ABC)^T = ((AB)C)^T = C^T(AB)^T$$
$$= C^T(B^T A^T) = C^T B^T A^T.$$

71. (b) After 20 years, the populations are given by
$$A^{20}\mathbf{p} = \begin{bmatrix} 205.668 \\ 994.332 \end{bmatrix}.$$
So the population of the city will be 205,668, and the population of the suburbs will be 994,332.
(c) After 50 years, the populations are given by
$$A^{50}\mathbf{p} = \begin{bmatrix} 200.015 \\ 999.985 \end{bmatrix}.$$
So the population of the city will be 200,015, and the population of the suburbs will be 999,985.
(d) As in (b) and (c), the populations after 100 years will be 200,000 in the city and 1,000,000 in the suburbs. Moreover, these numbers do not appear to change thereafter. Thus we conjecture that the population in the city will eventually be 200,000, and the population in the suburbs will eventually be 1,000,000.

2.2 APPLICATIONS OF MATRIX MULTIPLICATION

1. False, the population may be decreasing.
2. False, the population may continue to grow without bound.
3. False, this is only the case for $i = 1$.
4. True
5. True
6. False, $\mathbf{z} = BA\mathbf{x}$.
7. True
8. False, a $(0, 1)$-matrix need not be square, see Example 2.
9. False, a $(0, 1)$-matrix need not be symmetric, see Example 2.
13. (a) Using the notation on page 108, we have $p_1 = q$ and $p_2 = .5$. Also, $b_1 = 0$ because females under age 1 do not give birth. Likewise, $b_2 = 2$ and $b_3 = 1$. So the Leslie matrix is
$$\begin{bmatrix} b_1 & b_2 & b_3 \\ p_1 & 0 & 0 \\ 0 & p_2 & 0 \end{bmatrix} = \begin{bmatrix} 0 & 2 & 1 \\ q & 0 & 0 \\ 0 & .5 & 0 \end{bmatrix}.$$
(b) If $q = .8$, then
$$A = \begin{bmatrix} 0 & 2 & 1 \\ .8 & 0 & 0 \\ 0 & .5 & 0 \end{bmatrix}$$
and
$$\mathbf{x}_0 = \begin{bmatrix} 300 \\ 1180 \\ 130 \end{bmatrix}.$$

The population in 50 years is given by

$$A^{50}\mathbf{x}_0 \approx 10^9 \begin{bmatrix} 9.28 \\ 5.40 \\ 1.96 \end{bmatrix}.$$

So the population appears to grow without bound.

(c) If $q = .2$, then

$$A = \begin{bmatrix} 0 & 2 & 1 \\ .2 & 0 & 0 \\ 0 & .5 & 0 \end{bmatrix}.$$

As in (b), we compute

$$A^{50}\mathbf{x}_0 \approx 10^{-3} \begin{bmatrix} .3697 \\ 1.009 \\ .0689 \end{bmatrix}$$

and conclude that the population appears to approach zero.

(d) $q = .4$. The stable distribution is

$$\begin{bmatrix} 400 \\ 160 \\ 80 \end{bmatrix}.$$

(e) For $q = .4$ and $\mathbf{x}_0 = \begin{bmatrix} 210 \\ 240 \\ 180 \end{bmatrix}$, the respective vectors $A^5\mathbf{x}_0$, $A^{10}\mathbf{x}_0$, and $A^{30}\mathbf{x}_0$ equal

$$\begin{bmatrix} 513.60 \\ 144.96 \\ 114.00 \end{bmatrix}, \begin{bmatrix} 437.36 \\ 189.99 \\ 85.17 \end{bmatrix},$$

and

$$\begin{bmatrix} 499.98 \\ 180.01 \\ 89.99 \end{bmatrix}.$$

It appears that the populations approach

$$\begin{bmatrix} 450 \\ 180 \\ 90 \end{bmatrix}.$$

(f) The rank of $A - I_3$ must be less than 3 by Theorem 1.8. We can use elementary row operations to transform the matrix

$$A - I_3 = \begin{bmatrix} 0 & 2 & 1 \\ q & 0 & 0 \\ 0 & .5 & 0 \end{bmatrix}$$

to

$$\begin{bmatrix} 1 & -2 & -1 \\ 0 & 1 & -2 \\ 0 & 0 & -2(1-2q)+q \end{bmatrix}.$$

For rank $(A - I_3) < 3$, we need

$$-2(1-2q) + q = 0,$$

that is, $q = .4$. This is the value obtained in (d).

(g) For $q = .4$, the solution of the equation $(A - I_3)\mathbf{x} = \mathbf{0}$ is

$$\begin{bmatrix} x_1 \\ x_2 \\ x_3 \end{bmatrix} = x_3 \begin{bmatrix} 5 \\ 2 \\ 1 \end{bmatrix}.$$

For $x_3 = 90$, we obtain the solution in (e).

17. Let p and q be the amounts of donations and interest received by the foundation, and let n and a be the net income and fund raising costs, respectively. Then

$$n = .7p + .9q$$
$$a = .3p + .1q,$$

and hence

$$\begin{bmatrix} n \\ a \end{bmatrix} = \begin{bmatrix} .7 & .9 \\ .3 & .1 \end{bmatrix} \begin{bmatrix} p \\ q \end{bmatrix}.$$

Next, let r and c be the amounts of net income used for research and clinic maintenance, respectively. Then

$$r = .4n$$
$$c = .6n$$
$$a = a,$$

and hence

$$\begin{bmatrix} r \\ c \\ a \end{bmatrix} = \begin{bmatrix} .4 & 0 \\ .6 & 0 \\ 0 & 1 \end{bmatrix} \begin{bmatrix} n \\ a \end{bmatrix}.$$

Finally, let m and f be the material and personnel costs of the foundation, respectively. Then

$$m = .8r + .5c + .7a$$
$$f = .2r + .5c + .3a,$$

and hence

$$\begin{bmatrix} m \\ f \end{bmatrix} = \begin{bmatrix} .8 & .5 & .7 \\ .2 & .5 & .4 \end{bmatrix} \begin{bmatrix} r \\ c \\ a \end{bmatrix}.$$

Combining these matrix equations, we have

$$\begin{bmatrix} m \\ f \end{bmatrix} = \begin{bmatrix} .8 & .5 & .7 \\ .2 & .5 & .4 \end{bmatrix} \begin{bmatrix} .4 & 0 \\ .6 & 0 \\ 0 & 1 \end{bmatrix} \begin{bmatrix} n \\ a \end{bmatrix}$$

$$= \begin{bmatrix} .8 & .5 & .7 \\ .2 & .5 & .4 \end{bmatrix} \begin{bmatrix} .4 & 0 \\ .6 & 0 \\ 0 & 1 \end{bmatrix} \begin{bmatrix} .7 & .9 \\ .3 & .1 \end{bmatrix} \begin{bmatrix} p \\ q \end{bmatrix}$$

$$= \begin{bmatrix} 0.644 & 0.628 \\ 0.356 & 0.372 \end{bmatrix} \begin{bmatrix} p \\ q \end{bmatrix}.$$

21. (a) We need only find entries a_{ij} such that $a_{ij} = a_{ji} = 1$. The friends are 1 and 2, 1 and 4, 2 and 3, and 3 and 4.

(b) The (i,j)-entry of A^2 is

$$a_{i1}a_{1j} + a_{i2}a_{2j} + a_{i3}a_{3j} + a_{i4}a_{4j}.$$

The kth term equals 1 if and only if $a_{ik} = 1$ and $a_{kj} = 1$, that is, person i likes person k and person k likes person j. Otherwise, the term is 0. So the (i,j)-entry of A^2 equals the number of people who like person j and are liked by person i.

(c) We have

$$B = \begin{bmatrix} 0 & 1 & 0 & 1 \\ 1 & 0 & 1 & 0 \\ 0 & 1 & 0 & 1 \\ 1 & 0 & 1 & 0 \end{bmatrix},$$

which is symmetric because $B = B^T$.

(d) Because $B^3 = B^2 B$, the (i,i)-entry of B^3 equals a sum of terms of the form $c_{ik}b_{ki}$, where c_{ik} equals a sum of terms of the form $b_{ij}b_{jk}$. Therefore the (i,i)-entry of B^3 consists of terms of the form $b_{ij}b_{jk}b_{ki}$. The (i,i)-entry of B^3 is positive if and only if some term $b_{ij}b_{jk}b_{ki}$ is positive. This occurs if and only if $b_{ij} = 1 = b_{jk} = b_{ki}$, that is, there are friends k and j who are also friends of person i, that is, person i is in a clique.

(e) We have

$$B^3 = \begin{bmatrix} 0 & 4 & 0 & 4 \\ 4 & 0 & 4 & 0 \\ 0 & 4 & 0 & 4 \\ 4 & 0 & 4 & 0 \end{bmatrix};$$

so the (i,i)-entry is 0 for every i. Therefore there are no cliques.

25. (a) Using the notation in the example, we have

$$x_0 = \begin{bmatrix} 100 \\ 200 \\ 300 \\ 8000 \end{bmatrix}.$$

Using the equation $x_k = Ax_{k-1}$, we obtain the following table.

k	Sun	Noble	Hon.	MMQ
1	100	300	500	7700
2	100	400	800	7300
3	100	500	1200	6800

(b)

k	Sun	Noble	Hon.	MMQ
9	100	1100	5700	1700
10	100	1200	6800	500
11	100	1300	8000	-800

(c) The tribe will cease to exist because every member is required to marry a member of the MMQ, and the number of members of the MMQ decreases to zero.

(d) We must find k such that

$$s_k + n_k + h_k > m_k,$$

that is, there are enough members of the MMQ for the other classes to marry. From equation (6), this inequality is equivalent to

$$s_0 + (n_0 + ks_0)$$
$$+ \left(h_0 + kn_0 + \frac{k(k-1)}{2}s_0\right)$$
$$> m_0 - kn_0 - \frac{k(k+1)}{2}s_0.$$

Let $s_0 = 100$, $n_0 = 200$, $k_0 = 300$, and $m_0 = 8000$ in the preceding inequality. By simplifying the result, we obtain

$$k^2 + 5k - 74 > 0.$$

The smallest value of k that satisfies this inequality is $k = 7$.

2.3 INVERTIBILITY AND ELEMENTARY MATRICES

1. No, we must have $AB = BA = I_n$. In this case $AB \neq I_n$.

5. Since $AB = BA = I_n$, we have $B = A^{-1}$.

9. $(A^T)^{-1} = (A^{-1})^T = \begin{bmatrix} 1 & 2 & 1 \\ 2 & 0 & 1 \\ 3 & 1 & -1 \end{bmatrix}$

13. By Theorem 2.2, we have

$$(AB^T)^{-1} = (B^T)^{-1}A^{-1}$$
$$= (B^{-1})^T A^{-1}$$
$$= \begin{bmatrix} 5 & 7 & 3 \\ -3 & -4 & -1 \\ 12 & 7 & 12 \end{bmatrix}.$$

17. The given matrix is obtained from I_3 by adding -2 times row 1 to row 2. So adding 2 times row 1 to row 2 transforms the given matrix into I_3. Performing this elementary operation on I_3 produces $\begin{bmatrix} 1 & 0 & 0 \\ 2 & 1 & 0 \\ 0 & 0 & 1 \end{bmatrix}$, which is the inverse of the given matrix.

21. The given matrix is obtained from I_4 by interchanging rows 2 and 4. Interchanging these rows again transforms the given matrix into I_4. So the given matrix is its own inverse.

25. Matrix B is obtained by interchanging rows 1 and 2 of A. Performing this operation on I_2 produces the desired elementary matrix $\begin{bmatrix} 0 & 1 \\ 1 & 0 \end{bmatrix}$.

29. Since B is obtained from A by adding -5 times row 2 to row 3, the desired elementary matrix is $\begin{bmatrix} 1 & 0 & 0 \\ 0 & 1 & 0 \\ 0 & -5 & 1 \end{bmatrix}$, as in Exercise 17.

33. False, the $n \times n$ zero matrix is not invertible.

34. True   35. True

36. False, let

$$A = \begin{bmatrix} 1 & 0 & 0 \\ 0 & 1 & 0 \end{bmatrix} \text{ and } B = \begin{bmatrix} 1 & 0 \\ 0 & 1 \\ 0 & 0 \end{bmatrix}.$$

Then $AB = I_2$, but neither A nor B is square; so neither is invertible.

37. True   38. True

39. False, see the comment below the definition of *inverse*.

40. True   41. True   42. True

43. False, $(AB)^{-1} = B^{-1}A^{-1}$.

44. False, an elementary matrix is a matrix that can be obtained by performing one elementary row operation on an identity matrix.

45. True

46. False, $\begin{bmatrix} 2 & 0 \\ 0 & 1 \end{bmatrix}$ and $\begin{bmatrix} 1 & 0 \\ 0 & 3 \end{bmatrix}$ are elementary matrices, but

$$\begin{bmatrix} 2 & 0 \\ 0 & 1 \end{bmatrix} \begin{bmatrix} 1 & 0 \\ 0 & 3 \end{bmatrix} = \begin{bmatrix} 2 & 0 \\ 0 & 3 \end{bmatrix},$$

which is not an elementary matrix.

47. True   48. True   49. True

50. True

51. False, see Theorem 2.4(a).

52. True

55. By the boxed result on page 127, every elementary matrix is invertible. By the boxed result on page 125, the product of invertible matrices is invertible. Hence the product of elementary matrices is invertible.

59. Using Theorem 2.1(b) and Theorem 2.2(b), we have

$$\begin{aligned}(ABC)^{-1} &= [(AB)C]^{-1} \\ &= C^{-1}(AB)^{-1} \\ &= C^{-1}(B^{-1}A^{-1}) \\ &= C^{-1}B^{-1}A^{-1}.\end{aligned}$$

63. By Theorem 1.6, $B\mathbf{x} = \mathbf{b}$ has a solution for every $\mathbf{b}$ in $\mathcal{R}^n$. So, for every standard vector $\mathbf{e}_i$, there is a vector $\mathbf{u}_i$ that satisfies $B\mathbf{u}_i = \mathbf{e}_i$. Let

$$C = [\mathbf{u}_1 \ \mathbf{u}_2 \ \cdots \ \mathbf{u}_n].$$

Then

$$\begin{aligned}BC &= [B\mathbf{u}_1 \ B\mathbf{u}_2 \ \cdots \ B\mathbf{u}_n] \\ &= [\mathbf{e}_1 \ \mathbf{e}_2 \ \cdots \ \mathbf{e}_n] \\ &= I_n.\end{aligned}$$

67. From the column correspondence property, it follows that the third column of A equals

$$\mathbf{a}_1 + 2\mathbf{a}_2 = \begin{bmatrix} 3 \\ -1 \end{bmatrix} + 2\begin{bmatrix} 2 \\ 5 \end{bmatrix} = \begin{bmatrix} 7 \\ 9 \end{bmatrix}.$$

Therefore $A = \begin{bmatrix} 3 & 2 & 7 \\ -1 & 5 & 9 \end{bmatrix}$.

71. Because $\mathbf{r}_2 = 2\mathbf{r}_1$, we have $\mathbf{r}_1 = \frac{1}{2}\mathbf{r}_2$. Thus, by the column correspondence property, we have

$$\begin{aligned}\mathbf{a}_1 &= \frac{1}{2}\mathbf{a}_2 \\ &= \frac{1}{2}\begin{bmatrix} 2 \\ 4 \end{bmatrix} = \begin{bmatrix} 1 \\ 2 \end{bmatrix}.\end{aligned}$$

Also, $\mathbf{r}_4 = 4\mathbf{r}_1 + 3\mathbf{r}_3$. So

$$\begin{aligned}\mathbf{a}_4 &= 4\mathbf{a}_1 + 3\mathbf{a}_3 \\ &= 4\begin{bmatrix} 1 \\ 2 \end{bmatrix} + 3\begin{bmatrix} 3 \\ 5 \end{bmatrix} = \begin{bmatrix} 13 \\ 23 \end{bmatrix}.\end{aligned}$$

Thus $A = \begin{bmatrix} 1 & 2 & 3 & 13 \\ 2 & 4 & 5 & 23 \end{bmatrix}$.

75. The reduced row echelon form of A is
$$R = \begin{bmatrix} 1 & -2 & 0 & 2 & 3 \\ 0 & 0 & 1 & -3 & -5 \\ 0 & 0 & 0 & 0 & 0 \end{bmatrix}.$$

So columns 1 and 3 of A are its pivot columns. By the column correspondence property, the entries of the second column of R tell us how to write the second column of A as a linear combination of its pivot columns:
$$\mathbf{a}_2 = (-2)\mathbf{a}_1 + 0\mathbf{a}_3.$$

79. The reduced row echelon form of B is
$$R = \begin{bmatrix} 1 & 0 & 1 & -3 & 0 & 3 \\ 0 & 1 & -1 & 2 & 0 & -2 \\ 0 & 0 & 0 & 0 & 1 & -1 \\ 0 & 0 & 0 & 0 & 0 & 0 \end{bmatrix}$$

Proceeding as in the solution to Exercise 75, we have
$$\mathbf{b}_3 = 1\mathbf{b}_1 + (-1)\mathbf{b}_2 + 0\mathbf{b}_5.$$

83. Let R be the reduced row echelon form of A. Because $\mathbf{u}$ and $\mathbf{v}$ are linearly independent, $\mathbf{a}_1 = \mathbf{u} \neq \mathbf{0}$, and hence $\mathbf{a}_1$ is a pivot column. Thus $\mathbf{r}_1 = \mathbf{e}_1$. Since $\mathbf{a}_2 = 2\mathbf{u} = 2\mathbf{a}_1$, it follows that $\mathbf{r}_2 = 2\mathbf{r}_1 = 2\mathbf{e}_1$ by the column correspondence property. Since $\mathbf{u}$ and $\mathbf{v}$ are linearly independent, it is easy to show that $\mathbf{u}$ and $\mathbf{u} + \mathbf{v}$ are linearly independent, and hence $\mathbf{a}_3$ is not a linear combination of $\mathbf{a}_1$ and $\mathbf{a}_2$. Thus $\mathbf{a}_3$ is a pivot column, and so $\mathbf{r}_3 = \mathbf{e}_2$. Finally,
$$\mathbf{a}_4 = \mathbf{a}_3 - \mathbf{u} = \mathbf{a}_3 - \mathbf{a}_1,$$

and hence $\mathbf{r}_4 = \mathbf{r}_3 - \mathbf{r}_1$ by the column correspondence property. Therefore
$$R = \begin{bmatrix} 1 & 2 & 0 & -1 \\ 0 & 0 & 1 & 1 \\ 0 & 0 & 0 & 0 \end{bmatrix}.$$

87. (a) Note that, in the form described for R^T, the first r rows are the transposes of the standard vectors of $\mathcal{R}^m$, and the remaining rows are zero rows. As we learned on page 48, the standard vectors $\mathbf{e}'_1, \mathbf{e}'_2, \ldots, \mathbf{e}'_r$ of $\mathcal{R}^m$ must appear among the columns of R. Thus their transposes occur among the rows of R^T. By Theorem 2.4(b), every nonpivot column of R is a linear combination of $\mathbf{e}'_1, \mathbf{e}'_2, \ldots, \mathbf{e}'_r$. Thus, by appropriate row addition operations, the rows of R^T that correspond to the nonpivot columns of R can be changed to zero rows. Finally, by appropriate row interchanges, the first r rows of R^T can be changed to the transposes of $\mathbf{e}'_1, \mathbf{e}'_2, \ldots, \mathbf{e}'_r$. This is the form described for R^T.

(b) The reduced row echelon form of R^T given in (a) has r nonzero rows. Hence
$$\text{rank } R^T = r = \text{rank } R.$$

91. Let
$$A = \begin{bmatrix} a & b & c \\ p & q & r \end{bmatrix}.$$
We first prove the result for the operation of interchanging rows 1 and 2 of A. In this case, performing this operation on I_2 yields
$$E = \begin{bmatrix} 0 & 1 \\ 1 & 0 \end{bmatrix},$$

and

$$EA = \begin{bmatrix} 0 & 1 \\ 1 & 0 \end{bmatrix} \begin{bmatrix} a & b & c \\ p & q & r \end{bmatrix} = \begin{bmatrix} p & q & r \\ a & b & c \end{bmatrix},$$

which is the result of interchanging rows 1 and 2 of A.

Next, we prove the result for the operation of multiplying row 1 of A by the nonzero scalar k. In this case, performing this operation on I_2 yields

$$E = \begin{bmatrix} k & 0 \\ 0 & 1 \end{bmatrix},$$

and

$$EA = \begin{bmatrix} k & 0 \\ 0 & 1 \end{bmatrix} \begin{bmatrix} a & b & c \\ p & q & r \end{bmatrix}$$
$$= \begin{bmatrix} ka & kb & kc \\ p & q & r \end{bmatrix},$$

which is the result of multiplying row 1 of A by the k. The proof for the operation of multiplying row 2 of A by k is similar.

Finally, we prove the result for the operation of adding k times the second row of A to the first. In this case,

$$E = \begin{bmatrix} 1 & k \\ 0 & 1 \end{bmatrix}.$$

Then

$$EA = \begin{bmatrix} 1 & k \\ 0 & 1 \end{bmatrix} \begin{bmatrix} a & b & c \\ p & q & r \end{bmatrix}$$
$$= \begin{bmatrix} a+kp & b+kq & c+kr \\ p & q & r \end{bmatrix},$$

which is the result of adding k times the second row of A to the first. The proof for the operation of adding k times the first row to the second is similar.

95. (a) The reduced row echelon form of A is I_4; so A is invertible. In fact,

$$A^{-1} = \begin{bmatrix} -7 & 2 & 3 & -2 \\ 5 & -1 & -2 & 1 \\ 1 & 0 & 0 & 1 \\ -3 & 1 & 1 & -1 \end{bmatrix}.$$

(b) As in (a), both B and C are invertible, and

$$B^{-1} = \begin{bmatrix} 3 & 2 & -7 & -2 \\ -2 & -1 & 5 & 1 \\ 0 & 0 & 1 & 1 \\ 1 & 1 & -3 & -1 \end{bmatrix},$$

and

$$C^{-1} = \begin{bmatrix} -7 & -2 & 3 & 2 \\ 5 & 1 & -2 & -1 \\ 1 & 1 & 0 & 0 \\ -3 & -1 & 1 & 1 \end{bmatrix}.$$

(c) B^{-1} can be obtained by interchanging columns 1 and 3 of A^{-1}, and C^{-1} can be obtained by interchanging columns 2 and 4 of A^{-1}.

(d) B^{-1} can be obtained by interchanging columns i and j of A^{-1}.

(e) Let E be the elementary matrix that corresponds to interchanging rows i and j. Then $B = EA$. So by Theorem 2.2(b), we have $B^{-1} = (EA)^{-1} = A^{-1}E^{-1}$. It follows from Exercise 94 that B^{-1} is obtained from A^{-1} by performing the elementary column operation on A^{-1} that is associated with E^{-1}. But because E is associated with a row interchange, we have $E^2 = I_n$, and hence $E^{-1} = E$.

99. Observe that

$$A^{-1} = \begin{bmatrix} 10 & -2 & -1 & -1 \\ -6 & 1 & -1 & 2 \\ -2 & 0 & 1 & 0 \\ 1 & 0 & 1 & -1 \end{bmatrix}.$$

(a) The solution of $A\mathbf{x} = \mathbf{e}_1$ is $\begin{bmatrix} 10 \\ -6 \\ -2 \\ 1 \end{bmatrix}$.

(b) The solutions of $A\mathbf{x} = \mathbf{e}_i$ are the corresponding columns of A^{-1}.

2.4 THE INVERSE OF A MATRIX

1. We use the algorithm for matrix inversion to determine if A is invertible. First, form the matrix

$$\begin{bmatrix} 1 & 3 & | & 1 & 0 \\ 1 & 2 & | & 0 & 1 \end{bmatrix}.$$

Its reduced row echelon form is

$$\begin{bmatrix} 1 & 0 & | & -2 & 3 \\ 0 & 1 & | & 1 & -1 \end{bmatrix}.$$

Thus the algorithm implies that A is invertible and

$$A^{-1} = \begin{bmatrix} -2 & 3 \\ 1 & -1 \end{bmatrix}.$$

5. As in Exercise 1, the matrix is invertible, and its inverse is $\begin{bmatrix} 5 & -3 \\ -3 & 2 \end{bmatrix}$.

9. As in Exercise 1, the matrix is invertible, and its inverse is

$$\frac{1}{3}\begin{bmatrix} -7 & 2 & 3 \\ -6 & 0 & 3 \\ 8 & -1 & -3 \end{bmatrix}.$$

13. As in Exercise 1, the matrix is invertible, and its inverse is $\begin{bmatrix} -1 & -5 & 3 \\ 1 & 2 & -1 \\ 1 & 4 & -2 \end{bmatrix}.$

17. As in Exercise 1, the matrix is invertible, and its inverse is

$$\frac{1}{3}\begin{bmatrix} 1 & 1 & 1 & -2 \\ 1 & 1 & -2 & 1 \\ 1 & -2 & 1 & 1 \\ -2 & 1 & 1 & 1 \end{bmatrix}.$$

21. Using the algorithm for computing $A^{-1}B$, we form the matrix

$$\begin{bmatrix} 2 & 2 & | & 2 & 4 & 2 & 6 \\ 2 & 1 & | & 0 & -2 & 8 & -4 \end{bmatrix}$$

and compute its reduced row echelon form $[I_2 \ A^{-1}B]$, which equals

$$\begin{bmatrix} 1 & 0 & | & -1 & 4 & 7 & -7 \\ 0 & 1 & | & 2 & 6 & -6 & 10 \end{bmatrix}.$$

Thus $A^{-1}B = \begin{bmatrix} -1 & 4 & 7 & -7 \\ 2 & 6 & -6 & 10 \end{bmatrix}.$

25. As in Exercise 21, we obtain

$$A^{-1}B = \begin{bmatrix} -5 & -1 & -6 \\ -1 & 1 & 0 \\ 4 & 1 & 3 \\ 3 & 1 & 2 \end{bmatrix}.$$

29. $R = \begin{bmatrix} 1 & 0 & -2 & -1 \\ 0 & 1 & 1 & -1 \\ 0 & 0 & 0 & 0 \end{bmatrix}$

By the discussion on page 136, the reduced row echelon form of $[A \ I_n]$ is $[R \ P]$, where $PA = R$. Applying this procedure, we obtain

$$P = \begin{bmatrix} -1 & 0 & 0 \\ 0 & 1 & 0 \\ 2 & -3 & 1 \end{bmatrix}.$$

35. True

36. False, let

$$A = \begin{bmatrix} 1 & 0 & 0 \\ 0 & 1 & 0 \end{bmatrix} \text{ and } B = \begin{bmatrix} 1 & 0 \\ 0 & 1 \\ 0 & 0 \end{bmatrix}.$$

Then $AB = I_2$, but A is not square; so it is not invertible.

37. True **38.** True **39.** True

40. True **41.** True **42.** True

43. True **44.** True **45.** True

46. True **47.** True

48. False, if $A = I_2$ and $B = -I_2$, then $A + B = O$, which is not invertible.

49. True

50. False, $C = A^{-1}B$.

51. False, if $A = O$, then A^{-1} does not exist.

52. True **53.** True **54.** True

55. Let A be an $n \times n$ invertible matrix.

To prove that (a) implies (e), consider the system $A\mathbf{x} = \mathbf{b}$, where $\mathbf{b}$ is in $\mathcal{R}^n$. If we let $\mathbf{u} = A^{-1}\mathbf{b}$, then

$$A\mathbf{u} = A(A^{-1}\mathbf{b})$$
$$= (AA^{-1})\mathbf{b} = I_n\mathbf{b} = \mathbf{b},$$

and so $\mathbf{u}$ is a solution, that is, the system is consistent.

To prove that (a) implies (h), suppose that $\mathbf{u}$ is a solution of $A\mathbf{x} = \mathbf{0}$, that is, $A\mathbf{u} = \mathbf{0}$. Then

$$\mathbf{u} = (A^{-1}A)\mathbf{u}$$
$$= A^{-1}(A\mathbf{u}) = A^{-1}\mathbf{0} = \mathbf{0}.$$

59. (a) The matrix equation is

$$\begin{bmatrix} -1 & 0 & 1 \\ 1 & 2 & -2 \\ 2 & -1 & 1 \end{bmatrix} \begin{bmatrix} x_1 \\ x_2 \\ x_3 \end{bmatrix} = \begin{bmatrix} -4 \\ 3 \\ 1 \end{bmatrix}.$$

(b) When we use the algorithm for matrix inversion, we find that the reduced row echelon form of $[A \ I_3]$ is $[I_3 \ A^{-1}]$, where

$$A^{-1} = \frac{1}{5}\begin{bmatrix} 0 & 1 & 2 \\ 5 & 3 & 1 \\ 5 & 1 & 2 \end{bmatrix}.$$

(c) The solution is $A^{-1}\mathbf{b} = \begin{bmatrix} 1 \\ -2 \\ -3 \end{bmatrix}$.

63. (a) The matrix equation is

$$\begin{bmatrix} 1 & -2 & -1 & 1 \\ 1 & 1 & 0 & -1 \\ -1 & -1 & 1 & 1 \\ 3 & 1 & 2 & 0 \end{bmatrix} \begin{bmatrix} x_1 \\ x_2 \\ x_3 \\ x_4 \end{bmatrix}$$

$$= \begin{bmatrix} 4 \\ -2 \\ 1 \\ -1 \end{bmatrix}.$$

(b) Using the same reasoning as in Exercise 59(b), we have

$$A^{-1} = \begin{bmatrix} -1 & 0 & 1 & -1 \\ -3 & -2 & 1 & -2 \\ 0 & 1 & 1 & 0 \\ -4 & -3 & 2 & -3 \end{bmatrix}.$$

(c) The solution is $A^{-1}\mathbf{b} = \begin{bmatrix} -2 \\ -5 \\ -1 \\ -5 \end{bmatrix}$.

67. (a) If $k = 1$, then $A = I_n$, and the result is clear. If $k > 1$, rewrite the equation $A^k = I_n$ as $A(A^{k-1}) = I_n$. Then A is invertible by (j) of the Invertible Matrix Theorem.
(b) From (a), we have $A^{-1} = A^{k-1}$.

71. By Theorem 2.3, there exists an invertible matrix P such that $PA = R$, where R is the reduced row echelon form of A. Then by Theorem 2.2(c), Exercise 70, and Exercise 87(b) of Section 2.3, we have

$$\text{rank } A^T = \text{rank}(QR)^T \text{ rank } R^T Q^T$$
$$= \text{rank } R^T = \text{rank } R = \text{rank } A.$$

75. (a) Using Gaussian elimination, we obtain the solution

$$x_1 = -3 + x_3$$
$$x_2 = 4 - 2x_3$$
$$x_3 \text{ free.}$$

(b) It is not a solution because A is not invertible.

77. In Exercise 19(c) of Section 1.5, we have two sectors, oil and electricity. The input-output matrix is given by

$$C = \begin{bmatrix} .1 & .4 \\ .3 & .2 \end{bmatrix}.$$

As in Example 5, we need to compute 3 times the second column of $(I_2 - C)^{-1}$. Since

$$3(I_2 - C)^{-1} = 3 \begin{bmatrix} \frac{4}{3} & \frac{2}{3} \\ \frac{1}{2} & \frac{3}{2} \end{bmatrix} = \begin{bmatrix} 4.0 & 2.0 \\ 1.5 & 4.5 \end{bmatrix},$$

the amount required is $2 million of electricity and $4.5 million of oil.

81. Suppose that the net production of sector i must be increased by k units, where $k > 0$. The gross production vector is given by

$$(I_n - C)^{-1}\mathbf{d} + k\mathbf{p}_i,$$

where C is the input-output matrix, $\mathbf{d}$ is the original demand vector, and $\mathbf{p}_i$ is the ith column of $(I_n - C)^{-1}$. All the entries of $\mathbf{p}_i$ are positive. Hence the gross production of *every* sector of the economy must be increased.

85. (a) We are given that $I_n = P^{-1}AP$ for some invertible matrix P. It follows that

$$A = PI_nP^{-1} = PP^{-1} = I_n.$$

(b) We are given that $O = P^{-1}AP$ for some invertible matrix P. It follows that $A = POP^{-1} = O$.

(c) We are given that A is similar to $B = cI_n$. It follows that

$$A = PBP^{-1} = P(cI_n)P^{-1}$$
$$= cPP^{-1} = cI_n.$$

Thus $A = B$.

89. The reduced row echelon form of A is I_4.

2.5 PARTITIONED MATRICES AND BLOCK MULTIPLICATION

1. We have

$$\begin{bmatrix} -1 & 3 & 1 \end{bmatrix} \begin{bmatrix} 1 \\ -1 \\ 0 \end{bmatrix} = -4$$

and

$$\begin{bmatrix} -1 & 3 & 1 \end{bmatrix} \begin{bmatrix} 2 \\ 1 \\ 1 \end{bmatrix} = 2.$$

So the product is $[-4 \mid 2]$.

2.5 Partitioned Matrices and Block Multiplication

5. We have

$$\begin{bmatrix} 2 & 0 \\ 3 & 1 \end{bmatrix} \begin{bmatrix} -1 & 2 \\ 2 & 2 \end{bmatrix} = \begin{bmatrix} -2 & 4 \\ -1 & 8 \end{bmatrix},$$

$$\begin{bmatrix} 2 & 0 \\ 3 & 1 \end{bmatrix} \begin{bmatrix} 3 & 0 \\ -1 & 2 \end{bmatrix} = \begin{bmatrix} 6 & 0 \\ 8 & 2 \end{bmatrix},$$

$$\begin{bmatrix} -1 & 5 \\ 1 & 2 \end{bmatrix} \begin{bmatrix} -1 & 2 \\ 2 & 2 \end{bmatrix} = \begin{bmatrix} 11 & 8 \\ 3 & 6 \end{bmatrix},$$

and

$$\begin{bmatrix} -1 & 5 \\ 1 & 2 \end{bmatrix} \begin{bmatrix} 3 & 0 \\ -1 & 2 \end{bmatrix} = \begin{bmatrix} -8 & 10 \\ 1 & 4 \end{bmatrix}.$$

So the product is

$$\left[\begin{array}{cc|cc} -2 & 4 & 6 & 0 \\ -1 & 8 & 8 & 2 \\ \hline 11 & 8 & -8 & 10 \\ 3 & 6 & 1 & 4 \end{array}\right].$$

9. As in Exercise 5, the product is

$$\begin{bmatrix} 3 & 6 \\ 9 & 12 \\ \hline 2 & 4 \\ 6 & 8 \end{bmatrix}.$$

13. Multiplying row 1 of A times B, we obtain the first row of AB, which is $[16 \ -4]$.

17. Multiplying row 3 of B times C, we obtain the third row of BC, which is $[-2 \ -3 \ 1]$.

21. By multiplying each column of A times the corresponding row of B (as described on page 149), we obtain $AB =$

$$\begin{bmatrix} -1 & 0 \\ -2 & 0 \\ 3 & 0 \end{bmatrix} + \begin{bmatrix} 8 & 2 \\ -4 & -1 \\ -8 & -2 \end{bmatrix} + \begin{bmatrix} 9 & -6 \\ 12 & -8 \\ 0 & 0 \end{bmatrix}.$$

25. As in Exercise 21, we obtain $B^T A =$

$$\begin{bmatrix} -1 & -2 & -3 \\ 0 & 0 & 0 \end{bmatrix} + \begin{bmatrix} 8 & -4 & 16 \\ 2 & -1 & 4 \end{bmatrix}$$

$$+ \begin{bmatrix} -9 & -6 & 0 \\ 6 & 4 & 0 \end{bmatrix}.$$

29. True   **30.** True

31. False, for example $\mathbf{v}$ can be in $\mathcal{R}^2$ and $\mathbf{w}$ can be in $\mathcal{R}^3$, then $\mathbf{vw}^T$ is a 2×3 matrix.

32. True

33. False, if either $\mathbf{v}$ or $\mathbf{w}$ is $\mathbf{0}$, then $\mathbf{vw}^T$ is $\mathbf{0}$.

34. False, if the matrices have sizes 2×1 and 1×2, then their product can be written as a sum of two matrices of rank 1.

35. The product equals $A^{-1}A + I_n I_n = 2I_n$.

39. We have $\begin{bmatrix} A^T & C^T \\ B^T & D^T \end{bmatrix} \begin{bmatrix} A & B \\ C & D \end{bmatrix} =$

$$\begin{bmatrix} A^T A + C^T C & A^T B + C^T D \\ B^T A + D^T C & B^T B + D^T D \end{bmatrix}.$$

43. We have

$$\begin{bmatrix} O & A \\ D & O \end{bmatrix} \begin{bmatrix} O & D^{-1} \\ A^{-1} & O \end{bmatrix}$$

$$= \begin{bmatrix} AA^{-1} & O \\ O & DD^{-1} \end{bmatrix}$$

$$= \begin{bmatrix} I_n & O \\ O & I_n \end{bmatrix} = I_{2n}.$$

47. We have

$$\begin{bmatrix} I_n & B \\ C & I_n \end{bmatrix} \begin{bmatrix} P & -PB \\ -CP & I_n + CPB \end{bmatrix}$$

$$= \begin{bmatrix} P - BCP & -PB + B(I_n + CPB) \\ CP - CP & -CPB + I_n + CPB \end{bmatrix}.$$

$$= \begin{bmatrix} P(I_n - BC) & -PB + B + BCP \\ O & I_n \end{bmatrix}$$

$$= \begin{bmatrix} PP^{-1} & B - (PB - BCPB) \\ O & I_n \end{bmatrix}$$

$$= \begin{bmatrix} I_n & B - (I_n - BC)PB \\ O & I_n \end{bmatrix}$$

$$= \begin{bmatrix} I_n & B - P^{-1}PB \\ O & I_n \end{bmatrix}$$

$$= \begin{bmatrix} I_n & B - B \\ O & I_n \end{bmatrix} = \begin{bmatrix} I_n & O \\ O & I_n \end{bmatrix} = I_{2n}.$$

51. In order to guess the form of the inverse if the matrix is invertible, consider the matrix $\begin{bmatrix} a & 0 \\ 1 & b \end{bmatrix}$, where a and b are nonzero scalars. It is easy to show that this matrix is invertible with inverse $\begin{bmatrix} -a^{-1} & 0 \\ -(ab)^{-1} & b^{-1} \end{bmatrix}$. So a reasonable guess for the inverse of $\begin{bmatrix} A & O \\ I_n & B \end{bmatrix}$ is $\begin{bmatrix} A^{-1} & O \\ -B^{-1}A^{-1} & B^{-1} \end{bmatrix}$.

Now we must verify that this guess is correct. In the product

$$\begin{bmatrix} A & O \\ I_n & B \end{bmatrix} \begin{bmatrix} A^{-1} & O \\ -B^{-1}A^{-1} & B^{-1} \end{bmatrix},$$

the upper left submatrix is

$$AA^{-1} + O(-B^{-1}A^{-1}) = I_n + O = I_n,$$

the upper right submatrix is

$$AO + OB^{-1} = O + O = O,$$

the lower left submatrix is

$$I_n A^{-1} - B(B^{-1}A^{-1}) = A^{-1} - A^{-1} = O,$$

and the lower right submatrix is

$$I_n O + BB^{-1} = O + I_n = I_n.$$

Thus

$$\begin{bmatrix} A & O \\ I_n & B \end{bmatrix} \begin{bmatrix} A^{-1} & O \\ -B^{-1}A^{-1} & B^{-1} \end{bmatrix} = I_{2n},$$

and so $\begin{bmatrix} A & O \\ I_n & B \end{bmatrix}$ is invertible with inverse

$$\begin{bmatrix} A^{-1} & O \\ -B^{-1}A^{-1} & B^{-1} \end{bmatrix}.$$

2.6 THE LU DECOMPOSITION OF A MATRIX

3. We apply elementary row operations to transform the given matrix into an upper triangular matrix:

$$\begin{bmatrix} 1 & -1 & 2 & 1 \\ 2 & -3 & 5 & 4 \\ -3 & 2 & 5 & 4 \end{bmatrix} \xrightarrow{-2r_1 + r_2 \to r_2}$$

$$\begin{bmatrix} 1 & -1 & 2 & 1 \\ 0 & -1 & 1 & 2 \\ -3 & 2 & -4 & 0 \end{bmatrix} \xrightarrow{3r_1 + r_3 \to r_3}$$

$$\begin{bmatrix} 1 & -1 & 2 & 1 \\ 0 & -1 & 1 & 2 \\ 0 & -1 & 2 & 3 \end{bmatrix} \xrightarrow{-1r_2 + r_3 \to r_3}$$

$$\begin{bmatrix} 1 & -1 & 2 & 1 \\ 0 & -1 & 1 & 2 \\ 0 & 0 & 1 & 1 \end{bmatrix} = U.$$

Since U consists of 3 rows, L is a 3×3 matrix. As in Example 3, the entries of L below the diagonal are the multipliers, and these can be obtained directly from the labels above the arrows describing the transformation of the given matrix into an upper triangular matrix. In particular, a label of the form $c\mathbf{r}_i + \mathbf{r}_j$ indicates that the (i,j)-entry of L is $-c$. Thus

$$L = \begin{bmatrix} 1 & 0 & 0 \\ 2 & 1 & 0 \\ -3 & 1 & 1 \end{bmatrix}.$$

7. We apply elementary row operations to transform the given matrix into an upper triangular matrix:

$$\begin{bmatrix} 1 & 0 & -3 & -1 & -2 & 1 \\ 2 & -1 & -8 & -1 & -5 & 0 \\ -1 & 1 & 5 & 1 & 4 & 2 \\ 0 & 1 & 2 & 1 & 3 & 4 \end{bmatrix} \xrightarrow{-2\mathbf{r}_1+\mathbf{r}_2 \to \mathbf{r}_2}$$

$$\begin{bmatrix} 1 & 0 & -3 & -1 & -2 & 1 \\ 0 & -1 & -2 & 1 & -1 & -2 \\ -1 & 1 & 5 & 1 & 4 & 2 \\ 0 & 1 & 2 & 1 & 3 & 4 \end{bmatrix} \xrightarrow{1\mathbf{r}_1+\mathbf{r}_3 \to \mathbf{r}_3}$$

$$\begin{bmatrix} 1 & 0 & -3 & -1 & -2 & 1 \\ 0 & -1 & -2 & 1 & -1 & -2 \\ 0 & 1 & 2 & 0 & 2 & 3 \\ 0 & 1 & 2 & 1 & 3 & 4 \end{bmatrix} \xrightarrow{1\mathbf{r}_2+\mathbf{r}_3 \to \mathbf{r}_3}$$

$$\begin{bmatrix} 1 & 0 & -3 & -1 & -2 & 1 \\ 0 & -1 & -2 & 1 & -1 & -2 \\ 0 & 0 & 0 & 1 & 1 & 1 \\ 0 & 1 & 2 & 1 & 3 & 4 \end{bmatrix} \xrightarrow{1\mathbf{r}_2+\mathbf{r}_4 \to \mathbf{r}_4}$$

$$\begin{bmatrix} 1 & 0 & -3 & -1 & -2 & 1 \\ 0 & -1 & -2 & 1 & -1 & -2 \\ 0 & 0 & 0 & 1 & 1 & 1 \\ 0 & 0 & 0 & 2 & 2 & 2 \end{bmatrix} = U.$$

Since U consists of 4 rows, L is a 4×4 matrix. As in Example 3, the entries of L below the diagonal are the multipliers, and these can be obtained directly from the labels above the arrows describing the transformation of the given matrix into an upper triangular matrix. In particular, a label of the form $c\mathbf{r}_i + \mathbf{r}_j$ indicates that the (i,j)-entry of L is $-c$.

Thus

$$L = \begin{bmatrix} 1 & 0 & 0 & 0 \\ 2 & 1 & 0 & 0 \\ -1 & -1 & 1 & 0 \\ 0 & -1 & 0 & 1 \end{bmatrix}.$$

11. Let

$$A = \begin{bmatrix} 1 & -1 & 2 & 1 \\ 2 & -3 & 5 & 4 \\ -3 & 2 & -4 & 0 \end{bmatrix}$$

and

$$\mathbf{b} = \begin{bmatrix} 1 \\ 8 \\ 5 \end{bmatrix}.$$

Then the system of equations can be written as $A\mathbf{x} = \mathbf{b}$. By Exercise 3, $A = LU$ is an LU decomposition of A, where

$$L = \begin{bmatrix} 1 & 0 & 0 \\ 2 & 1 & 0 \\ -3 & 1 & 1 \end{bmatrix}$$

and

$$U = \begin{bmatrix} 1 & -1 & 2 & 1 \\ 0 & -1 & 1 & 2 \\ 0 & 0 & 1 & 1 \end{bmatrix}.$$

We first solve the system $L\mathbf{y} = \mathbf{b}$, which is

$$\begin{aligned} y_1 &= 1 \\ 2y_1 + y_2 &= 8 \\ -3y_1 + y_2 + y_3 &= 5. \end{aligned}$$

Clearly $y_1 = 1$. Substitututing this value into the second equation, we obtain $y_2 = 6$. Substituting these values into the third equation we obtain $y_3 = 2$. Thus we obtain

$$\mathbf{y} = \begin{bmatrix} y_1 \\ y_2 \\ y_3 \end{bmatrix} = \begin{bmatrix} 1 \\ 6 \\ 2 \end{bmatrix}.$$

Next, using back substitution, we solve the system $U\mathbf{x} = \mathbf{y}$, which is

$$\begin{aligned} x_1 - x_2 + 2x_3 + x_4 &= 1 \\ -x_2 + x_3 + 2x_4 &= 6 \\ x_3 + x_4 &= 2. \end{aligned}$$

We solve the third equation for x_3, while treating x_4 as a free variable to obtain

$$x_3 = 2 - x_4.$$

Similarly, we substitute the value of x_3 obtained in second equation to solve for x_2, and we substitute both these values into the first eqution and solve for x_1. Thus we have

$$x_2 = -4 + x_4 \quad \text{and} \quad x_1 = -7 + 2x_4.$$

Therefore we obtain the general solution

$$\begin{bmatrix} x_1 \\ x_2 \\ x_3 \\ x_4 \end{bmatrix} = \begin{bmatrix} -7 \\ -4 \\ 2 \\ 0 \end{bmatrix} + x_4 \begin{bmatrix} 2 \\ 1 \\ -1 \\ 1 \end{bmatrix}.$$

15. Let

$$A = \begin{bmatrix} 1 & 0 & -3 & -1 & -2 & 1 \\ 2 & -1 & -8 & -1 & -5 & 0 \\ -1 & 1 & 5 & 1 & 4 & 2 \\ 0 & 1 & 2 & 1 & 3 & 4 \end{bmatrix}$$

and

$$\mathbf{b} = \begin{bmatrix} 1 \\ 8 \\ -5 \\ -2 \end{bmatrix}.$$

Then the system of equations can be written as $A\mathbf{x} = \mathbf{b}$. By Exercise 7, $A = LU$ is an LU decomposition of A, where

$$L = \begin{bmatrix} 1 & 0 & 0 & 0 \\ 2 & 1 & 0 & 0 \\ -1 & -1 & 1 & 0 \\ 0 & -1 & 0 & 1 \end{bmatrix}$$

and

$$U = \begin{bmatrix} 1 & 0 & -3 & -1 & -2 & 1 \\ 0 & -1 & -2 & 1 & -1 & -2 \\ 0 & 0 & 0 & 1 & 1 & 1 \\ 0 & 0 & 0 & 2 & 2 & 2 \end{bmatrix}.$$

We first solve the system $L\mathbf{y} = \mathbf{b}$, which becomes

$$\begin{aligned} y_1 &= 1 \\ 2y_1 + y_2 &= 8 \\ -y_1 - y_2 + y_3 &= -5 \\ -y_2 + y_4 &= -2. \end{aligned}$$

Clearly $y_1 = 1$. Substituting this value into the second equation, we obtain $y_2 = 6$. Continuing in this manner we can solve for the other values to obtain

$$\mathbf{y} = \begin{bmatrix} y_1 \\ y_2 \\ y_3 \\ y_4 \end{bmatrix} = \begin{bmatrix} 1 \\ 6 \\ 2 \\ 4 \end{bmatrix}.$$

Next, using back substitution, we solve the system $U\mathbf{x} = \mathbf{y}$, which becomes

$$\begin{aligned} x_1 - 3x_3 - x_4 - 2x_5 + x_6 &= 1 \\ -x_2 - 2x_3 + x_4 - x_5 - 2x_6 &= 6 \\ x_4 + x_5 + x_6 &= 2 \\ 2x_4 + 2x_5 + 2x_6 &= 4. \end{aligned}$$

We solve the fourth equation for x_4, while treating x_5 and x_5 as free variables to obtain

$$x_4 = 2 - x_5 - x_6.$$

Since the third equation is equivalent to the fourth equation, it can be ignored. Now solve the third equation for x_2, using the value for x_4 obtained in the fourth equation. Notice that x_3 is a free variable, and hence we obtain

$$x_2 = -4 - 2x_3 - 2x_5 - 3x_6.$$

Finally, we solve the first equation for x_1 to obtain

$$x_1 = 3 + 3x_3 + x_5 - 2x_6.$$

This produces the general solution

$$\begin{bmatrix} x_1 \\ x_2 \\ x_3 \\ x_4 \\ x_5 \\ x_6 \end{bmatrix} = \begin{bmatrix} 3 \\ -4 \\ 0 \\ 2 \\ 0 \\ 0 \end{bmatrix} + x_3 \begin{bmatrix} 3 \\ -2 \\ 1 \\ 0 \\ 0 \\ 0 \end{bmatrix} + x_5 \begin{bmatrix} 1 \\ -2 \\ 0 \\ -1 \\ 1 \\ 0 \end{bmatrix} + x_6 \begin{bmatrix} -2 \\ -3 \\ 0 \\ -1 \\ 0 \\ 1 \end{bmatrix}$$

19. Using the procedure described in the solution of Exercise 23, which follows, we obtain

$$P = \begin{bmatrix} 1 & 0 & 0 \\ 0 & 0 & 1 \\ 0 & 1 & 0 \end{bmatrix}, \quad L = \begin{bmatrix} 1 & 0 & 0 \\ -1 & 1 & 0 \\ 2 & 0 & 1 \end{bmatrix},$$

and

$$U = \begin{bmatrix} 1 & 1 & -2 & -1 \\ 0 & 1 & -3 & 0 \\ 0 & 0 & 1 & 1 \end{bmatrix}.$$

23. We use Examples 5 and 6 as a model, placing the multipliers in parentheses in appropriate matrix entries:

$$A = \begin{bmatrix} 1 & 2 & 1 & -1 \\ 2 & 4 & 1 & 1 \\ 3 & 2 & -1 & -2 \\ 2 & 5 & 3 & 0 \end{bmatrix} \xrightarrow{-2r_1+r_2 \to r_2}$$

$$\begin{bmatrix} 1 & 2 & 1 & -1 \\ (2) & 0 & -1 & 3 \\ 3 & 2 & -1 & -2 \\ 2 & 5 & 3 & 0 \end{bmatrix} \xrightarrow{-3r_1+r_3 \to r_3}$$

$$\begin{bmatrix} 1 & 2 & 1 & -1 \\ (2) & 0 & -1 & 3 \\ (3) & -4 & -4 & 1 \\ 2 & 5 & 3 & 0 \end{bmatrix} \xrightarrow{-2r_1+r_4 \to r_4}$$

$$\begin{bmatrix} 1 & 2 & 1 & -1 \\ (2) & 0 & -1 & 3 \\ (3) & -4 & -4 & 1 \\ (2) & 1 & 1 & 2 \end{bmatrix} \xrightarrow{r_2 \leftrightarrow r_4}$$

$$\begin{bmatrix} 1 & 2 & 1 & -1 \\ (2) & 1 & 1 & 2 \\ (3) & -4 & -4 & 1 \\ (2) & 0 & -1 & 3 \end{bmatrix} \xrightarrow{4r_2+r_3 \to r_3}$$

$$\begin{bmatrix} 1 & 2 & 1 & -1 \\ (2) & 1 & 1 & 2 \\ (3) & (-4) & 0 & 9 \\ (2) & 0 & -1 & 3 \end{bmatrix} \xrightarrow{r_3 \leftrightarrow r_4}$$

$$\begin{bmatrix} 1 & 2 & 1 & -1 \\ (2) & 1 & 1 & 2 \\ (2) & 0 & -1 & 3 \\ (3) & (-4) & 0 & 9 \end{bmatrix}.$$

The last matrix in the sequence contains the information necessary to construct the matrices L and U in an LU decomposition of A. Matrix L is the unit lower triangular matrix whose subdiagonal entries are the same as the subdiagonal entries of the final matrix, where parentheses are removed, if necessary. U is the upper triangular matrix obtained from the final matrix in the sequence by replacing all subdiagonal entries by zeros. Thus we obtain

$$L = \begin{bmatrix} 1 & 0 & 0 & 0 \\ 2 & 1 & 0 & 0 \\ 2 & 0 & 1 & 0 \\ 3 & -4 & 0 & 1 \end{bmatrix}$$

and
$$U = \begin{bmatrix} 1 & 2 & 1 & -1 \\ 0 & 1 & 1 & 2 \\ 0 & 0 & -1 & 3 \\ 0 & 0 & 0 & 9 \end{bmatrix}.$$

Finally, we obtain P by applying to I_4 the row interchanges that occur in the preceding sequence of elementary row operations. Thus

$$I_4 \xrightarrow{r_2 \leftrightarrow r_4} \begin{bmatrix} 1 & 0 & 0 & 0 \\ 0 & 0 & 0 & 1 \\ 0 & 0 & 1 & 0 \\ 0 & 1 & 0 & 0 \end{bmatrix}$$

$$\xrightarrow{r_3 \leftrightarrow r_4} \begin{bmatrix} 1 & 0 & 0 & 0 \\ 0 & 0 & 0 & 1 \\ 0 & 1 & 0 & 0 \\ 0 & 0 & 1 & 0 \end{bmatrix} = P.$$

27. Let
$$A = \begin{bmatrix} 1 & 1 & -2 & -1 \\ 2 & 2 & -3 & -1 \\ -1 & -2 & -1 & 1 \end{bmatrix}$$

and
$$\mathbf{b} = \begin{bmatrix} 1 \\ 5 \\ -1 \end{bmatrix}.$$

Then the system can be written as a matrix equation $A\mathbf{x} = \mathbf{b}$. By Exercise 19, $PA = LU$, where

$$P = \begin{bmatrix} 1 & 0 & 0 \\ 0 & 0 & 1 \\ 0 & 1 & 0 \end{bmatrix},$$

$$L = \begin{bmatrix} 1 & 0 & 0 \\ -1 & 1 & 0 \\ 2 & 0 & 1 \end{bmatrix},$$

and
$$U = \begin{bmatrix} 1 & 1 & -2 & -1 \\ 0 & -1 & -3 & 0 \\ 0 & 0 & 1 & 1 \end{bmatrix}.$$

Since P is invertible, $A\mathbf{x} = \mathbf{b}$ is equivalent to

$$PA\mathbf{x} = P\mathbf{b} = \begin{bmatrix} 1 & 0 & 0 \\ 0 & 0 & 1 \\ 0 & 1 & 0 \end{bmatrix} \begin{bmatrix} 1 \\ 5 \\ -1 \end{bmatrix}$$

$$= \begin{bmatrix} 1 \\ -1 \\ 5 \end{bmatrix} = \mathbf{b}'.$$

We can solve this system using the LU decomposition of PA given above. As in Example 4, set $\mathbf{y} = U\mathbf{x}$, and use forward substitution to solve the system $L\mathbf{y} = \mathbf{b}'$, which can be written as

$$\begin{aligned} y_1 &= 1 \\ -y_1 + y_2 &= -1 \\ 2y_1 \phantom{{}+{}} + y_3 &= 5. \end{aligned}$$

The resulting solution is

$$\mathbf{y} = \begin{bmatrix} y_1 \\ y_2 \\ y_3 \end{bmatrix} = \begin{bmatrix} 1 \\ 0 \\ 3 \end{bmatrix}.$$

Finally to obtain the original solution, use back substitution to solve the system $U\mathbf{x} = \mathbf{y}$, which can be written as

$$\begin{aligned} x_1 + x_2 - 2x_3 - x_4 &= 1 \\ -x_2 - 3x_3 \phantom{{}-x_4} &= 0 \\ x_3 + x_4 &= 3. \end{aligned}$$

The solution is

$$\begin{bmatrix} x_1 \\ x_2 \\ x_3 \\ x_4 \end{bmatrix} = \begin{bmatrix} 16 \\ -9 \\ 3 \\ 0 \end{bmatrix} + x_4 \begin{bmatrix} -4 \\ 3 \\ -1 \\ 1 \end{bmatrix}.$$

2.6 The LU Decomposition of a Matrix

31. Let
$$A = \begin{bmatrix} 1 & 2 & 1 & -1 \\ 2 & 4 & 1 & 1 \\ 3 & 2 & -1 & -2 \\ 2 & 5 & 3 & 0 \end{bmatrix}$$

and
$$\mathbf{b} = \begin{bmatrix} 3 \\ 2 \\ -4 \\ 7 \end{bmatrix}.$$

Then the system can be written as the matrix equation $A\mathbf{x} = \mathbf{b}$. By Exercise 23, $PA = LU$, where

$$P = \begin{bmatrix} 1 & 0 & 0 & 0 \\ 0 & 0 & 0 & 1 \\ 0 & 1 & 0 & 0 \\ 0 & 0 & 1 & 0 \end{bmatrix},$$

$$L = \begin{bmatrix} 1 & 0 & 0 & 0 \\ 2 & 1 & 0 & 0 \\ 2 & 0 & 1 & 0 \\ 3 & -4 & 0 & 1 \end{bmatrix},$$

and
$$U = \begin{bmatrix} 1 & 2 & 1 & -1 \\ 0 & 1 & 1 & 2 \\ 0 & 0 & -1 & 3 \\ 0 & 0 & 0 & 9 \end{bmatrix}.$$

Since P is invertible, $A\mathbf{x} = \mathbf{b}$ is equivalent to

$$PA\mathbf{x} = P\mathbf{b}$$

$$= \begin{bmatrix} 1 & 0 & 0 & 0 \\ 0 & 0 & 0 & 1 \\ 0 & 1 & 0 & 0 \\ 0 & 0 & 1 & 0 \end{bmatrix} \begin{bmatrix} 3 \\ 2 \\ -4 \\ 7 \end{bmatrix}$$

$$= \begin{bmatrix} 3 \\ 7 \\ -2 \\ 4 \end{bmatrix} = \mathbf{b}'.$$

We can solve this system using the LU decomposition of PA given above. As in Example 4, set $\mathbf{y} = U\mathbf{x}$, and use forward substitution to solve the system $L\mathbf{y} = \mathbf{b}'$, which can be written

$$\begin{aligned} y_1 &= 3 \\ 2y_1 + y_2 &= 7 \\ 2y_1 + y_3 &= 2 \\ 3y_1 - 4y_2 + y_4 &= -4. \end{aligned}$$

The resulting solution is

$$\mathbf{y} = \begin{bmatrix} y_1 \\ y_2 \\ y_3 \\ y_4 \end{bmatrix} = \begin{bmatrix} 3 \\ 1 \\ -4 \\ -9 \end{bmatrix}.$$

Finally, to obtain the solution of the original system, use back substitution to solve $U\mathbf{x} = \mathbf{y}$, which can be written as

$$\begin{aligned} x_1 + 2x_2 + x_3 - x_4 &= 3 \\ 2x_2 + x_3 + 2x_4 &= 1 \\ -x_3 + 3x_4 &= -4 \\ 9x_4 &= 9. \end{aligned}$$

This solution is

$$\begin{bmatrix} x_1 \\ x_2 \\ x_3 \\ x_4 \end{bmatrix} = \begin{bmatrix} -3 \\ 2 \\ 1 \\ -1 \end{bmatrix}.$$

33. False, the matrices in Exercises 17–24 do not have LU decompositions.

34. True

35. False, the entries below and to the left of the diagonal entries are zeros.

36. False, consider the LU decomposition of the matrix in Exercise 1.

37. False, for example, if A is the $m \times n$ zero matrix and $U = A$, then $A = LU$, where U is any $m \times m$ unit lower triangular matrix.

48   Chapter 2   Matrices and Linear Transformations

38. True

39. False, the (i,j)-entry of L is $-c$.

40. True   **41.** True

45. By means of elementary row operations, L can be transformed into a unit lower triangular matrix L_1 whose first column is $\mathbf{e}_1$. Additional elementary row operations can be applied to transform L_1 into a unit lower triangular matrix whose first two columns are $\mathbf{e}_1$ and $\mathbf{e}_2$. This process can be continued until L is transformed into I_n, which is in reduced row echelon form. Hence L has rank n, and so L is invertible. Thus L^T is invertible upper triangular matrix whose diagonal entries all equal 1. So it follows from Exercise 43 that $(L^{-1})^T = (L^T)^{-1}$ is an upper triangular matrix with diagonal entries equal to $1/1 = 1$. Therefore, L^{-1} is a lower triangular matrix whose diagonal entries are all equal to 1.

49. Each entry of AB requires $n-1$ additions and n multiplications for a total of $2n-1$ flops. Since AB has mp entries, a total of $(2n-1)mp$ flops are required to compute all the entries of AB.

53. Use the imported MATLAB function elu2 (see Appendix D) to compute the answer. Enter $[L\ U\ P] = \text{elu2}(A)$ to obtain

$$P = \begin{bmatrix} 0 & 1 & 0 & 0 & 0 \\ 1 & 0 & 0 & 0 & 0 \\ 0 & 0 & 1 & 0 & 0 \\ 0 & 0 & 0 & 1 & 0 \\ 0 & 0 & 0 & 0 & 1 \end{bmatrix},$$

$$L = \begin{bmatrix} 1.0 & 0 & 0 & 0 & 0 \\ 0.0 & 1 & 0 & 0 & 0 \\ 0.5 & 2 & 1 & 0 & 0 \\ -0.5 & -1 & -3 & 1 & 0 \\ 1.5 & 7 & 9 & -9 & 1 \end{bmatrix},$$

and

$$U = \begin{bmatrix} 2 & -2 & -1.0 & 3.0 & 4 \\ 0 & 1 & 2.0 & -1.0 & 1 \\ 0 & 0 & -1.5 & -0.5 & -2 \\ 0 & 0 & 0.0 & -1.0 & -2 \\ 0 & 0 & 0.0 & 0.0 & -9 \end{bmatrix}.$$

2.7   LINEAR TRANSFORMATIONS AND MATRICES

1. Since A is a 2×3 matrix, the domain is $\mathcal{R}^3$ and the codomain is $\mathcal{R}^2$.

5. Since B^T is a 3×3 matrix, the domain is $\mathcal{R}^3$ and the codomain is $\mathcal{R}^3$.

9. $T_C\left(\begin{bmatrix} 2 \\ 3 \end{bmatrix}\right) = C\begin{bmatrix} 2 \\ 3 \end{bmatrix} = \begin{bmatrix} 8 \\ -6 \\ 11 \end{bmatrix}$

13. $T_A\left(\begin{bmatrix} 4 \\ 0 \\ -3 \end{bmatrix}\right) = A\begin{bmatrix} 4 \\ 0 \\ -3 \end{bmatrix} = \begin{bmatrix} 5 \\ 22 \end{bmatrix}$

17. $T_B\left(\begin{bmatrix} -3 \\ 0 \\ -1 \end{bmatrix}\right) = B\begin{bmatrix} -3 \\ 0 \\ -1 \end{bmatrix} = \begin{bmatrix} -3 \\ -9 \\ 2 \end{bmatrix}$

21. The domain consists of vectors with 3 components; so $n=3$. The codomain consists of vectors with 2 components; so $m=2$.

25. The standard matrix of T is

$$[T(\mathbf{e}_1)\ \ T(\mathbf{e}_2)] = \begin{bmatrix} 0 & 1 \\ 1 & 1 \end{bmatrix}.$$

29. The standard matrix of T is

$$[T(\mathbf{e}_1)\ \ T(\mathbf{e}_2)] = \begin{bmatrix} 1 & -1 \\ 2 & -3 \\ 0 & 0 \\ 0 & 1 \end{bmatrix}.$$

33. The standard matrix of T is
$$[T(\mathbf{e}_1)\ T(\mathbf{e}_2)\ T(\mathbf{e}_3)] = \begin{bmatrix} 1 & 0 & 0 \\ 0 & 1 & 0 \\ 0 & 0 & 1 \end{bmatrix}.$$

35. False, only a linear transformation has a standard matrix.

36. True

37. False, the function must also preserve vector addition.

38. True

39. False, the standard matrix is a 2×3 matrix.

40. True

41. False, the function must be linear.

42. True

43. False, the range of a function is the set of all images.

44. False, the range is contained in the codomain.

45. False, the function must be one-to-one.

46. True   47. True   48. True

49. True   50. True   51. True

52. False, $f: \mathcal{R} \to \mathcal{R}$ defined by $f(x) = x^2$ does not preserve scalar multiplication.

53. False, the functions must be linear.

54. True

57.
$$T\left(\begin{bmatrix} 16 \\ 4 \end{bmatrix}\right) = T\left(2\begin{bmatrix} 8 \\ 2 \end{bmatrix}\right) = 2T\left(\begin{bmatrix} 8 \\ 2 \end{bmatrix}\right)$$

$$= 2\begin{bmatrix} 2 \\ -4 \\ 6 \end{bmatrix} = \begin{bmatrix} 4 \\ -8 \\ 12 \end{bmatrix}$$

and
$$T\left(\begin{bmatrix} -4 \\ -1 \end{bmatrix}\right) = T\left(-\frac{1}{2}\begin{bmatrix} 8 \\ 2 \end{bmatrix}\right)$$
$$= -\frac{1}{2}T\left(\begin{bmatrix} 8 \\ 2 \end{bmatrix}\right)$$
$$= -\frac{1}{2}\begin{bmatrix} 2 \\ -4 \\ 6 \end{bmatrix} = \begin{bmatrix} -1 \\ 2 \\ -3 \end{bmatrix}$$

61. Write $\begin{bmatrix} -2 \\ 6 \end{bmatrix} = a\begin{bmatrix} -3 \\ 0 \end{bmatrix} + b\begin{bmatrix} 0 \\ 4 \end{bmatrix}$ and solve for a and b. We obtain
$$\begin{bmatrix} -2 \\ 6 \end{bmatrix} = \frac{2}{3}\begin{bmatrix} -3 \\ 0 \end{bmatrix} + \frac{3}{2}\begin{bmatrix} 0 \\ 4 \end{bmatrix}.$$

Hence
$$T\left(\begin{bmatrix} -2 \\ 6 \end{bmatrix}\right) = T\left(\frac{2}{3}\begin{bmatrix} -3 \\ 0 \end{bmatrix} + \frac{3}{2}\begin{bmatrix} 0 \\ 4 \end{bmatrix}\right)$$
$$= \frac{2}{3}T\left(\begin{bmatrix} -3 \\ 0 \end{bmatrix}\right) + \frac{3}{2}T\left(\begin{bmatrix} 0 \\ 4 \end{bmatrix}\right)$$
$$= \frac{2}{3}\begin{bmatrix} 6 \\ 3 \\ 9 \end{bmatrix} + \frac{3}{2}\begin{bmatrix} 8 \\ 0 \\ -4 \end{bmatrix} = \begin{bmatrix} 16 \\ 2 \\ 0 \end{bmatrix}.$$

65.
$$T\left(\begin{bmatrix} x_1 \\ x_2 \end{bmatrix}\right) = T(x_1\mathbf{e}_1 + x_2\mathbf{e}_2)$$
$$= x_1 T(\mathbf{e}_1) + x_2 T\mathbf{e}_2)$$
$$= x_1\begin{bmatrix} 2 \\ 3 \end{bmatrix} + x_2\begin{bmatrix} 4 \\ 1 \end{bmatrix}$$
$$= \begin{bmatrix} 2x_1 \\ 3x_1 \end{bmatrix} + \begin{bmatrix} 4x_2 \\ x_2 \end{bmatrix} = \begin{bmatrix} 2x_1 + 4x_2 \\ 3x_1 + x_2 \end{bmatrix}$$

69. We begin by finding the standard matrix of T. This requires that we express each of the standard vectors of $\mathcal{R}^2$ as a linear combination of $\begin{bmatrix} 1 \\ -2 \end{bmatrix}$ and $\begin{bmatrix} -1 \\ 3 \end{bmatrix}$. For the equation

$$\begin{bmatrix} 1 \\ 0 \end{bmatrix} = a \begin{bmatrix} 1 \\ -2 \end{bmatrix} + b \begin{bmatrix} -1 \\ 3 \end{bmatrix},$$

we obtain

$$\mathbf{e}_1 = 3 \begin{bmatrix} 1 \\ -2 \end{bmatrix} + 2 \begin{bmatrix} -1 \\ 3 \end{bmatrix}.$$

Likewise

$$\mathbf{e}_2 = 1 \begin{bmatrix} 1 \\ -2 \end{bmatrix} + 1 \begin{bmatrix} -1 \\ 3 \end{bmatrix}.$$

So

$$T(\mathbf{e}_1) = 3T\left(\begin{bmatrix} 1 \\ -2 \end{bmatrix}\right) + 2T\left(\begin{bmatrix} -1 \\ 3 \end{bmatrix}\right)$$

$$= 3 \begin{bmatrix} 2 \\ 1 \end{bmatrix} + 2 \begin{bmatrix} 3 \\ 0 \end{bmatrix} = \begin{bmatrix} 12 \\ 3 \end{bmatrix}.$$

Similarly, $T(\mathbf{e}_2) = \begin{bmatrix} 5 \\ 1 \end{bmatrix}$. Therefore the standard matrix of T is

$$A = \begin{bmatrix} 12 & 5 \\ 3 & 1 \end{bmatrix}.$$

So

$$T\left(\begin{bmatrix} x_1 \\ x_2 \end{bmatrix}\right) = A \begin{bmatrix} x_1 \\ x_2 \end{bmatrix} = \begin{bmatrix} 12x_1 + 5x_2 \\ 3x_1 + x_2 \end{bmatrix}.$$

73. T is linear. Let $A = \begin{bmatrix} 0 & 0 \\ 2 & 0 \end{bmatrix}$. Then

$$A\mathbf{x} = \begin{bmatrix} 0 & 0 \\ 2 & 0 \end{bmatrix} \begin{bmatrix} x_1 \\ x_2 \end{bmatrix} = \begin{bmatrix} 0 \\ 2x_1 \end{bmatrix} = T(\mathbf{x}).$$

So $T = T_A$, and hence T is a linear transformation by Theorem 2.7.

ALTERNATE PROOF. We can use the definition of a linear transformation by proving that T preserves vector addition and scalar multiplication.

Let $\mathbf{u}$ and $\mathbf{v}$ be vectors in $\mathcal{R}^2$. Then we have

$$T(\mathbf{u} + \mathbf{v}) = T\left(\begin{bmatrix} u_1 \\ u_2 \end{bmatrix} + \begin{bmatrix} v_1 \\ v_2 \end{bmatrix}\right)$$

$$= T\left(\begin{bmatrix} u_1 + v_1 \\ u_2 + v_2 \end{bmatrix}\right)$$

$$= \begin{bmatrix} 0 \\ 2(u_1 + v_1) \end{bmatrix} = \begin{bmatrix} 0 \\ 2u_1 + 2v_1 \end{bmatrix}.$$

Also

$$T(\mathbf{u}) + T(\mathbf{v}) = T\left(\begin{bmatrix} u_1 \\ u_2 \end{bmatrix}\right) + T\left(\begin{bmatrix} v_1 \\ v_2 \end{bmatrix}\right)$$

$$= \begin{bmatrix} 0 \\ 2u_1 \end{bmatrix} + \begin{bmatrix} 0 \\ 2v_1 \end{bmatrix}$$

$$= \begin{bmatrix} 0 \\ 2u_1 + 2v_1 \end{bmatrix}.$$

So $T(\mathbf{u} + \mathbf{v}) = T(\mathbf{u}) + T(\mathbf{v})$, and hence T preserves vector addition.

Now suppose c is any scalar. Then

$$T(c\mathbf{u}) = T\left(c \begin{bmatrix} u_1 \\ u_2 \end{bmatrix}\right) = T\left(\begin{bmatrix} cu_1 \\ cu_2 \end{bmatrix}\right)$$

$$= \begin{bmatrix} 0 \\ 2(cu_1) \end{bmatrix} = \begin{bmatrix} 0 \\ c(2u_1) \end{bmatrix}.$$

Also

$$cT(\mathbf{u}) = cT\left(\begin{bmatrix} u_1 \\ u_2 \end{bmatrix}\right)$$

$$= c \begin{bmatrix} 0 \\ 2u_1 \end{bmatrix} = \begin{bmatrix} 0 \\ c(2u_1) \end{bmatrix}.$$

So $T(c\mathbf{u}) = cT(\mathbf{u})$. Therefore T preserves scalar multiplication. Hence T is linear.

75. T is not linear. We must show that either T does not preserve vector addition or T does not preserve scalar multiplication. For example, let $\mathbf{u} = \mathbf{e}_1$ and $\mathbf{v} = \mathbf{e}_2$. Then

$$T(\mathbf{u} + \mathbf{v}) = T(\mathbf{e}_1 + \mathbf{e}_2) = T\left(\begin{bmatrix} 1 \\ 1 \\ 0 \end{bmatrix}\right)$$

$$= 1 + 1 + 0 - 1 = 1.$$

On the other hand,

$$T(\mathbf{u}) + T(\mathbf{v}) = T(\mathbf{e}_1) + T(\mathbf{e}_2)$$

$$= T\left(\begin{bmatrix} 1 \\ 0 \\ 0 \end{bmatrix}\right) + T\left(\begin{bmatrix} 0 \\ 1 \\ 0 \end{bmatrix}\right)$$

$$= (1 + 0 + 0 - 1)$$
$$+ (0 + 1 + 0 - 1)$$

$$= 0.$$

So $T(\mathbf{u}+\mathbf{v}) \neq T(\mathbf{u})+T(\mathbf{v})$ for the given vectors. Therefore T does not preserve vector addition and hence is not linear.

ALTERNATE PROOF. Let $c = 4$ and $\mathbf{u} = \mathbf{e}_1$. Then

$$T(4\mathbf{u}) = T(4\mathbf{e}_1) = T\left(\begin{bmatrix} 4 \\ 0 \\ 0 \end{bmatrix}\right)$$

$$= 4 + 0 + 0 - 1 = 3.$$

On the other hand,

$$4T(\mathbf{u}) = 4T(\mathbf{e}_1) = 4T\left(\begin{bmatrix} 1 \\ 0 \\ 0 \end{bmatrix}\right)$$

$$= 4(1 + 0 + 0 - 1) = 0.$$

So $T(4\mathbf{u}) \neq 4T(\mathbf{u})$ and hence T does not preserve scalar multiplication. Therefore T is not linear.

COMMENT. For this example, we can also show that T is not linear by noting that

$$T(\mathbf{0}) = 0 + 0 + 0 - 1 = -1 \neq 0.$$

So T is not linear by Theorem 2.8(a).

79. Since

$$2 \cdot T\left(\frac{\pi}{2}\right) = 2 \cdot \begin{bmatrix} 1 \\ \frac{\pi}{2} \end{bmatrix} = \begin{bmatrix} 2 \\ \pi \end{bmatrix},$$

but

$$T\left(2 \cdot \frac{\pi}{2}\right) = T(\pi) = \begin{bmatrix} 0 \\ \pi \end{bmatrix},$$

T does not preserve scalar multiplication and so is not linear.

83. We must show that the transformation cT preserves vector addition and scalar multiplication. Let $\mathbf{u}$ and $\mathbf{v}$ be in $\mathcal{R}^n$. Because T is linear,

$$(cT)(\mathbf{u} + \mathbf{v}) = cT(\mathbf{u} + \mathbf{v})$$
$$= c(T(\mathbf{u}) + T(\mathbf{v}))$$
$$= cT(\mathbf{u}) + cT(\mathbf{v})$$
$$= (cT)(\mathbf{u}) + (cT)(\mathbf{v}).$$

Also

$$(cT)(\mathbf{u}) + (cT)(\mathbf{v}) = cT(\mathbf{u}) + cT(\mathbf{v}).$$

So cT preserves vector addition. Now suppose k is a scalar. Because T is linear,

$$(cT)(k\mathbf{u}) = cT(k\mathbf{u})$$
$$= c(kT(\mathbf{u})) = ckT(\mathbf{u}).$$

Also

$$k((cT)(\mathbf{u})) = k(cT(\mathbf{u}))$$
$$= kcT(\mathbf{u}) = ckT(\mathbf{u}).$$

So cT preserves scalar multiplication. Hence cT is linear.

87. By Theorem 2.9, there exists a unique matrix A such that $T(\mathbf{v}) = A\mathbf{v}$ for all $\mathbf{v}$ in $\mathcal{R}^2$. Let $A = \begin{bmatrix} a & b \\ c & d \end{bmatrix}$. Then

$$T\left(\begin{bmatrix} x_1 \\ x_2 \end{bmatrix}\right) = \begin{bmatrix} a & b \\ c & d \end{bmatrix}\begin{bmatrix} x_1 \\ x_2 \end{bmatrix}$$

$$= \begin{bmatrix} ax_1 + bx_2 \\ cx_1 + dx_2 \end{bmatrix}.$$

91. (a) Because it is given that T is linear, it follows from Theorem 2.9 that T is a matrix transformation.
ALTERNATE PROOF. Let

$$A = \begin{bmatrix} -1 & 0 \\ 0 & 1 \end{bmatrix}.$$

Then

$$T_A\left(\begin{bmatrix} x_1 \\ x_2 \end{bmatrix}\right) = \begin{bmatrix} -1 & 0 \\ 0 & 1 \end{bmatrix}\begin{bmatrix} x_1 \\ x_2 \end{bmatrix}$$

$$= \begin{bmatrix} -x_1 \\ x_2 \end{bmatrix}$$

$$= T\left(\begin{bmatrix} x_1 \\ x_2 \end{bmatrix}\right),$$

and hence $T = T_A$.

(b) Every vector $\mathbf{v} = \begin{bmatrix} v_1 \\ v_2 \end{bmatrix}$ in $\mathcal{R}^2$ is an image because

$$T\left(\begin{bmatrix} -v_1 \\ v_2 \end{bmatrix}\right) = \begin{bmatrix} v_1 \\ v_2 \end{bmatrix} = \mathbf{v}.$$

Thus the range of T is $\mathcal{R}^2$.

95. We have $T(\mathbf{u}) = T(\mathbf{v})$ if and only if $T(\mathbf{u}) - T(\mathbf{v}) = \mathbf{0}$. Because T is linear, the preceding equation is true if and only if $T(\mathbf{u} - \mathbf{v}) = \mathbf{0}$.

99. A vector $\mathbf{v}$ is in the range of T if and only if $\mathbf{v} = T(\mathbf{u}) = A\mathbf{u}$ for some $\mathbf{u}$ in $\mathcal{R}^n$, which is true if and only if $\mathbf{v}$ is in the span of the columns of A.

103. The given vector $\mathbf{v} = \begin{bmatrix} 2 \\ -1 \\ 0 \\ 3 \end{bmatrix}$ is in the range of T if and only there is a vector $\mathbf{u}$ such that $T(\mathbf{u}) = \mathbf{v}$. If A is the standard matrix of T, then this condition is equivalent to the system $A\mathbf{x} = \mathbf{v}$ being consistent, where

$$A = \begin{bmatrix} 1 & 1 & 1 & 2 \\ 1 & 2 & -3 & 4 \\ 0 & 1 & 0 & 2 \\ 1 & 5 & -1 & 0 \end{bmatrix}.$$

If we solve this system, we obtain

$$\mathbf{u} = \frac{1}{4}\begin{bmatrix} 5 \\ 2 \\ 3 \\ -1 \end{bmatrix}.$$

So $T(\mathbf{u}) = \mathbf{v}$, and thus $\mathbf{v}$ is in the range of T.

Alternatively, we can show that the reduced row echelon form of A is I_4 and conclude from (b) and (e) of Theorem 2.6 that the system $A\mathbf{x} = \mathbf{b}$ is consistent for every $\mathbf{b}$ in $\mathcal{R}^4$.

2.8 COMPOSITION AND INVERTIBILITY OF LINEAR TRANSFORMATIONS

1. The columns of the standard matrix
$$\begin{bmatrix} 2 & 3 \\ 4 & 5 \end{bmatrix}$$
of T form a generating set for the range of T. So $\left\{ \begin{bmatrix} 2 \\ 4 \end{bmatrix}, \begin{bmatrix} 3 \\ 5 \end{bmatrix} \right\}$ is one possible generating set.

5. As in Exercise 1, one generating set is
$$\left\{ \begin{bmatrix} 2 \\ 2 \\ 4 \end{bmatrix}, \begin{bmatrix} 1 \\ 2 \\ 1 \end{bmatrix}, \begin{bmatrix} 1 \\ 3 \\ 0 \end{bmatrix} \right\}.$$

9. As in Exercise 1, one generating set is
$$\left\{ \begin{bmatrix} 1 \\ 0 \\ 0 \end{bmatrix}, \begin{bmatrix} 0 \\ 1 \\ 0 \end{bmatrix} \right\}.$$

13. The null space of T is the solution set of $A\mathbf{x} = \mathbf{0}$, where
$$A = \begin{bmatrix} 0 & 1 \\ 1 & 1 \end{bmatrix}$$
is the standard matrix of T. Thus the general solution of $A\mathbf{x} = \mathbf{0}$ is
$$x_1 = 0$$
$$x_2 = 0.$$
So a generating set is $\{\mathbf{0}\}$. By Theorem 2.11, T is one-to-one.

17. The null space of T is the solution set of $A\mathbf{x} = \mathbf{0}$, where
$$A = \begin{bmatrix} 1 & 2 & 1 \\ 1 & 3 & 2 \\ 2 & 5 & 3 \end{bmatrix}$$
is the standard matrix of T. The general solution of $A\mathbf{x} = \mathbf{0}$ is
$$x_1 = x_3$$
$$x_2 = -x_3$$
$$x_3 \text{ free,}$$
or
$$\begin{bmatrix} x_1 \\ x_2 \\ x_3 \end{bmatrix} = \begin{bmatrix} x_3 \\ -x_3 \\ x_3 \end{bmatrix} = x_3 \begin{bmatrix} 1 \\ -1 \\ 1 \end{bmatrix}.$$
So a generating set is
$$\left\{ \begin{bmatrix} 1 \\ -1 \\ 1 \end{bmatrix} \right\}.$$
By Theorem 2.11, T is not one-to-one.

21. As in Exercise 17, a generating set for the null space of T is $\{\mathbf{e}_2\}$, and so T is not one-to-one.

25. The standard matrix of T is
$$[T(\mathbf{e}_1) \ T(\mathbf{e}_2)] = \begin{bmatrix} 2 & 3 \\ 4 & 5 \end{bmatrix}.$$
The reduced row echelon form of this matrix is
$$\begin{bmatrix} 1 & 0 \\ 0 & 1 \end{bmatrix},$$
which has rank 2. So by Theorem 2.11, T is one-to-one.

29. As in Exercise 25, the standard matrix of T is
$$\begin{bmatrix} 1 & -1 & 0 \\ 0 & 1 & -1 \\ 1 & 0 & -1 \end{bmatrix},$$
which has rank 2. So T is not one-to-one.

33. The standard matrix of T is

54 Chapter 2 Matrices and Linear Transformations

$$A = \begin{bmatrix} 2 & 3 \\ 4 & 5 \end{bmatrix}.$$

Because rank $A = 2$, we see that T is onto by Theorem 2.10.

37. The standard matrix of T is

$$\begin{bmatrix} 0 & 1 & -2 \\ 1 & 0 & -1 \\ -1 & 2 & -3 \end{bmatrix}.$$

Its reduced row echelon form is

$$\begin{bmatrix} 1 & 0 & -1 \\ 0 & 1 & -2 \\ 0 & 0 & 0 \end{bmatrix},$$

and so its rank is 2. So T is not onto by Theorem 2.10.

41. True

42. False, the span of the columns must equal the codomain for the transformation to be onto.

43. False, $A = \begin{bmatrix} 1 & 0 \\ 0 & 1 \\ 0 & 0 \end{bmatrix}$ has linearly independent columns, but the vector $\begin{bmatrix} 0 \\ 0 \\ 1 \end{bmatrix}$ is not in the range of T_A.

44. True   45. True   46. True

47. False, T_A must be onto.

48. True

49. False, the range must equal its codomain.

50. True

51. False, the function must be linear.

52. False, the rank must equal m.

53. True

54. False, the function must be linear.

55. False, the rank must equal n.

56. True   57. True

58. False, the standard matrix of TU is AB.

59. True   60. True

63. (a) Span $\{\mathbf{e}_1\}$. The only vectors that are projected to $\mathbf{0}$ are the multiples of $\mathbf{e}_1$.
    (b) No, because $T(\mathbf{e}_1) = \mathbf{0}$.
    (c) Span $\{\mathbf{e}_2\}$. Clearly every vector is projected onto the y-axis and hence is a multiple of $\mathbf{e}_2$.
    (d) No, from (c), it follows that $\mathbf{e}_1$ is not in the range of T.

67. (a) T is one-to-one. The columns of the standard matrix of T are $T(\mathbf{e}_1)$ and $T(\mathbf{e}_2)$, which are linearly independent because neither is a multiple of the other. So by Theorem 2.11, T is one-to-one.
    (b) T is onto. The standard matrix of T is $\begin{bmatrix} 3 & 4 \\ 1 & 2 \end{bmatrix}$. Since its reduced row echelon form is I_2, its rank is 2. Thus, by Theorem 2.10, T is onto.

71. The standard matrices of T and U are

$$A = \begin{bmatrix} 1 & 1 \\ 1 & -3 \\ 4 & 0 \end{bmatrix} \text{ and } B = \begin{bmatrix} 1 & -1 & 4 \\ 1 & 3 & 0 \end{bmatrix},$$

respectively.

75. The standard matrix of TU found in Exercise 74 equals

$$AB = \begin{bmatrix} 2 & 2 & 4 \\ -2 & -10 & 4 \\ 4 & -4 & 16 \end{bmatrix},$$

2.8 Composition and Invertibility of Linear Transformations

where A and B are the matrices in the solution to Exercise 71.

79. The standard matrix of UT found in Exercise 77 equals

$$BA = \begin{bmatrix} -1 & 5 \\ 15 & -5 \end{bmatrix},$$

where A and B are the matrices in the solution to Exercise 78.

83. The standard matrix of T is

$$A = \begin{bmatrix} 2 & -1 \\ 1 & 1 \end{bmatrix}.$$

This matrix is invertible, and

$$A^{-1} = \frac{1}{3}\begin{bmatrix} 1 & 1 \\ -1 & 2 \end{bmatrix}.$$

Thus, by Theorem 2.13, T is invertible, and the standard matrix of its inverse is A^{-1}. Therefore

$$T^{-1}\left(\begin{bmatrix} x_1 \\ x_2 \end{bmatrix}\right) = \begin{bmatrix} \frac{1}{3}x_1 + \frac{1}{3}x_2 \\ -\frac{1}{3}x_1 + \frac{2}{3}x_2 \end{bmatrix}.$$

87. As in Exercise 83, we have

$$T^{-1}\left(\begin{bmatrix} x_1 \\ x_2 \\ x_3 \end{bmatrix}\right) = \begin{bmatrix} x_1 - 2x_2 + x_3 \\ -x_1 + x_2 - x_3 \\ 2x_1 - 7x_2 + 3x_3 \end{bmatrix}.$$

91. The result is true for arbitrary functions. Suppose that $f: \mathcal{R}^n \to \mathcal{R}^m$ and $g: \mathcal{R}^m \to \mathcal{R}^p$ are one-to-one functions. To show $gf: \mathcal{R}^n \to \mathcal{R}^p$ is one-to-one, assume that $(gf)(\mathbf{u}) = (gf)(\mathbf{v})$. Then $g(f(\mathbf{u})) = g(f(\mathbf{v}))$. Because g is one-to-one, we have $f(\mathbf{u}) = f(\mathbf{v})$, and since f is also one-to-one, we have $\mathbf{u} = \mathbf{v}$.

95. Let T be the projection on the x-axis, and let U be the reflection about the y-axis. Then

$$T\left(\begin{bmatrix} x_1 \\ x_2 \end{bmatrix}\right) = \begin{bmatrix} x_1 \\ 0 \end{bmatrix}$$

and

$$U\left(\begin{bmatrix} x_1 \\ x_2 \end{bmatrix}\right) = \begin{bmatrix} -x_1 \\ x_2 \end{bmatrix}.$$

So

$$(UT)\left(\begin{bmatrix} x_1 \\ x_2 \end{bmatrix}\right) = \begin{bmatrix} -x_1 \\ 0 \end{bmatrix}$$

$$= (TU)\left(\begin{bmatrix} x_1 \\ x_2 \end{bmatrix}\right).$$

99. (a) We have $A = \begin{bmatrix} 1 & 3 & -2 & 1 \\ 3 & 0 & 4 & 1 \\ 2 & -1 & 0 & 2 \\ 0 & 0 & 1 & 1 \end{bmatrix}$

and $B = \begin{bmatrix} 0 & 1 & 0 & -3 \\ 2 & 0 & 1 & -1 \\ 1 & -2 & 0 & 4 \\ 0 & 5 & 1 & 0 \end{bmatrix}.$

(b) $AB = \begin{bmatrix} 4 & 10 & 4 & -14 \\ 4 & 0 & 1 & 7 \\ -2 & 12 & 1 & -5 \\ 1 & 3 & 1 & 4 \end{bmatrix}$

(c) By Theorem 2.12, the rule for TU is $TU\left(\begin{bmatrix} x_1 \\ x_2 \\ x_3 \\ x_4 \end{bmatrix}\right) =$

$$\begin{bmatrix} 4x_1 + 10x_2 + 4x_3 - 14x_4 \\ 4x_1 + x_3 + 7x_4 \\ -2x_1 + 12x_2 + x_3 - 5x_4 \\ x_1 + 3x_2 + x_3 + 4x_4 \end{bmatrix}.$$

CHAPTER 2 REVIEW

1. True

2. False, consider $\begin{bmatrix} 1 & 2 \\ 2 & 4 \end{bmatrix}$.

3. False, the product of a 2×2 and 3×3 matrix is not defined.

4. True

5. False, see page 122.

6. False, consider $I_2 + I_2 = 2I_2 \neq O$.

7. True   8. True   9. True

10. False, $O\mathbf{x} = \mathbf{0}$ is consistent for $\mathbf{b} = \mathbf{0}$, but O is not invertible.

11. True

12. False, the null space is contained in the domain.

13. True

14. False, let T be the projection on the x-axis. Then $\{\mathbf{e}_2\}$ is a linearly independent set but $\{T(\mathbf{e}_2)\} = \{\mathbf{0}\}$ is a linearly dependent set.

15. True   16. True

17. False, the transformation
$$T_A\left(\begin{bmatrix} x_1 \\ x_2 \end{bmatrix}\right) = \begin{bmatrix} x_1 \\ x_2 \\ 0 \end{bmatrix}$$
is one-to-one, but not onto.

18. True

19. False, the null space consists exactly of the zero vector.

20. False, the columns of its standard matrix form a generating set for its codomain.

21. True

23. (a) BA is defined if and only if $q = m$.
    (b) If BA is defined, it is a $p \times n$ matrix.

27. $C\mathbf{w} = \begin{bmatrix} 2 \\ 29 \\ 4 \end{bmatrix}$

31.
$$A^{-1}B^T = \frac{1}{6}\begin{bmatrix} 1 & 1 \\ 4 & -2 \end{bmatrix}\begin{bmatrix} 2 & 4 \\ 3 & 6 \end{bmatrix}$$
$$= \frac{1}{6}\begin{bmatrix} 5 & 10 \\ 2 & 4 \end{bmatrix}$$

35. Because $\mathbf{u}$ is a 3×1 matrix, the product $\mathbf{u}^2 = \mathbf{u}\mathbf{u}$ is not defined.

39. The reduced row echelon form of the given matrix is I_3; so it is invertible. Its inverse is $\dfrac{1}{50}\begin{bmatrix} 22 & 14 & -2 \\ -42 & -2 & 11 \\ -5 & -10 & 5 \end{bmatrix}$.

43. The inverse of the coefficient matrix is
$$\begin{bmatrix} 1 & -1 \\ -1 & 2 \end{bmatrix}.$$
So the solution is
$$\begin{bmatrix} 1 & -1 \\ -1 & 2 \end{bmatrix}\begin{bmatrix} 3 \\ 5 \end{bmatrix} = \begin{bmatrix} -2 \\ 7 \end{bmatrix}.$$

47. Since $T_B \colon \mathcal{R}^2 \to \mathcal{R}^3$, the codomain is $\mathcal{R}^3$. The range equals
$$\text{Span}\left\{\begin{bmatrix} 4 \\ 1 \\ 0 \end{bmatrix}, \begin{bmatrix} 2 \\ -3 \\ 1 \end{bmatrix}\right\}.$$

51. The standard matrix of T is
$$[T(\mathbf{e}_1)\ T(\mathbf{e}_2)\ T(\mathbf{e}_3)] = \begin{bmatrix} 2 & 0 & -1 \\ 4 & 0 & 0 \end{bmatrix}.$$

53. The standard matrix of T is
$$A = [T(\mathbf{e}_1) \ T(\mathbf{e}_2)].$$
Now
$$T(\mathbf{e}_1) = 2\mathbf{e}_1 + U(\mathbf{e}_1)$$
$$= 2\begin{bmatrix}1\\0\end{bmatrix} + U\left(\begin{bmatrix}1\\0\end{bmatrix}\right)$$
$$= \begin{bmatrix}2\\0\end{bmatrix} + \begin{bmatrix}2\\3\end{bmatrix} = \begin{bmatrix}4\\3\end{bmatrix}.$$
Also
$$T(\mathbf{e}_2) = 2\mathbf{e}_2 + U(\mathbf{e}_2)$$
$$= 2\begin{bmatrix}0\\1\end{bmatrix} + U\left(\begin{bmatrix}0\\1\end{bmatrix}\right)$$
$$= \begin{bmatrix}0\\2\end{bmatrix} + \begin{bmatrix}1\\0\end{bmatrix} = \begin{bmatrix}1\\2\end{bmatrix}.$$
So
$$A = \begin{bmatrix}4 & 1\\3 & 2\end{bmatrix}.$$

57. The given function is the matrix transformation induced by $\begin{bmatrix}1 & 1 & 0\\0 & 0 & 1\end{bmatrix}$; so this function is linear by Theorem 2.7.

61. The null space is the solution set of $A\mathbf{x} = \mathbf{0}$, where A is the standard matrix of T. The general solution is
$$\begin{array}{rl} x_1 = & -2x_3\\ x_2 = & x_3\\ x_3 & \text{free.}\end{array}$$
Thus
$$\begin{bmatrix}x_1\\x_2\\x_3\end{bmatrix} = \begin{bmatrix}-2x_3\\x_3\\x_3\end{bmatrix} = x_3\begin{bmatrix}-2\\1\\1\end{bmatrix}.$$
So the generating set is $\left\{\begin{bmatrix}-2\\1\\1\end{bmatrix}\right\}$. By Theorem 2.11, T is not one-to-one.

65. The standard matrix of T is $\begin{bmatrix}3 & -1\\0 & 1\\1 & 1\end{bmatrix}$.
Because its rank is 2 and the codomain of T is $\mathcal{R}^3$, T is not onto by Theorem 2.10.

69. The matrix $BA = \begin{bmatrix}5 & -1 & 4\\1 & 1 & -1\\3 & 1 & 0\end{bmatrix}$ is the standard matrix of the the linear transformation UT.

73. The standard matrix of T is
$$A = \begin{bmatrix}1 & 2\\-1 & 3\end{bmatrix}.$$
By Theorem 2.13, T is invertible and $A^{-1} = \frac{1}{5}\begin{bmatrix}3 & -2\\1 & 1\end{bmatrix}$ is the standard matrix of T^{-1}. Therefore
$$T^{-1}\left(\begin{bmatrix}x_1\\x_2\end{bmatrix}\right) = \frac{1}{5}\begin{bmatrix}3x_1 - 2x_2\\x_1 + x_2\end{bmatrix}.$$

CHAPTER 2 MATLAB EXERCISES

1. (a) $AD = \begin{bmatrix}4 & 10 & 9\\1 & 2 & 9\\5 & 8 & 15\\5 & 8 & -8\\-4 & -8 & 1\end{bmatrix}$

(b) $DB = \begin{bmatrix}6 & -2 & 5 & 11 & 9\\-3 & -1 & 10 & 7 & -3\\-3 & 1 & 2 & -1 & -3\\2 & -2 & 7 & 9 & 3\\0 & -1 & 10 & 10 & 2\end{bmatrix}$

(c), (d) $(AB^T)C = A(B^TC) = \begin{bmatrix}38 & -22 & 14 & 38 & 57\\10 & -4 & 4 & 10 & 11\\-12 & -9 & -11 & -12 & 12\\9 & -5 & 4 & 9 & 14\\28 & 10 & 20 & 28 & -9\end{bmatrix}$

(e) $D(B - 2C) =$
$$\begin{bmatrix} -2 & 10 & 5 & 3 & -17 \\ -31 & -7 & -8 & -21 & -1 \\ -11 & -5 & -4 & -9 & 7 \\ -14 & 2 & -1 & -7 & -11 \\ -26 & -1 & -4 & -16 & -6 \end{bmatrix}$$

(f) $A\mathbf{v} = \begin{bmatrix} 11 \\ 8 \\ 20 \\ -3 \\ -9 \end{bmatrix}$

(g), (h) $C(A\mathbf{v}) = (CA)\mathbf{v} = \begin{bmatrix} 1 \\ -18 \\ 81 \end{bmatrix}$

(i) $A^3 =$
$$\begin{bmatrix} 23 & 14 & 9 & -7 & 46 \\ 2 & 11 & 6 & -2 & 10 \\ 21 & 26 & -8 & -17 & 11 \\ -6 & 18 & 53 & 24 & -36 \\ -33 & -6 & 35 & 25 & -12 \end{bmatrix}$$

5. (a) First, note that if the matrix $[C \ D]$ is in reduced row echelon form, then C is also in reduced row echelon form.

Given an $m \times n$ matrix A, let $B = [A \ I_m]$, the $m \times (m+n)$ matrix whose first n columns are the columns of A, and whose last m columns are those of I_m. By Theorem 2.3, there is an invertible $m \times m$ matrix P such that PB is in reduced row echelon form. Furthermore, $PB = P[A \ I_m] = [PA \ P]$. Thus PA is in reduced echelon form, and the final m columns of PB are the columns of P.

(b) $P =$
$$\begin{bmatrix} 0.0 & -0.8 & -2.2 & -1.8 & 1.0 \\ 0.0 & -0.8 & -1.2 & -1.8 & 1.0 \\ 0.0 & 0.4 & 1.6 & 2.4 & -1.0 \\ 0.0 & 1.0 & 2.0 & 2.0 & -1.0 \\ 1.0 & 0.0 & -1.0 & -1.0 & 0.0 \end{bmatrix}$$

7. The calculations in (a) and (b) both produce the matrix $A^{-1}B =$
$$\begin{bmatrix} 6 & -4 & 3 & 19 & 5 & -2 & -5 \\ -1 & 2 & -4 & -1 & 4 & -3 & -2 \\ -2 & 0 & 2 & 6 & -1 & 6 & 3 \\ 0 & 1 & -3 & -8 & 2 & -3 & 1 \\ -1 & 0 & 2 & -6 & -5 & 2 & 2 \end{bmatrix}.$$

Chapter 3

Determinants

3.1 COFACTOR EXPANSION

1. $\det \begin{bmatrix} 6 & 2 \\ -3 & -1 \end{bmatrix} = 6(-1) - 2(-3)$
$= -6 + 6 = 0$

5. $\det \begin{bmatrix} -5 & -6 \\ 10 & 12 \end{bmatrix} = (-5)(12) - (-6)(10)$
$= -60 + 60 = 0$

7. $\det \begin{bmatrix} 4 & 3 \\ -2 & -1 \end{bmatrix} = 4(-1) - 3(-2)$
$= -4 + 6 = 2$

11. The $(3, 1)$-cofactor of A is
$$(-1)^{3+1} \det \begin{bmatrix} -2 & 4 \\ 6 & 3 \end{bmatrix}$$
$$= 1[(-2)(3) - 4(6)]$$
$$= 1(-30) = -30.$$

15. The cofactor expansion along the third row is
$$0(-1)^{3+1} \det \begin{bmatrix} -2 & 2 \\ -1 & 3 \end{bmatrix}$$
$$+ 1(-1)^{3+2} \det \begin{bmatrix} 1 & 2 \\ 2 & 3 \end{bmatrix}$$
$$+ (-1)(-1)^{3+3} \det \begin{bmatrix} 1 & -2 \\ 2 & -1 \end{bmatrix}$$
$$= 0 + (-1)[1(3) - 2(2)]$$
$$+ (-1)[1(-1) - (-2)(2)]$$
$$= -2.$$

19. The cofactor expansion along the second row is
$$0 + (-1)(-1)^{2+2} \det \begin{bmatrix} 1 & 1 & -1 \\ 4 & 2 & -1 \\ 0 & 0 & -2 \end{bmatrix} + 0$$
$$+ 1(-1)^{2+4} \det \begin{bmatrix} 1 & 2 & 1 \\ 4 & -3 & 2 \\ 0 & 3 & 0 \end{bmatrix}$$
$$= (-1)(-2)(-1)^{3+3} \det \begin{bmatrix} 1 & 1 \\ 4 & 2 \end{bmatrix}$$
$$+ 1(3)(-1)^{3+2} \det \begin{bmatrix} 1 & 1 \\ 4 & 2 \end{bmatrix}$$
$$= 2[1(2) - 1(4)] - 3[1(2) - 1(4)]$$
$$= 2(-2) - 3(-2)$$
$$= 2.$$

23. We have
$$\det \begin{bmatrix} -6 & 0 & 0 \\ 7 & -3 & 2 \\ 2 & 9 & 4 \end{bmatrix}$$
$$= -6(-1)^{1+1} \det \begin{bmatrix} -3 & 2 \\ 9 & 4 \end{bmatrix}$$
$$= -6[(-3)4 - 2(9)] = 180.$$

27. We have
$$\det \begin{bmatrix} -2 & -1 & -5 & 1 \\ 0 & 0 & 0 & 4 \\ 0 & -2 & 0 & 5 \\ 3 & 1 & 6 & -2 \end{bmatrix}$$

60   Chapter 3   Determinants

$$= 4(-1)^{2+4} \det \begin{bmatrix} -2 & -1 & -5 \\ 0 & -2 & 0 \\ 3 & 1 & 6 \end{bmatrix}$$

$$= 4(-2)(-1)^{2+2} \det \begin{bmatrix} -2 & -5 \\ 3 & 6 \end{bmatrix}$$

$$= -8[-2(6) - (-5)(3)]$$

$$= -24.$$

31. The area of the parallelogram determined by **u** and **v** is

$$|\det [\mathbf{u}\ \mathbf{v}]| = \left|\det \begin{bmatrix} 6 & 3 \\ 4 & 2 \end{bmatrix}\right|$$

$$= |6(2) - 3(4)| = |0| = 0.$$

35. The area of the parallelogram determined by **u** and **v** is

$$|\det [\mathbf{u}\ \mathbf{v}]| = \left|\det \begin{bmatrix} 6 & 4 \\ -1 & 3 \end{bmatrix}\right|$$

$$= |6(3) - 4(-1)|$$

$$= |22| = 22.$$

37. The matrix is not invertible if its determinant equals 0. Now

$$\det \begin{bmatrix} 3 & 6 \\ c & 4 \end{bmatrix} = 3(4) - 6(c) = 12 - 6c.$$

Therefore the matrix is not invertible if $12 - 6c = 0$, that is, if $c = 2$.

41. We have

$$\det \begin{bmatrix} c & -2 \\ -8 & c \end{bmatrix} = c^2 - (-2)(-8)$$

$$= c^2 - 16.$$

Therefore the matrix is not invertible if $c^2 - 16 = 0$, that is, if $c = \pm 4$.

45. False, the determinant of a matrix is a *scalar*.

46. False, $\det \begin{bmatrix} a & b \\ c & d \end{bmatrix} = ad - bc.$

47. False, if the determinant of a 2×2 matrix is *nonzero*, then the matrix is invertible.

48. False, if a 2×2 matrix is invertible, then its determinant is *nonzero*.

49. True

50. False, the (i,j)-cofactor of A equals $(-1)^{i+j}$ times the determinant of the $(n-1) \times (n-1)$ matrix obtained by deleting row i and column j from A.

51. True   **52.** True

53. False, cofactor expansion is very inefficient. (See pages 204–205.)

54. True

55. False, consider $\begin{bmatrix} 1 & 2 \\ 2 & 1 \end{bmatrix}$.

56. True   **57.** True

57. False, see Example 1 on page 154.

58. False, a matrix in which all the entries to the left and below the diagonal entries equal zero is called an *upper* triangular matrix.

59. True   **60.** True

61. False, the determinant of an upper triangular or a lower triangular square matrix equals the *product* of its diagonal entries.

62. True

63. False, the area of the parallelogram determined by **u** and **v** is $|\det [\mathbf{u}\ \mathbf{v}]|$.

64. False, if A is the standard matrix of T, then

$$|\det [T(\mathbf{u})\ T(\mathbf{v})]| = |\det A| \cdot |\det [\mathbf{u}\ \mathbf{v}]|.$$

65. We have

$$\det A_\theta = \det \begin{bmatrix} \cos\theta & -\sin\theta \\ \sin\theta & \cos\theta \end{bmatrix}$$
$$= \cos^2\theta - (-\sin^2\theta)$$
$$= \cos^2\theta + \sin^2\theta = 1.$$

69. Let $A = \begin{bmatrix} a & b \\ c & d \end{bmatrix}$. Then

$$\det A^T = \det \begin{bmatrix} a & c \\ b & d \end{bmatrix}$$
$$= ad - cb = \det \begin{bmatrix} a & b \\ c & d \end{bmatrix}$$
$$= \det A.$$

73. We have $\det E = k$ and $\det A = ad - bc$; so $(\det E)(\det A) = k(ad - bc)$. Also

$$EA = \begin{bmatrix} a & b \\ kc & kd \end{bmatrix}.$$

Thus

$$\det EA = a(kd) - b(kc)$$
$$= k(ad - bc) = (\det E)(\det A).$$

77. We have

$$\det \begin{bmatrix} a & b \\ c+kp & d+kq \end{bmatrix}$$
$$= a(d+kq) - b(c+kp)$$
$$= ad + akq - bc - bkp$$

and

$$\det \begin{bmatrix} a & b \\ c & d \end{bmatrix} + k \det \begin{bmatrix} a & b \\ p & q \end{bmatrix}$$
$$= (ad - bc) + k(aq - bp)$$
$$= ad - bc + akq - bkq$$
$$= ad + akq - bc - bkp.$$

Thus

$$\det \begin{bmatrix} a & b \\ c+kp & d+kq \end{bmatrix}$$
$$= \det \begin{bmatrix} a & b \\ c & d \end{bmatrix} + k \det \begin{bmatrix} a & b \\ p & q \end{bmatrix}.$$

81. (c) No, $\det(A+B) \neq \det A + \det B$ for all $n \times n$ matrices A and B.

3.2 PROPERTIES OF DETERMINANTS

3. The cofactor expansion along the second column yields

$$-1(-1)^{1+2} \det \begin{bmatrix} 1 & -2 \\ -1 & 1 \end{bmatrix}$$
$$+ 4(-1)^{2+2} \det \begin{bmatrix} 2 & 3 \\ -1 & 1 \end{bmatrix}$$
$$+ 0(-1)^{3+2} \det \begin{bmatrix} 2 & 3 \\ 1 & -2 \end{bmatrix}$$
$$= 1[1(1) - (-2)(-1)]$$
$$+ 4[2(1) - 3(-1)] + 0$$
$$= 1(-1) + 4(5) + 0$$
$$= 19.$$

7. The cofactor expansion along the first column yields

$$0(-1)^{1+1} \det \begin{bmatrix} 1 & 2 \\ -1 & 1 \end{bmatrix}$$
$$+ 1(-1)^{2+1} \det \begin{bmatrix} 2 & 0 \\ -1 & 1 \end{bmatrix}$$
$$+ 0(-1)^{3+1} \det \begin{bmatrix} 2 & 0 \\ 1 & 2 \end{bmatrix}$$
$$= 0 + (-1)[2(1) - 0(-1)] + 0$$
$$= 0 + (-2) + 0 = -2.$$

62 Chapter 3 Determinants

11. We have
$$\det \begin{bmatrix} 0 & 0 & 5 \\ 0 & 3 & 7 \\ 4 & -1 & -2 \end{bmatrix}$$
$$= -\det \begin{bmatrix} 4 & -1 & 2 \\ 0 & 3 & 7 \\ 0 & 0 & 5 \end{bmatrix}$$
$$= -(4)(3)(5) = -60.$$

15. We have
$$\det \begin{bmatrix} 3 & -2 & 1 \\ 0 & 0 & 5 \\ -9 & 4 & 2 \end{bmatrix}$$
$$= -\det \begin{bmatrix} 3 & -2 & 1 \\ -9 & 4 & 2 \\ 0 & 0 & 5 \end{bmatrix}$$
$$= -\det \begin{bmatrix} 3 & -2 & 1 \\ 0 & -2 & 5 \\ 0 & 0 & 5 \end{bmatrix}$$
$$= -(3)(-2)(5) = 30.$$

19. We have
$$\det \begin{bmatrix} 1 & 2 & 1 \\ 1 & 1 & 2 \\ 3 & 4 & 8 \end{bmatrix} = \det \begin{bmatrix} 1 & 2 & 1 \\ 0 & -1 & 1 \\ 0 & -2 & 5 \end{bmatrix}$$
$$= \det \begin{bmatrix} 1 & 2 & 1 \\ 0 & -1 & 1 \\ 0 & 0 & 3 \end{bmatrix}$$
$$= 1(-1)(3) = -3.$$

23. We have
$$\det \begin{bmatrix} 0 & 4 & -1 & 1 \\ -3 & 1 & 1 & 2 \\ 1 & 0 & -2 & 3 \\ 2 & 3 & 0 & 1 \end{bmatrix}$$

$$= -\det \begin{bmatrix} 1 & 0 & -2 & 3 \\ -3 & 1 & 1 & 2 \\ 0 & 4 & -1 & 1 \\ 2 & 3 & 0 & 1 \end{bmatrix}$$
$$= -\det \begin{bmatrix} 1 & 0 & -2 & 3 \\ 0 & 1 & -5 & 11 \\ 0 & 4 & -1 & 1 \\ 0 & 3 & 4 & -5 \end{bmatrix}$$
$$= -\det \begin{bmatrix} 1 & 0 & -2 & 3 \\ 0 & 1 & -5 & 11 \\ 0 & 0 & 19 & -43 \\ 0 & 0 & 19 & -38 \end{bmatrix}$$
$$= -\det \begin{bmatrix} 1 & 0 & -2 & 3 \\ 0 & 1 & -5 & 11 \\ 0 & 0 & 19 & -43 \\ 0 & 0 & 0 & 5 \end{bmatrix}$$
$$= -1(1)(19)(5) = -95.$$

27. We have
$$\det \begin{bmatrix} c & 6 \\ 2 & c+4 \end{bmatrix} = c(c+4) - 12$$
$$= c^2 + 4c - 12$$
$$= (c+6)(c-2).$$

The matrix is not invertible if its determinant equals 0; so the matrix is not invertible if $c = -6$ or $c = 2$.

31. We have
$$\det \begin{bmatrix} 1 & -1 & 2 \\ -1 & 0 & 4 \\ 2 & 1 & c \end{bmatrix}$$
$$= \det \begin{bmatrix} 1 & -1 & 2 \\ 0 & -1 & 6 \\ 0 & 3 & c-4 \end{bmatrix}$$
$$= \det \begin{bmatrix} 1 & -1 & 2 \\ 0 & -1 & 6 \\ 0 & 0 & c+14 \end{bmatrix}$$

$= 1(-1)(c+14) = -(c+14).$

The matrix is not invertible if its determinant equals 0; so the matrix is not invertible if $c = -14$.

35. The given matrix A is not invertible if its determinant equals 0. Because $\det A = -(c+1)$, we see that A is not invertible if $c = -1$.

39. False, $\det \begin{bmatrix} 1 & 2 \\ 3 & 4 \end{bmatrix} \neq 1 \cdot 4$.

40. True

41. False, multiplying a row of a square matrix by a scalar c changes the determinant by a factor of c.

42. True

43. False, consider $A = [\mathbf{e}_1 \ \mathbf{0}]$ and $B = [\mathbf{0} \ \mathbf{e}_2]$.

44. True

45. False, if A is an invertible matrix, then $\det A \neq 0$.

46. False, for any square matrix A, $\det A^T = \det A$.

47. True

48. False, the determinant of $2I_2$ is 4, but its reduced row echelon form is I_2.

49. True   50. True

51. False, if A is an $n \times n$ matrix, then $\det cA = c^n \det A$.

52. False, Cramer's rule can be used to solve only systems that have an invertible coefficient matrix.

53. True   54. True   55. True

56. False, if A is a 5×5 matrix, then $\det(-A) = -\det A$.

57. True

58. False, if an $n \times n$ matrix A is transformed into an upper triangular matrix U using only row interchanges and row addition operations, then

$$\det A = (-1)^r u_{11} u_{22} \cdots u_{nn},$$

where r is the number of row interchanges performed.

59. We have

$$x_1 = \frac{\det \begin{bmatrix} 6 & 2 \\ -3 & 4 \end{bmatrix}}{\det \begin{bmatrix} 1 & 2 \\ 3 & 4 \end{bmatrix}} = \frac{6(4) - 2(-3)}{1(4) - 2(3)}$$

$$= -15$$

and

$$x_2 = \frac{\det \begin{bmatrix} 1 & 6 \\ 3 & -3 \end{bmatrix}}{\det \begin{bmatrix} 1 & 2 \\ 3 & 4 \end{bmatrix}} = \frac{1(-3) - 6(3)}{-2}$$

$$= 10.5.$$

63. We have

$$x_1 = \frac{\det \begin{bmatrix} 6 & 0 & -2 \\ -5 & 1 & 3 \\ 4 & 2 & 1 \end{bmatrix}}{\det \begin{bmatrix} 1 & 0 & -2 \\ -1 & 1 & 3 \\ 0 & 2 & 1 \end{bmatrix}}$$

$$= \frac{\det \begin{bmatrix} 6 & 0 & -2 \\ -5 & 1 & 3 \\ 14 & 0 & -5 \end{bmatrix}}{\det \begin{bmatrix} 1 & 0 & -2 \\ 0 & 1 & 1 \\ 0 & 2 & 1 \end{bmatrix}}$$

$$x_2 = \frac{\det \begin{bmatrix} 1 & 6 & -2 \\ -1 & -5 & 3 \\ 0 & 4 & 1 \end{bmatrix}}{\det \begin{bmatrix} 1 & 0 & -2 \\ -1 & 1 & 3 \\ 0 & 2 & 1 \end{bmatrix}}$$

$$= \frac{\det \begin{bmatrix} 1 & 6 & -2 \\ 0 & 1 & 1 \\ 0 & 4 & 1 \end{bmatrix}}{-1}$$

$$= \frac{1(1) - 1(4)}{-1} = 3,$$

and

$$x_3 = \frac{\det \begin{bmatrix} 1 & 0 & 6 \\ -1 & 1 & -5 \\ 0 & 2 & 4 \end{bmatrix}}{\det \begin{bmatrix} 1 & 0 & -2 \\ -1 & 1 & 3 \\ 0 & 2 & 1 \end{bmatrix}}$$

$$= \frac{\det \begin{bmatrix} 1 & 0 & 6 \\ 0 & 1 & 1 \\ 0 & 2 & 4 \end{bmatrix}}{-1}$$

$$= \frac{1(4) - 1(2)}{-1} = -2.$$

67. Take $A = I_2$ and $k = 3$. Then

$$\det kA = \det \begin{bmatrix} 3 & 0 \\ 0 & 3 \end{bmatrix} = 3 \cdot 3 = 9,$$

whereas $k \cdot \det A = 3 \cdot 1 = 3$.

71. By (b) and (d) of Theorem 3.4, we have

$$\det (B^{-1}AB)$$

$$= (\det B^{-1})(\det A)(\det B)$$
$$= (\det A)(\det B^{-1})(\det B)$$
$$= (\det A)\left(\frac{1}{\det B}\right)(\det B)$$
$$= \det A.$$

75. We have

$$\det \begin{bmatrix} 1 & a & a^2 \\ 1 & b & b^2 \\ 1 & c & c^2 \end{bmatrix}$$

$$= \det \begin{bmatrix} 1 & a & a^2 \\ 0 & b-a & b^2-a^2 \\ 0 & c-a & c^2-a^2 \end{bmatrix}$$

$$= (b-a)(c-a) \det \begin{bmatrix} 1 & a & a^2 \\ 0 & 1 & b+a \\ 0 & 1 & c+a \end{bmatrix}$$

$$= (b-a)(c-a) \det \begin{bmatrix} 1 & a & a^2 \\ 0 & 1 & b+a \\ 0 & 0 & c-b \end{bmatrix}$$

$$= (b-a)(c-a) \cdot (1)(1)(c-b)$$
$$= (b-a)(c-a)(c-b).$$

79. Let A be an $n \times n$ matrix, and let B be obtained by multiplying each entry of row r of A by the scalar k. Suppose that c_{ij} is the (i,j)-cofactor of A. Because the entries of A and B differ only in row r, the (r,j)-cofactor of B is c_{rj} for $j = 1, 2, \ldots, n$. Evaluating $\det B$ by cofactor expansion along row r gives

$$\det B = b_{r1}c_{r1} + \cdots + b_{rn}c_{rn}$$
$$= (ka_{r1})c_{r1} + \cdots + (ka_{rn})c_{rn}$$
$$= k(a_{r1}c_{r1} + \cdots + a_{rn}c_{rn})$$
$$= k(\det A).$$

83. (a) We have

$$\begin{bmatrix} 0.0 & -3.0 & -2 & -5 \\ 2.4 & 3.0 & -6 & 9 \\ -4.8 & 6.3 & 4 & -2 \\ 9.6 & 1.5 & 5 & 9 \end{bmatrix}$$

$$\xrightarrow{r_1 \leftrightarrow r_2}$$

$$\begin{bmatrix} 2.4 & 3.0 & -6 & 9 \\ 0.0 & -3.0 & -2 & -5 \\ -4.8 & 6.3 & 4 & -2 \\ 9.6 & 1.5 & 5 & 9 \end{bmatrix}$$

$$\xrightarrow[-4r_1+r_4 \to r_4]{2r_1+r_3 \to r_3}$$

$$\begin{bmatrix} 2.4 & 3.0 & -6 & 9 \\ 0.0 & -3.0 & -2 & -5 \\ 0.0 & 12.3 & -8 & 16 \\ 0.0 & -10.5 & 29 & -27 \end{bmatrix}$$

$$\xrightarrow[-3.5r_2+r_4 \to r_4]{4.1r_2+r_3 \to r_3}$$

$$\begin{bmatrix} 2.4 & 3 & -6.0 & 9.0 \\ 0.0 & -3 & -2.0 & -5.0 \\ 0.0 & 0 & -16.2 & -4.5 \\ 0.0 & 0 & 36.0 & -9.5 \end{bmatrix}$$

$$\xrightarrow{-\frac{20}{9}r_3+r_4 \to r_4}$$

$$\begin{bmatrix} 2.4 & 3 & -6.0 & 9.0 \\ 0.0 & -3 & -2.0 & -5.0 \\ 0.0 & 0 & -16.2 & -4.5 \\ 0.0 & 0 & 0.0 & -19.5 \end{bmatrix}.$$

(b) $\det A$
$= (-1)^1(2.4)(-3)(-16.2)(-19.5)$
$= 2274.48$

CHAPTER 3 REVIEW

1. False, $\det \begin{bmatrix} a & b \\ c & d \end{bmatrix} = ad - bc$.

2. False, for $n \geq 2$, the (i,j)-cofactor of an $n \times n$ matrix A equals $(-1)^{i+j}$ times the determinant of the $(n-1) \times (n-1)$ matrix obtained by deleting row i and column j from A.

3. True

4. False, consider $A = [\mathbf{e}_1 \ \mathbf{0}]$ and $B = [\mathbf{0} \ \mathbf{e}_2]$.

5. True

6. False, if B is obtained by interchanging two rows of an $n \times n$ matrix A, then $\det B = -\det A$.

7. False, an $n \times n$ matrix is invertible if and only if its determinant is *nonzero*.

8. True

9. False, for any invertible matrix A, $\det A^{-1} = \dfrac{1}{\det A}$.

10. False, for any $n \times n$ matrix A and scalar c, $\det cA = c^n(\det A)$.

11. False, the determinant of an upper triangular or a lower triangular square matrix equals the *product* of its diagonal entries.

15. The $(3,1)$-cofactor of the matrix is

$$(-1)^{3+1} \det \begin{bmatrix} -1 & 2 \\ 2 & -1 \end{bmatrix}$$
$$= 1[(-1)(-1) - 2(2)] = -3.$$

19. The determinant of the given matrix is

$$1(-1)^{1+1} \det \begin{bmatrix} 2 & -1 \\ 1 & 3 \end{bmatrix}$$
$$+ (-1)(-1)^{1+2} \det \begin{bmatrix} -1 & 2 \\ 1 & 3 \end{bmatrix}$$
$$+ 2(-1)^{1+3} \det \begin{bmatrix} -1 & 2 \\ 2 & -1 \end{bmatrix}$$

$$= 1[2(3) - (-1)(1)]$$
$$+ [(-1)(3) - 2(1)]$$
$$+ 2[(-1)(-1) - 2(2)]$$
$$= -4.$$

23. We have

$$\det \begin{bmatrix} 1 & -3 & 1 \\ 4 & -2 & 1 \\ 2 & 5 & -1 \end{bmatrix} = \det \begin{bmatrix} 1 & -3 & 1 \\ 0 & 10 & -3 \\ 0 & 11 & -3 \end{bmatrix}$$

$$= 1(-1)^{1+1} \det \begin{bmatrix} 10 & -3 \\ 11 & -3 \end{bmatrix}$$

$$= 1[10(-3) - (-3)(11)] = 3.$$

27. We have

$$\det \begin{bmatrix} c+4 & -1 & c+5 \\ -3 & 3 & -4 \\ c+6 & -3 & c+7 \end{bmatrix}$$

$$= \det \begin{bmatrix} c+4 & -1 & c+5 \\ 3c+9 & 0 & 3c+11 \\ -2c-6 & 0 & -2c-8 \end{bmatrix}$$

$$= (-1)(-1)^{1+2} \det \begin{bmatrix} 3c+9 & 3c+11 \\ -2c-6 & -2c-8 \end{bmatrix}$$

$$= (c+3) \det \begin{bmatrix} 3 & 3c+11 \\ -2 & -2c-8 \end{bmatrix}$$

$$= (c+3)[3(-2c-8) - (3c+11)(-2)]$$

$$= -2(c+3).$$

So the matrix is not invertible if and only if $c = -3$.

29. The area of the parallelogram in $\mathcal{R}^2$ determined by $\begin{bmatrix} 3 \\ 7 \end{bmatrix}$ and $\begin{bmatrix} 4 \\ 1 \end{bmatrix}$ is

$$\left| \det \begin{bmatrix} 3 & 4 \\ 7 & 1 \end{bmatrix} \right| = |3(1) - 4(7)| = 25.$$

31. We have

$$x_1 = \frac{\det \begin{bmatrix} 5 & 1 \\ -6 & 3 \end{bmatrix}}{\det \begin{bmatrix} 2 & 1 \\ -4 & 3 \end{bmatrix}}$$

$$= \frac{5(3) - 1(-6)}{2(3) - 1(-4)} = \frac{21}{10} = 2.1$$

and

$$x_2 = \frac{\det \begin{bmatrix} 2 & 5 \\ -4 & -6 \end{bmatrix}}{\det \begin{bmatrix} 2 & 1 \\ -4 & 3 \end{bmatrix}}$$

$$= \frac{2(-6) - 5(-4)}{10} = \frac{8}{10} = 0.8.$$

35. If $\det A = 5$, then

$$\det 2A = 2^3 (\det A) = 8(5) = 40$$

because we can remove a factor of 2 from each row of $2A$ and apply Theorem 3.3(b).

37. If we add 3 times row 2 of the given matrix to row 1, we obtain A. Since, by Theorem 3.3(b), row addition operations do not change the value of the determinant, the determinant of the given matrix equals $\det A = 5$.

41. If $B^2 = B$, then

$$\det B = \det B^2 = \det BB$$
$$= (\det B)(\det B) = (\det B)^2.$$

Thus $\det B$ is a solution of the equation $x^2 = x$, so that $\det B = 1$ or $\det B = 0$.

CHAPTER 3 MATLAB EXERCISES

1. Matrix A can be transformed into an upper triangular matrix U using only row addition operations. The diagonal entries of U (rounded to 4 places after the decimal point) are -0.8000, -30.4375, 1.7865, -0.3488, -1.0967, and 0.3749. Thus det A equals the product of these numbers, which is 6.2400.

Chapter 4

Subspaces and Their Properties

4.1 SUBSPACES

1. A vector in the subspace has the form $\begin{bmatrix} 0 \\ s \end{bmatrix} = s \begin{bmatrix} 0 \\ 1 \end{bmatrix}$ for some scalar s. Hence

$$\left\{ \begin{bmatrix} 0 \\ 1 \end{bmatrix} \right\}$$

is a generating set for the subspace.

5. Since

$$\begin{bmatrix} -s + t \\ 2s - t \\ s + 3t \end{bmatrix} = s \begin{bmatrix} -1 \\ 2 \\ 1 \end{bmatrix} + t \begin{bmatrix} 1 \\ -1 \\ 3 \end{bmatrix},$$

a generating set for the subspace is

$$\left\{ \begin{bmatrix} -1 \\ 2 \\ 1 \end{bmatrix}, \begin{bmatrix} 1 \\ -1 \\ 3 \end{bmatrix} \right\}.$$

9. Since

$$\begin{bmatrix} 2s - 5t \\ 3r + s - 2t \\ r - 4s + 3t \\ -r + 2s \end{bmatrix} = r \begin{bmatrix} 0 \\ 3 \\ 1 \\ -1 \end{bmatrix} + s \begin{bmatrix} 2 \\ 1 \\ -4 \\ 2 \end{bmatrix} + t \begin{bmatrix} -5 \\ -2 \\ 3 \\ 0 \end{bmatrix},$$

a generating set for the subspace is

$$\left\{ \begin{bmatrix} 0 \\ 3 \\ 1 \\ -1 \end{bmatrix}, \begin{bmatrix} 2 \\ 1 \\ -4 \\ 2 \end{bmatrix}, \begin{bmatrix} -5 \\ -2 \\ 3 \\ 0 \end{bmatrix} \right\}.$$

13. For the given vector $\mathbf{v}$, we have $A\mathbf{v} \neq \mathbf{0}$. Hence $\mathbf{v}$ does not belong to Null A.

17. Because

$$A \begin{bmatrix} 3 \\ 1 \\ 1 \\ 2 \end{bmatrix} = \begin{bmatrix} 0 \\ 0 \\ 0 \end{bmatrix},$$

the vector $\begin{bmatrix} 3 \\ 1 \\ 1 \\ 2 \end{bmatrix}$ is in Null A.

21. Vector $\mathbf{u} = \begin{bmatrix} 1 \\ -4 \\ 2 \end{bmatrix}$ belongs to Col A if and only if $A\mathbf{x} = \mathbf{u}$ is consistent. Since

$$\begin{bmatrix} -7 \\ -4 \\ 0 \\ 0 \end{bmatrix}$$

is a solution of this system, $\mathbf{u}$ belongs to Col A.

68

25. Let $\mathbf{u} = \begin{bmatrix} 5 \\ -4 \\ -6 \end{bmatrix}$. The equation $A\mathbf{x} = \mathbf{u}$ is consistent; in fact, $\begin{bmatrix} -3 \\ -4 \\ 0 \\ 0 \end{bmatrix}$ is a solution. Hence $\mathbf{u}$ is in the column space of A.

29. The reduced row echelon form of the given matrix A is
$$\begin{bmatrix} 1 & 0 & -2 & 1 \\ 0 & 1 & 1 & 3 \\ 0 & 0 & 0 & 0 \end{bmatrix}.$$

Hence the general solution of $A\mathbf{x} = \mathbf{0}$ is
$$\begin{aligned} x_1 &= 2x_3 - x_4 \\ x_2 &= -x_3 - 3x_4 \\ x_3 &\quad \text{free} \\ x_4 &\quad \text{free}, \end{aligned}$$

and so the vector form of the general solution is
$$\begin{bmatrix} x_1 \\ x_2 \\ x_3 \\ x_4 \end{bmatrix} = \begin{bmatrix} 2x_3 - x_4 \\ -x_3 - 3x_4 \\ x_3 \\ x_4 \end{bmatrix}$$
$$= x_3 \begin{bmatrix} 2 \\ -1 \\ 1 \\ 0 \end{bmatrix} + x_4 \begin{bmatrix} -1 \\ -3 \\ 0 \\ 1 \end{bmatrix}.$$

Hence a generating set for Null A is
$$\left\{ \begin{bmatrix} 2 \\ -1 \\ 1 \\ 0 \end{bmatrix}, \begin{bmatrix} -1 \\ -3 \\ 0 \\ 1 \end{bmatrix} \right\}.$$

33. The reduced row echelon form of the given matrix A is
$$\begin{bmatrix} 1 & -3 & 0 & 1 & 0 & 2 \\ 0 & 0 & 1 & 2 & 0 & 3 \\ 0 & 0 & 0 & 0 & 1 & 2 \end{bmatrix}.$$

Hence the general solution of $A\mathbf{x} = \mathbf{0}$ is
$$\begin{aligned} x_1 &= 3x_2 - x_4 - 2x_6 \\ x_2 &\quad \text{free} \\ x_3 &= -2x_4 - 3x_6 \\ x_4 &\quad \text{free} \\ x_5 &= - 2x_6 \\ x_6 &\quad \text{free}, \end{aligned}$$

and the vector form of the general solution of $A\mathbf{x} = \mathbf{0}$ is
$$\begin{bmatrix} x_1 \\ x_2 \\ x_3 \\ x_4 \\ x_5 \\ x_6 \end{bmatrix} = x_2 \begin{bmatrix} 3 \\ 1 \\ 0 \\ 0 \\ 0 \\ 0 \end{bmatrix} + x_4 \begin{bmatrix} -1 \\ 0 \\ -2 \\ 1 \\ 0 \\ 0 \end{bmatrix} + x_6 \begin{bmatrix} -2 \\ 0 \\ -3 \\ 0 \\ -2 \\ 1 \end{bmatrix},$$

Thus a generating set for Null A is
$$\left\{ \begin{bmatrix} 3 \\ 1 \\ 0 \\ 0 \\ 0 \\ 0 \end{bmatrix}, \begin{bmatrix} -1 \\ 0 \\ -2 \\ 1 \\ 0 \\ 0 \end{bmatrix}, \begin{bmatrix} -2 \\ 0 \\ -3 \\ 0 \\ -2 \\ 1 \end{bmatrix} \right\}.$$

35. The standard matrix of T is
$$A = \begin{bmatrix} 1 & 2 & -1 \end{bmatrix}.$$

The range of T equals the column space of A; so $\{1, 2, -1\}$ is a generating set for the range of T. Note that A is in reduced row echelon form. The general solution of $A\mathbf{x} = \mathbf{0}$ is
$$\begin{aligned} x_1 &= -2x_2 + x_3 \\ x_2 &\quad \text{free} \\ x_3 &\quad \text{free}, \end{aligned}$$

and its vector form is
$$\begin{bmatrix} x_1 \\ x_2 \\ x_3 \end{bmatrix} = x_2 \begin{bmatrix} -2 \\ 1 \\ 0 \end{bmatrix} + x_3 \begin{bmatrix} 1 \\ 0 \\ 1 \end{bmatrix}.$$

Hence
$$\left\{ \begin{bmatrix} -2 \\ 1 \\ 0 \end{bmatrix}, \begin{bmatrix} 1 \\ 0 \\ 1 \end{bmatrix} \right\}$$
is a generating set for the null spaces of both A and T.

39. The standard matrix of T is
$$A = \begin{bmatrix} 1 & 1 & -1 \\ 0 & 0 & 0 \\ 2 & 0 & -1 \end{bmatrix}.$$

Since the range of T equals the column space of A, a generating set for the range of T is
$$\left\{ \begin{bmatrix} 1 \\ 0 \\ 2 \end{bmatrix}, \begin{bmatrix} 1 \\ 0 \\ 0 \end{bmatrix}, \begin{bmatrix} -1 \\ 0 \\ -1 \end{bmatrix} \right\}.$$

Since the reduced row echelon form of A is
$$\begin{bmatrix} 1 & 0 & -.5 \\ 0 & 1 & -.5 \\ 0 & 0 & 0 \end{bmatrix},$$
the vector form of the genereal solution of $A\mathbf{x} = \mathbf{0}$ is
$$\begin{bmatrix} x_1 \\ x_2 \\ x_3 \end{bmatrix} = x_3 \begin{bmatrix} .5 \\ .5 \\ 1 \end{bmatrix}.$$

Hence a generating set for Null A is
$$\left\{ \begin{bmatrix} .5 \\ .5 \\ 1 \end{bmatrix} \right\} \quad \text{or} \quad \left\{ \begin{bmatrix} 1 \\ 1 \\ 2 \end{bmatrix} \right\}.$$

43. True   44. True

45. False, $\{\mathbf{0}\}$ is called the *zero subspace*.

46. True   47. True

48. False, the column space of an $m \times n$ matrix is contained in $\mathcal{R}^m$.

49. False, the row space of an $m \times n$ matrix is contained in $\mathcal{R}^n$.

50. False, the *column space* of an $m \times n$ matrix equals $\{A\mathbf{v} : \mathbf{v} \text{ is in } \mathcal{R}^n\}$.

51. True   52. True   53. True

54. True

55. False, the range of a linear transformation equals the *column space* of its standard matrix.

56. True   57. True   58. True

59. True   60. True   61. True

62. True

65. From the reduced row echelon form of the matrix A in Exercise 32, we see that the pivot columns of A are columns 1, 2, and 4. Choosing each of these columns and exactly one of the other columns gives a generating set for the column space of A that contains exactly four vectors. (See Theorems 2.4(b) and 1.7.) One such generating set is the set containing the first four columns of A.

69. The reduced row echelon form of the given matrix A is
$$\begin{bmatrix} 1 & 0 & 2 & 0 \\ 0 & 1 & -3 & 0 \\ 0 & 0 & 0 & 1 \\ 0 & 0 & 0 & 0 \end{bmatrix}.$$

The columns of A form a generating set for the column space of A. But the matrix above shows that the third column of A is a linear combination of columns 1 and 2. Hence, by Theorem 1.7, a generating set for Col A containing exactly

3 vectors is
$$\left\{\begin{bmatrix}-2\\4\\5\\-1\end{bmatrix}, \begin{bmatrix}-1\\1\\2\\0\end{bmatrix}, \begin{bmatrix}3\\-4\\-5\\1\end{bmatrix}\right\}.$$

73. Consider $A = \begin{bmatrix}1 & 1\\2 & 2\end{bmatrix}$. The reduced row echelon form of A is $R = \begin{bmatrix}1 & 1\\0 & 0\end{bmatrix}$. Since $\begin{bmatrix}1\\2\end{bmatrix}$ belongs to Col A but not to Col R, we see that Col $A \neq$ Col R.

77. Let V and W be subspaces of $\mathcal{R}^n$. Since $\mathbf{0}$ is contained in both V and W, $\mathbf{0}$ is contained in $V \cap W$. Let $\mathbf{v}$ and $\mathbf{w}$ be contained in $V \cap W$. Then $\mathbf{v}$ and $\mathbf{w}$ are contained in both V and W, and thus $\mathbf{v} + \mathbf{w}$ is contained in both V and W. Hence $\mathbf{v} + \mathbf{w}$ is contained in $V \cap W$. Finally, for any scalar c, $c\mathbf{v}$ is contained in both V and W; so $c\mathbf{v}$ is in $V \cap W$. It follows that $V \cap W$ is a subspace of $\mathcal{R}^n$.

81. The vectors $\begin{bmatrix}1\\0\end{bmatrix}$ and $\begin{bmatrix}0\\1\end{bmatrix}$ are in the set, but their sum is not.

83. If $\mathbf{0}$ were in the set, then there would be scalars s and t such that
$$\begin{aligned}3s &= 2\\2s + 4t &= 0\\-t &= 0.\end{aligned}$$
Since the system has no solutions, $\mathbf{0}$ is not in the given set.

87. Consider
$$\mathbf{v} = \begin{bmatrix}2\\1\\2\end{bmatrix} \text{ and } 3\mathbf{v} = \begin{bmatrix}6\\3\\6\end{bmatrix}.$$
The vector $\mathbf{v}$ belongs to the given set because $2 = 1(2)$. However, $3\mathbf{v}$ does not belong to the given set because $6 \neq 3(6)$.

91. Denote the given set by V. Since
$$2u_1 + 5u_2 - 4u_3 = 0$$
for $u_1 = u_2 = u_3 = 0$, we see that $\mathbf{0}$ is in V. Let $\mathbf{u}$ and $\mathbf{v}$ be in V. Then
$$2u_1 + 5u_2 - 4u_3 = 0$$
and
$$2v_1 + 5v_2 - 4v_3 = 0.$$
Now
$$\mathbf{u} + \mathbf{v} = \begin{bmatrix}u_1 + v_1\\u_2 + v_2\\u_3 + v_3\end{bmatrix},$$
and
$$\begin{aligned}2(u_1 + v_1) + 5(u_2 + v_2) &- 4(u_3 + v_3)\\= (2u_1 + 5u_2 &- 4u_3)\\+ (2v_1 + 5v_2 &- 4v_3)\\= 0 + 0&\\= 0.&\end{aligned}$$
So $\mathbf{u} + \mathbf{v}$ belongs to V. Thus V is closed under vector addition.

For any scalar c,
$$c\mathbf{u} = \begin{bmatrix}cu_1\\cu_2\\cu_3\end{bmatrix},$$
and
$$\begin{aligned}2(cu_1) + 5(cu_2) - 4(cu_3)\\= c(2u_1 + 5u_2 - 4u_3)\\= c(0)\\= 0.\end{aligned}$$
Thus $c\mathbf{u}$ belongs to V, and hence V is also closed under scalar multiplication.

Since V is a subset of $\mathcal{R}^3$ that contains $\mathbf{0}$ and is closed under both vector addition and scalar multiplication, V is a subspace of $\mathcal{R}^3$.

95. Let V denote the null space of T. Since $T(\mathbf{0}) = \mathbf{0}$, $\mathbf{0}$ is in V. If $\mathbf{u}$ and $\mathbf{v}$ are in V, then $T(\mathbf{u}) = T(\mathbf{v}) = \mathbf{0}$. Hence

$$T(\mathbf{u}+\mathbf{v}) = T(\mathbf{u}) + T(\mathbf{v}) = \mathbf{0} + \mathbf{0} = \mathbf{0};$$

so $\mathbf{u} + \mathbf{v}$ is in V. Finally, for any scalar c and any vector $\mathbf{u}$ in V, we have

$$T(c\mathbf{u}) = cT(\mathbf{u}) = c(\mathbf{0}) = \mathbf{0};$$

so $c\mathbf{u}$ is in V. Thus V is also closed under scalar multiplication. Since V is a subset of $\mathcal{R}^n$ that contains $\mathbf{0}$ and is closed under both vector addition and scalar multiplication, V is a subspace of $\mathcal{R}^n$.

99. Because $A\mathbf{0} = \mathbf{0} = B\mathbf{0}$, the zero vector is in V. Assume that $\mathbf{u}$ and $\mathbf{v}$ are in V. Then $A\mathbf{u} = B\mathbf{u}$ and $A\mathbf{v} = B\mathbf{v}$. Hence

$$\begin{aligned} A(\mathbf{u}+\mathbf{v}) &= A\mathbf{u} + A\mathbf{v} \\ &= B\mathbf{u} + B\mathbf{v} = B(\mathbf{u}+\mathbf{v}). \end{aligned}$$

Thus $\mathbf{u}+\mathbf{v}$ is in V, and so V is closed under vector addition. Also, for any scalar c,

$$A(c\mathbf{u}) = cA\mathbf{u} = cB\mathbf{u} = B(c\mathbf{u}).$$

Hence $c\mathbf{u}$ is in V, and V is closed under scalar multiplication. Since V is a subset of $\mathcal{R}^n$ that contains $\mathbf{0}$ and is closed under both vector addition and scalar multiplication, V is a subspace of $\mathcal{R}^n$.

101. (a) The system $A\mathbf{x} = \mathbf{u}$ is consistent since the reduced row echelon form of $[A\ \mathbf{u}]$ contains no row whose only nonzero entry lies in the last column. Hence $\mathbf{u}$ belongs to Col A.
(b) On the other hand, $A\mathbf{x} = \mathbf{v}$ is not consistent, and so $\mathbf{v}$ does not belong to Col A.

4.2 BASIS AND DIMENSION

1. The reduced row echelon form of the given matrix A is

$$\begin{bmatrix} 1 & -3 & 4 & -2 \\ 0 & 0 & 0 & 0 \end{bmatrix}.$$

(a) The pivot columns of A form a basis for Col A. Hence

$$\left\{ \begin{bmatrix} 1 \\ -1 \end{bmatrix} \right\}$$

is a basis for the column space of A.

(b) The null space of A is the solution set of $A\mathbf{x} = \mathbf{0}$. Now the general solution of $A\mathbf{x} = \mathbf{0}$ is

$$\begin{aligned} x_1 &= 3x_2 - 4x_3 + 2x_4 \\ x_2 &\quad \text{free} \\ x_3 &\quad \text{free} \\ x_4 &\quad \text{free.} \end{aligned}$$

Thus the vector form of the general solution is

$$\begin{bmatrix} x_1 \\ x_2 \\ x_3 \\ x_4 \end{bmatrix} = \begin{bmatrix} 3x_2 - 4x_3 + 2x_4 \\ x_2 \\ x_3 \\ x_4 \end{bmatrix}$$

$$= x_2 \begin{bmatrix} 3 \\ 1 \\ 0 \\ 0 \end{bmatrix} + x_3 \begin{bmatrix} -4 \\ 0 \\ 1 \\ 0 \end{bmatrix} + x_4 \begin{bmatrix} 2 \\ 0 \\ 0 \\ 1 \end{bmatrix}.$$

Hence

$$\left\{ \begin{bmatrix} 3 \\ 1 \\ 0 \\ 0 \end{bmatrix}, \begin{bmatrix} -4 \\ 0 \\ 1 \\ 0 \end{bmatrix}, \begin{bmatrix} 2 \\ 0 \\ 0 \\ 1 \end{bmatrix} \right\}$$

is a basis for the null space of A.

5. The reduced row echelon form of the given matrix A is
$$\begin{bmatrix} 1 & -2 & 0 & 2 \\ 0 & 0 & 1 & -1 \\ 0 & 0 & 0 & 0 \end{bmatrix}.$$

(a) Hence the first and third columns of the given matrix are its pivot columns, and so
$$\left\{ \begin{bmatrix} 1 \\ -1 \\ 2 \end{bmatrix}, \begin{bmatrix} 0 \\ 1 \\ 3 \end{bmatrix} \right\}$$
is a basis for the column space of A.

(b) The general solution of $A\mathbf{x} = \mathbf{0}$ is
$$\begin{aligned} x_1 &= 2x_2 - 2x_4 \\ x_2 & \text{ free} \\ x_3 &= \phantom{2x_2 -{}} x_4 \\ x_4 & \text{ free.} \end{aligned}$$

Thus the vector form of the general solution is
$$\begin{bmatrix} x_1 \\ x_2 \\ x_3 \\ x_4 \end{bmatrix} = \begin{bmatrix} 2x_2 - 2x_4 \\ x_2 \\ x_4 \\ x_4 \end{bmatrix}$$
$$= x_2 \begin{bmatrix} 2 \\ 1 \\ 0 \\ 0 \end{bmatrix} + x_4 \begin{bmatrix} -2 \\ 0 \\ 1 \\ 1 \end{bmatrix}.$$

Hence
$$\left\{ \begin{bmatrix} 2 \\ 1 \\ 0 \\ 0 \end{bmatrix}, \begin{bmatrix} -2 \\ 0 \\ 1 \\ 1 \end{bmatrix} \right\}$$
is a basis for the null space of A.

7. The reduced row echelon form of the given matrix A is
$$\begin{bmatrix} 1 & 0 & 0 & 4 \\ 0 & 1 & 0 & 4 \\ 0 & 0 & 1 & 1 \\ 0 & 0 & 0 & 0 \end{bmatrix}.$$

(a) Hence the first three columns of the given matrix are its pivot columns, and so
$$\left\{ \begin{bmatrix} -1 \\ 2 \\ 1 \\ 0 \end{bmatrix}, \begin{bmatrix} 1 \\ 0 \\ -1 \\ 1 \end{bmatrix}, \begin{bmatrix} 2 \\ -5 \\ -1 \\ -2 \end{bmatrix} \right\}$$
is a basis for the column space of A.

(b) The null space of A is the solution set of $A\mathbf{x} = \mathbf{0}$. Since the vector form of the general solution of $A\mathbf{x} = \mathbf{0}$ is
$$\begin{bmatrix} x_1 \\ x_2 \\ x_3 \\ x_4 \end{bmatrix} = x_4 \begin{bmatrix} -4 \\ -4 \\ -1 \\ 1 \end{bmatrix},$$
the set
$$\left\{ \begin{bmatrix} -4 \\ -4 \\ -1 \\ 1 \end{bmatrix} \right\}$$
is a basis for the null space of A.

11. The standard matrix of T is
$$A = \begin{bmatrix} 1 & -2 & 1 & 1 \\ 2 & -5 & 1 & 3 \\ 1 & -3 & 0 & 2 \end{bmatrix}.$$

(a) The range of T equals the column space of A; so we proceed as in Exercise 7. The reduced row echelon form of A is
$$R = \begin{bmatrix} 1 & 0 & 3 & -1 \\ 0 & 1 & 1 & -1 \\ 0 & 0 & 0 & 0 \end{bmatrix}.$$

Hence the set of pivot columns of A,

$$\left\{ \begin{bmatrix} 1 \\ 2 \\ 1 \end{bmatrix}, \begin{bmatrix} -2 \\ -5 \\ -3 \end{bmatrix} \right\},$$

is a basis for the range of T.

(b) Since the null space of T is the same as the null space of A, we must determine the vector form of the general solution of $A\mathbf{x} = \mathbf{0}$. This representation is:

$$\begin{bmatrix} x_1 \\ x_2 \\ x_3 \\ x_4 \end{bmatrix} = x_3 \begin{bmatrix} -3 \\ -1 \\ 1 \\ 0 \end{bmatrix} + x_4 \begin{bmatrix} 1 \\ 1 \\ 0 \\ 1 \end{bmatrix}.$$

Hence

$$\left\{ \begin{bmatrix} -3 \\ -1 \\ 1 \\ 0 \end{bmatrix}, \begin{bmatrix} 1 \\ 1 \\ 0 \\ 1 \end{bmatrix} \right\}$$

is a basis for the null space of T.

15. The standard matrix of T is

$$A = \begin{bmatrix} 1 & 2 & 3 & 0 & 4 \\ 3 & 1 & -1 & 0 & -3 \\ 7 & 4 & 1 & 0 & -2 \end{bmatrix},$$

and the reduced row echelon form of A is

$$\begin{bmatrix} 1 & 0 & -1 & 0 & -2 \\ 0 & 1 & 2 & 0 & 3 \\ 0 & 0 & 0 & 0 & 0 \end{bmatrix}.$$

(a) As in Exercise 9, the set of pivot columns of A,

$$\left\{ \begin{bmatrix} 1 \\ 3 \\ 7 \end{bmatrix}, \begin{bmatrix} 2 \\ 1 \\ 4 \end{bmatrix} \right\},$$

is a basis for the range of T.

(b) The vector form of the general solution of $A\mathbf{x} = \mathbf{0}$ is

$$\begin{bmatrix} x_1 \\ x_2 \\ x_3 \\ x_4 \\ x_5 \end{bmatrix} =$$

$$x_3 \begin{bmatrix} 1 \\ -2 \\ 1 \\ 0 \\ 0 \end{bmatrix} + x_4 \begin{bmatrix} 0 \\ 0 \\ 0 \\ 1 \\ 0 \end{bmatrix} + x_5 \begin{bmatrix} 2 \\ -3 \\ 0 \\ 0 \\ 1 \end{bmatrix}.$$

Thus

$$\left\{ \begin{bmatrix} 1 \\ -2 \\ 1 \\ 0 \\ 0 \end{bmatrix}, \begin{bmatrix} 0 \\ 0 \\ 0 \\ 1 \\ 0 \end{bmatrix}, \begin{bmatrix} 2 \\ -3 \\ 0 \\ 0 \\ 1 \end{bmatrix} \right\}$$

is a basis for the null space of T.

17. Since

$$\begin{bmatrix} s \\ -2s \end{bmatrix} = s \begin{bmatrix} 1 \\ -2 \end{bmatrix}$$

and $\left\{ \begin{bmatrix} 1 \\ -2 \end{bmatrix} \right\}$ is linearly independent, this set is a basis for the given subspace.

21. The general solution of $x_1 - 3x_2 + 5x_3 = 0$ is

$$x_1 = 3x_2 - 5x_3$$
$$x_2 \quad \text{free}$$
$$x_3 \quad \text{free}.$$

Thus the vector form of the general solution is

$$\begin{bmatrix} x_1 \\ x_2 \\ x_3 \end{bmatrix} = \begin{bmatrix} 3x_2 - 5x_3 \\ x_2 \\ x_3 \end{bmatrix}$$

$$= x_2 \begin{bmatrix} 3 \\ 1 \\ 0 \end{bmatrix} + x_3 \begin{bmatrix} -5 \\ 0 \\ 1 \end{bmatrix}.$$

Hence

$$\left\{ \begin{bmatrix} 3 \\ 1 \\ 0 \end{bmatrix}, \begin{bmatrix} -5 \\ 0 \\ 1 \end{bmatrix} \right\}$$

is a basis for the given subspace.

25. Let
$$A = \begin{bmatrix} 1 & 2 & 1 \\ 2 & 1 & -4 \\ 1 & 3 & 3 \end{bmatrix}.$$

Then the given subspace is Col A, and so a basis for the given subspace can be obtained by choosing the pivot columns of A. Since the reduced row echelon form of A is
$$\begin{bmatrix} 1 & 0 & -3 \\ 0 & 1 & 2 \\ 0 & 0 & 0 \end{bmatrix},$$
this basis is
$$\left\{ \begin{bmatrix} 1 \\ 2 \\ 1 \end{bmatrix}, \begin{bmatrix} 2 \\ 1 \\ 3 \end{bmatrix} \right\}.$$

29. As in Exercise 25, form a 4×5 matrix whose columns are the vectors in the given set. The pivot columns of this matrix form a basis for the given subspace. Since the reduced row echelon form of this matrix is
$$\begin{bmatrix} 1 & 0 & -5 & 2 & 0 \\ 0 & 1 & 3 & -1 & 0 \\ 0 & 0 & 0 & 0 & 1 \\ 0 & 0 & 0 & 0 & 0 \end{bmatrix},$$
one basis for the given subspace is
$$\left\{ \begin{bmatrix} 1 \\ 0 \\ -1 \\ 2 \end{bmatrix}, \begin{bmatrix} 1 \\ 1 \\ -2 \\ 1 \end{bmatrix}, \begin{bmatrix} 0 \\ 1 \\ -1 \\ 2 \end{bmatrix} \right\}.$$

33. False, every nonzero subspace of $\mathcal{R}^n$ has infinitely many bases.

34. True

35. False, a basis for a subspace is a generating set that is as *small* as possible.

36. True   37. True   38. True

39. True

40. False, the *pivot columns* of any matrix form a basis for its column space.

41. False, if $A = \begin{bmatrix} 1 & 2 \\ 1 & 2 \end{bmatrix}$, then the pivot columns of the reduced row echelon form of A do not form a basis for Col A.

42. True

43. False, every generating set for V contains *at least k* vectors.

44. True   45. True   46. True

47. True   48. True   49. True

50. False, neither standard vector is in the subspace $\left\{ \begin{bmatrix} u_1 \\ u_2 \end{bmatrix} \in \mathcal{R}^2 : u_1 + u_2 = 0 \right\}$.

51. True   52. True

53. A generating set for $\mathcal{R}^n$ must contain at least n vectors. Because the given set is a set of 3 vectors from $\mathcal{R}^4$, it cannot be a generating set for $\mathcal{R}^4$.

55. It follows from Theorem 4.5 that every basis for $\mathcal{R}^n$ must contain exactly n vectors. Hence the given set of 2 vectors cannot be a basis for $\mathcal{R}^3$.

57. By property 4 of linearly dependent and independent sets in Section 1.7, a set of more than 2 vectors from $\mathcal{R}^2$ must be linearly dependent.

61. We proceed as in Example 4. Let W be the subspace in Exercise 29, A denote the matrix whose columns are the vectors in Exercise 29, and $\mathcal{B}$ be the set of three vectors given in Exercise 61. For each vector $\mathbf{v}$ in $\mathcal{B}$, the equation $A\mathbf{x} = \mathbf{v}$ is consistent. Hence $\mathcal{B}$ is contained in W. Moreover, the reduced row echelon form of the matrix whose columns are the vectors in $\mathcal{B}$ is $[\mathbf{e}_1\ \mathbf{e}_2\ \mathbf{e}_3]$; so $\mathcal{B}$ is linearly independent. Finally, Exercise 29 shows that the dimension of W is 3, which equals the number of vectors in $\mathcal{B}$. Thus the 3 conditions of the boxed statement on pages 248–249 are satisfied, and so $\mathcal{B}$ is a basis for W.

65. Let A denote the matrix in Exercise 7 and $\mathcal{B}$ be the set of three vectors given in Exercise 65. Because the column space of a matrix equals the span of its columns, Exercise 65 can be worked in the same way as Exercise 61. For each vector $\mathbf{v}$ in $\mathcal{B}$, the equation $A\mathbf{x} = \mathbf{v}$ is consistent, and so $\mathcal{B}$ is contained in Col A. Moreover, the reduced row echelon form of the matrix whose columns are the vectors in $\mathcal{B}$ is $[\mathbf{e}_1\ \mathbf{e}_2\ \mathbf{e}_3]$; so $\mathcal{B}$ is linearly independent. Finally, Exercise 7 shows that the dimension of Col A is 3, which equals the number of vectors in $\mathcal{B}$. Thus the 3 conditions of the boxed statement on pages 248–249 are satisfied, and so $\mathcal{B}$ is a basis for W.

69. Let V denote the given subspace of $\mathcal{R}^n$. Clearly $\mathcal{B} = \{\mathbf{e}_3, \mathbf{e}_4, \ldots, \mathbf{e}_n\}$ is a subset of V, and $\mathcal{B}$ is linearly independent because every column of $[\mathbf{e}_3\ \mathbf{e}_4\ \cdots\ \mathbf{e}_n]$ is a pivot column. Moreover, $\mathcal{B}$ is a generating set for V, for if $\mathbf{v}$ is in V, then

$$\mathbf{v} = \begin{bmatrix} 0 \\ 0 \\ v_3 \\ \vdots \\ v_n \end{bmatrix} = v_3 \mathbf{e}_3 + \cdots + v_n \mathbf{e}_n.$$

Since $\mathcal{B}$ is a linearly independent generating set for V, we see that $\mathcal{B}$ is a basis for V. Hence the dimension of V equals the number of vectors in $\mathcal{B}$, which is $n - 2$.

73. Vector $\mathbf{v}$ belongs to $V = \text{Span}\,\mathcal{A}$. Thus $\mathcal{B} = \{\mathbf{v}, \mathbf{u}_2, \mathbf{u}_3, \ldots, \mathbf{u}_k\}$ is a subset of V, because $\mathbf{u}_2, \mathbf{u}_3, \ldots, \mathbf{u}_k$ belong to $\mathcal{A}$, which is a subset of V.

We claim that $\mathcal{B}$ is linearly independent. Suppose that $c_1, c_2, \ldots, c_k$ are scalars such that

$$c_1 \mathbf{v} + c_2 \mathbf{u}_2 + \cdots + c_k \mathbf{u}_k = \mathbf{0}.$$

Then

$$c_1(\mathbf{u}_1 + \mathbf{u}_2 + \cdots + \mathbf{u}_k) + c_2 \mathbf{u}_2 + \cdots + c_k \mathbf{u}_k = \mathbf{0}$$

that is,

$$c_1 \mathbf{u}_1 + (c_1 + c_2)\mathbf{u}_2 + \cdots + (c_1 + c_k)\mathbf{u}_k = \mathbf{0}.$$

Since $\mathcal{A}$ is linearly independent, it follows that $c_1 = 0$, $c_1 + c_2 = 0$, $\cdots$, $c_1 + c_k = 0$. Hence $c_1 = c_2 = \cdots = c_k = 0$, proving that $\mathcal{B}$ is linearly independent. Since $\mathcal{B}$ contains k vectors, it follows from Theorem 4.7 that $\mathcal{B}$ is a basis for V.

77. (a) Because V and W are subspaces of $\mathcal{R}^n$, $\mathbf{0}$ is in both V and W. Assume that $\mathbf{u}$ is in both V and W. Then

$\mathbf{u} = \mathbf{v}_1 + \mathbf{w}_1$, where $\mathbf{v}_1 = \mathbf{u}$ and $\mathbf{w}_1 = \mathbf{0}$, and also $\mathbf{u} = \mathbf{v}_2 + \mathbf{w}_2$, where $\mathbf{v}_2 = \mathbf{0}$ and $\mathbf{w}_2 = \mathbf{u}$. The uniqueness of the representation of $\mathbf{u}$ in the form $\mathbf{v} + \mathbf{w}$ for some $\mathbf{v}$ in V and some $\mathbf{w}$ in W implies that $\mathbf{u} = \mathbf{v}_1 = \mathbf{v}_2 = \mathbf{0}$. Hence the only vector in both V and W is $\mathbf{0}$.

(b) Let $\mathcal{B}_1 = \{\mathbf{v}_1, \mathbf{v}_2, \ldots, \mathbf{v}_k\}$ be a basis for V, $\mathcal{B}_2 = \{\mathbf{w}_1, \mathbf{w}_2, \ldots, \mathbf{w}_m\}$ be a basis for W, and $\mathcal{B} = \{\mathbf{v}_1, \mathbf{v}_2, \ldots, \mathbf{v}_k, \mathbf{w}_1, \mathbf{w}_2, \ldots, \mathbf{w}_m\}$. Note that $\dim V = k$ and $\dim W = m$. We will show that $\mathcal{B}$ is a basis for $\mathcal{R}^n$ so that, by Theorem 4.5, the number of vectors in $\mathcal{B}$ must be n, that is, $\dim V + \dim W = k + m = n$.

First we show that $\mathcal{B}$ is linearly independent. Let $a_1, a_2, \ldots, a_k$ and $b_1, b_2, \ldots, b_m$ be scalars such that

$$a_1\mathbf{v}_1 + a_2\mathbf{v}_2 + \cdots + a_k\mathbf{v}_k$$
$$+ b_1\mathbf{w}_1 + b_2\mathbf{w}_2 + \cdots + b_m\mathbf{w}_m$$
$$= \mathbf{0}.$$

Let
$$\mathbf{v} = a_1\mathbf{v}_1 + a_2\mathbf{v}_2 + \cdots + a_k\mathbf{v}_k$$
and
$$\mathbf{w} = b_1\mathbf{w}_1 + b_2\mathbf{w}_2 + \cdots + b_m\mathbf{w}_m.$$

Then $\mathbf{v} = -\mathbf{w}$. Because $\mathbf{0}$ is the only vector in both V and W, it follows that $\mathbf{v} = \mathbf{0}$ and $\mathbf{w} = \mathbf{0}$. But if
$$\mathbf{v} = a_1\mathbf{v}_1 + a_2\mathbf{v}_2 + \cdots + a_k\mathbf{v}_k$$
$$= \mathbf{0},$$
then $a_1 = a_2 = \ldots = a_k = 0$ because $\mathcal{B}_1$ is linearly independent.

Similarly, $b_1 = b_2 = \ldots = b_m = 0$. Thus $\mathcal{B}$ is linearly independent.

Next, we show that $\mathcal{B}$ is a generating set for $\mathcal{R}^n$. For any $\mathbf{u}$ in $\mathcal{R}^n$, there exist $\mathbf{v}$ in V and $\mathbf{w}$ in W such that $\mathbf{u} = \mathbf{v} + \mathbf{w}$. Because $\mathcal{B}_1$ is a basis for V, there exist scalars $a_1, a_2, \ldots, a_k$ such that

$$\mathbf{v} = a_1\mathbf{v}_1 + a_2\mathbf{v}_2 + \cdots + a_k\mathbf{v}_k.$$

Similarly, there exist scalars $b_1, b_2, \ldots, b_m$ such that

$$\mathbf{w} = b_1\mathbf{w}_1 + b_2\mathbf{w}_2 + \cdots + b_m\mathbf{w}_m.$$

Hence $\mathbf{u} = \mathbf{v} + \mathbf{w}$ is a linear combination of the vectors in $\mathcal{B}$, and so $\mathcal{B}$ is a generating set for $\mathcal{R}^n$. Because $\mathcal{B}$ is a linearly independent generating set for $\mathcal{R}^n$, $\mathcal{B}$ is a basis for $\mathcal{R}^n$, completing the proof.

81. Let
$$A = \begin{bmatrix} 1 & -1 & 2 & 1 \\ 2 & -2 & 4 & 2 \\ -3 & 3 & -6 & -3 \end{bmatrix}.$$

Since the reduced row echelon form of A is
$$\begin{bmatrix} 1 & -1 & 2 & 1 \\ 0 & 0 & 0 & 0 \\ 0 & 0 & 0 & 0 \end{bmatrix},$$
the vector form of the general solution of $A\mathbf{x} = \mathbf{0}$ is

$$\begin{bmatrix} x_1 \\ x_2 \\ x_3 \\ x_4 \end{bmatrix} = x_2 \begin{bmatrix} 1 \\ 1 \\ 0 \\ 0 \end{bmatrix} + x_3 \begin{bmatrix} -2 \\ 0 \\ 1 \\ 0 \end{bmatrix} + x_4 \begin{bmatrix} -1 \\ 0 \\ 0 \\ 1 \end{bmatrix}.$$

Hence
$$\left\{ \begin{bmatrix} 1 \\ 1 \\ 0 \\ 0 \end{bmatrix}, \begin{bmatrix} -2 \\ 0 \\ 1 \\ 0 \end{bmatrix}, \begin{bmatrix} -1 \\ 0 \\ 0 \\ 1 \end{bmatrix} \right\}$$

is a basis for Null A.

Since the reduced row echelon form of

$$\begin{bmatrix} 0 & 1 & -2 & -1 \\ 2 & 1 & 0 & 0 \\ 1 & 0 & 1 & 0 \\ 0 & 0 & 0 & 1 \end{bmatrix}$$

is

$$\begin{bmatrix} 1 & 0 & 1 & 0 \\ 0 & 1 & -2 & 0 \\ 0 & 0 & 0 & 1 \\ 0 & 0 & 0 & 0 \end{bmatrix},$$

it follows from Exercise 78 that

$$\left\{ \begin{bmatrix} 0 \\ 2 \\ 1 \\ 0 \end{bmatrix}, \begin{bmatrix} 1 \\ 1 \\ 0 \\ 0 \end{bmatrix}, \begin{bmatrix} -1 \\ 0 \\ 0 \\ 1 \end{bmatrix} \right\}$$

is a basis for Null A that contains $\mathcal{L}$.

85. The reduced row echelon form of A is

$$\begin{bmatrix} 1 & 0 & -1.2 & 0 & 1.4 \\ 0 & 1 & 2.3 & 0 & -2.9 \\ 0 & 0 & 0.0 & 1 & 0.7 \end{bmatrix}.$$

(a) As in Exercise 7,

$$\left\{ \begin{bmatrix} 0.1 \\ 0.7 \\ -0.5 \end{bmatrix}, \begin{bmatrix} 0.2 \\ 0.9 \\ 0.5 \end{bmatrix}, \begin{bmatrix} 0.5 \\ -0.5 \\ -0.5 \end{bmatrix} \right\}$$

is a basis for the column space of A.

(b) The vector form of the general solution of $A\mathbf{x} = \mathbf{0}$ is

$$\begin{bmatrix} x_1 \\ x_2 \\ x_3 \\ x_4 \\ x_5 \end{bmatrix} = x_3 \begin{bmatrix} 1.2 \\ -2.3 \\ 1.0 \\ 0.0 \\ 0.0 \end{bmatrix} + x_5 \begin{bmatrix} -1.4 \\ 2.9 \\ 0.0 \\ -0.7 \\ 1.0 \end{bmatrix}.$$

Hence

$$\left\{ \begin{bmatrix} 1.2 \\ -2.3 \\ 1.0 \\ 0.0 \\ 0.0 \end{bmatrix}, \begin{bmatrix} -1.4 \\ 2.9 \\ 0.0 \\ -0.7 \\ 1.0 \end{bmatrix} \right\}$$

is a basis for Null A.

4.3 THE DIMENSION OF SUBSPACES ASSOCIATED WITH A MATRIX

1. (a) The dimension of Col A equals rank A, which is 2.
(b) The dimension of Null A equals the nullity of A, which is $4 - 2 = 2$.
(c) The dimension of Row A equals rank A, which is 2.
(d) The dimension of Null A^T equals the nullity of A^T. Because A^T is a 4×3 matrix, the nullity of A^T equals

$$3 - \text{rank } A^T = 3 - \text{rank } A$$
$$= 3 - 2 = 1.$$

5. Clearly rank $A = 1$. So, as in Exercise 1, the answers are:
(a) 1   (b) 3   (c) 1   (d) 0.

9. The reduced row echelon form of A is

$$\begin{bmatrix} 1 & 0 & 6 & 0 \\ 0 & 1 & -4 & 1 \\ 0 & 0 & 0 & 0 \end{bmatrix}.$$

Hence rank $A = 2$. As in Exercise 1, the answers are:
(a) 2   (b) 2   (c) 2   (d) 1.

13. Every vector in the given subspace V has the form $\begin{bmatrix} -2s \\ s \end{bmatrix} = s \begin{bmatrix} -2 \\ 1 \end{bmatrix}$ for some scalar s. Hence $\mathcal{B} = \left\{ \begin{bmatrix} -2 \\ 1 \end{bmatrix} \right\}$ is a generating set for V. But a set containing a single nonzero vector is linearly independent, and so $\mathcal{B}$ is a basis for V. Thus the dimension of V equals 1, the number of vectors in $\mathcal{B}$.

17. The reduced row echelon form of A is

$$R = \begin{bmatrix} 1 & 0 & 3 \\ 0 & 1 & 2 \end{bmatrix}.$$

By Theorem 4.8, a basis for Row A is formed from the nonzero rows of R:

$$\left\{ \begin{bmatrix} 1 \\ 0 \\ 3 \end{bmatrix}, \begin{bmatrix} 0 \\ 1 \\ 2 \end{bmatrix} \right\}.$$

21. Proceeding as in Exercise 17, we see that a basis for Row A is

$$\left\{ \begin{bmatrix} 1 \\ 0 \\ 0 \\ -3 \\ 1 \\ 3 \end{bmatrix}, \begin{bmatrix} 0 \\ 1 \\ 0 \\ 2 \\ -1 \\ -2 \end{bmatrix}, \begin{bmatrix} 0 \\ 0 \\ 1 \\ 0 \\ 0 \\ -1 \end{bmatrix} \right\}.$$

25. It follows from Exercise 17 that the dimension of Row A equals 2. Hence a basis for Row A must consist of 2 vectors, and so

$$\left\{ \begin{bmatrix} 1 \\ -1 \\ 1 \end{bmatrix}, \begin{bmatrix} 0 \\ 1 \\ 2 \end{bmatrix} \right\}$$

is the only basis for Row A consisting of rows of A.

29. Exercise 21 shows that the dimension of Row A equals 3, and so a basis for Row A consists of 3 linearly independent rows of A. The reduced row echelon form of A^T is

$$\begin{bmatrix} 1 & 0 & 1 & 0 \\ 0 & 1 & -1 & 0 \\ 0 & 0 & 0 & 1 \\ 0 & 0 & 0 & 0 \\ 0 & 0 & 0 & 0 \\ 0 & 0 & 0 & 0 \end{bmatrix}.$$

Because the pivot columns of A^T are columns 1, 2, and 4, it follows that rows 1, 2, and 4 of A are linearly independent. Hence

$$\left\{ \begin{bmatrix} 1 \\ 0 \\ -1 \\ -3 \\ 1 \\ 4 \end{bmatrix}, \begin{bmatrix} 2 \\ -1 \\ -1 \\ -8 \\ 3 \\ 9 \end{bmatrix}, \begin{bmatrix} 0 \\ 1 \\ 1 \\ 2 \\ -1 \\ -3 \end{bmatrix} \right\}$$

is a basis for Row A consisting of rows of A.

33. The standard matrix of T is $\begin{bmatrix} 1 & 2 \\ 2 & 1 \end{bmatrix}$, and its reduced row echelon form is I_2.

   (a) Since the range of T equals the column space of A, the dimension of the range of T equals the rank of A, which is 2. Thus, by Theorem 2.10, T is onto.

   (b) The null space of T equals the null space of A. Hence the dimension of the null space of T equals the nullity of A, which is 0. Thus T is one-to-one by Theorem 2.11.

37. The standard matrix of T is

$$\begin{bmatrix} 1 & 0 \\ 2 & 1 \\ 0 & -1 \end{bmatrix},$$

and its reduced row echelon form is

$$\begin{bmatrix} 1 & 0 \\ 0 & 1 \\ 0 & 0 \end{bmatrix}.$$

(a) As in Exercise 33, the dimension of the range of T is 2. Since the codomain of T is $\mathcal{R}^3$, T is not onto.
(b) As in Exercise 33, the dimension of the null space of T is 0. Hence T is one-to-one.

41. False, the dimensions of the subspaces $V = \text{Span } \{\mathbf{e}_1\}$ and $W = \text{Span } \{\mathbf{e}_2\}$ of $\mathcal{R}^2$ are both 1, but $V \neq W$.

42. True   43. True

44. False, the dimension of the null space of a matrix equals the *nullity* of the matrix.

45. False, the dimension of the column space of a matrix equals the *rank* of the matrix.

46. True   47. True

48. False, consider $A = \begin{bmatrix} 1 & 2 \\ 1 & 2 \end{bmatrix}$ and the reduced row echelon form of A, which is $\begin{bmatrix} 1 & 2 \\ 0 & 0 \end{bmatrix}$.

49. True

50. False, the nonzero rows of *the reduced row echelon form* of a matrix form a basis for its row space.

51. False, consider $\begin{bmatrix} 1 & 0 & 0 \\ 0 & 0 & 0 \\ 0 & 1 & 0 \end{bmatrix}$.

52. False, consider $\begin{bmatrix} 1 & 0 & 0 \\ 0 & 1 & 0 \end{bmatrix}$.

53. True   54. True

55. False, consider any nonsquare matrix.

56. False, the dimension of the null space of any $m \times n$ matrix A plus the dimension of its column space equals

$$\text{rank } A + \text{nullity } A$$
$$= \text{rank } A + (n - \text{rank } A) = n.$$

57. True   58. True   59. True

60. True

61. Taking $s = \frac{3}{7}$ and $t = -\frac{1}{7}$, we have

$$\begin{bmatrix} 2s - t \\ s + 3t \end{bmatrix} = \begin{bmatrix} 1 \\ 0 \end{bmatrix};$$

and taking $s = \frac{1}{7}$ and $t = \frac{2}{7}$, we have

$$\begin{bmatrix} 2s - t \\ s + 3t \end{bmatrix} = \begin{bmatrix} 0 \\ 1 \end{bmatrix}.$$

Hence $\mathcal{B}$ is contained in V. Moreover, $\mathcal{B}$ is linearly independent. Since the vectors in V have the form

$$\begin{bmatrix} 2s - t \\ s + 3t \end{bmatrix} = s \begin{bmatrix} 2 \\ 1 \end{bmatrix} + t \begin{bmatrix} -1 \\ 3 \end{bmatrix},$$

the set

$$\left\{ \begin{bmatrix} 2 \\ 1 \end{bmatrix}, \begin{bmatrix} -1 \\ 3 \end{bmatrix} \right\}$$

is a basis for V. Hence $\dim V = 2$. Therefore $\mathcal{B}$ is a basis for V because the 3 conditions of the boxed statement on pages 248–249 are satisfied.

65. Taking $r = 2$, $s = 1$, and $t = 1$, we have

$$\begin{bmatrix} -r + 3s \\ 0 \\ s - t \\ r - 2t \end{bmatrix} = \begin{bmatrix} 1 \\ 0 \\ 0 \\ 0 \end{bmatrix};$$

taking $r = 5$, $s = 2$, and $t = 3$, we have

$$\begin{bmatrix} -r + 3s \\ 0 \\ s - t \\ r - 2t \end{bmatrix} = \begin{bmatrix} 1 \\ 0 \\ -1 \\ -1 \end{bmatrix};$$

and taking $r = 2$, $s = 1$, and $t = 0$, we have
$$\begin{bmatrix} -r+3s \\ 0 \\ s-t \\ r-2t \end{bmatrix} = \begin{bmatrix} 1 \\ 0 \\ 1 \\ 2 \end{bmatrix}.$$

Hence $\mathcal{B}$ is contained in V.

Since the reduced row echelon form of
$$\begin{bmatrix} 1 & 1 & 1 \\ 0 & 0 & 0 \\ 0 & -1 & 1 \\ 0 & -1 & 2 \end{bmatrix}$$

is
$$\begin{bmatrix} 1 & 0 & 0 \\ 0 & 1 & 0 \\ 0 & 0 & 1 \\ 0 & 0 & 0 \end{bmatrix},$$

$\mathcal{B}$ is linearly independent. The vectors in V have the form
$$\begin{bmatrix} -r+3s \\ 0 \\ s-t \\ r-2t \end{bmatrix} = r\begin{bmatrix} -1 \\ 0 \\ 0 \\ 1 \end{bmatrix} + s\begin{bmatrix} 3 \\ 0 \\ 1 \\ 0 \end{bmatrix} + t\begin{bmatrix} 0 \\ 0 \\ -1 \\ -2 \end{bmatrix}.$$

It is easily checked that the set
$$\left\{ \begin{bmatrix} -1 \\ 0 \\ 0 \\ 1 \end{bmatrix}, \begin{bmatrix} 3 \\ 0 \\ 1 \\ 0 \end{bmatrix}, \begin{bmatrix} 0 \\ 0 \\ -1 \\ -2 \end{bmatrix} \right\}$$

is linearly independent, and so it is a basis for V. Hence $\dim V = 3$. It follows, as in Exercise 61, that $\mathcal{B}$ is a basis for V.

69. (a) Refer to the solution to Exercise 9. By Theorem 4.8,
$$\left\{ \begin{bmatrix} 1 \\ 0 \\ 6 \\ 0 \end{bmatrix}, \begin{bmatrix} 0 \\ 1 \\ -4 \\ 1 \end{bmatrix} \right\}$$

is a basis for Row A. Also, the vector form of the general solution of $A\mathbf{x} = \mathbf{0}$ is
$$\begin{bmatrix} x_1 \\ x_2 \\ x_3 \\ x_4 \end{bmatrix} = x_3 \begin{bmatrix} -6 \\ 4 \\ 1 \\ 0 \end{bmatrix} + x_4 \begin{bmatrix} 0 \\ -1 \\ 0 \\ 1 \end{bmatrix}.$$

Thus
$$\left\{ \begin{bmatrix} -6 \\ 4 \\ 1 \\ 0 \end{bmatrix}, \begin{bmatrix} 0 \\ -1 \\ 0 \\ 1 \end{bmatrix} \right\}$$

is a basis for Null A.

(b) It is easily checked that set
$$\left\{ \begin{bmatrix} 1 \\ 0 \\ 6 \\ 0 \end{bmatrix}, \begin{bmatrix} 0 \\ 1 \\ -4 \\ 1 \end{bmatrix}, \begin{bmatrix} -6 \\ 4 \\ 1 \\ 0 \end{bmatrix}, \begin{bmatrix} 0 \\ -1 \\ 0 \\ 1 \end{bmatrix} \right\}$$

is linearly independent. Since it contains 4 vectors, it is a basis for $\mathcal{R}^4$ by Theorem 4.7.

73. Let $\mathbf{v}$ be in the column space of AB. Then $\mathbf{v} = (AB)\mathbf{u}$ for some $\mathbf{u}$ in $\mathcal{R}^p$. Consider $\mathbf{w} = B\mathbf{u}$. Since $A\mathbf{w} = A(B\mathbf{u}) = (AB)\mathbf{u} = \mathbf{v}$, $\mathbf{v}$ is in the column space of A.

77. Since the ranks of a matrix and its transpose are equal, we have
$$\text{rank } AB = \text{rank } (AB)^T = \text{rank } B^T A^T.$$

By Exercise 75,
$$\text{rank } B^T A^T \leq \text{rank } B^T = \text{rank } B.$$

Combining the preceding results yields $\text{rank } AB \leq \text{rank } B$.

81. (a) Let $\mathbf{v}$ and $\mathbf{w}$ be in $\mathcal{R}^k$. Then

$$T(\mathbf{v}+\mathbf{w}) = T\left(\begin{bmatrix} v_1+w_1 \\ v_2+w_2 \\ \vdots \\ v_k+w_k \end{bmatrix}\right)$$

$$= (v_1+w_1)\mathbf{u}_1 + \cdots$$
$$\quad + (v_k+w_k)\mathbf{u}_k$$
$$= (v_1\mathbf{u}_1 + \cdots + v_k\mathbf{u}_k)$$
$$\quad + (w_1\mathbf{u}_1 + \cdots + w_k\mathbf{u}_k)$$
$$= T(\mathbf{v}) + T(\mathbf{u}).$$

Also, for any scalar c,

$$T(c\mathbf{v}) = T\left(\begin{bmatrix} cv_1 \\ \vdots \\ cv_k \end{bmatrix}\right)$$

$$= (cv_1)\mathbf{u}_1 + \cdots + (cv_k)\mathbf{u}_k$$
$$= c(v_1\mathbf{u}_1 + \cdots + v_k\mathbf{u}_k)$$
$$= cT(\mathbf{v}).$$

Thus T is a linear transformation.

(b) Since $\{\mathbf{u}_1, \mathbf{u}_2, \ldots, \mathbf{u}_k\}$ is linearly independent,

$$x_1\mathbf{u}_1 + x_2\mathbf{u}_2 + \cdots + x_k\mathbf{u}_k = \mathbf{0}$$

implies $x_1 = x_2 = \cdots = x_k = 0$. Thus $T(\mathbf{x}) = \mathbf{0}$ implies $\mathbf{x} = \mathbf{0}$, so that the null space of T is $\{\mathbf{0}\}$. It follows from Theorem 2.11 that T is one-to-one.

(c) For every $\mathbf{x}$ in $\mathcal{R}^k$, $T(\mathbf{x})$ is a linear combination of $\mathbf{u}_1, \mathbf{u}_2, \ldots, \mathbf{u}_k$ and hence is a vector in V. Conversely, if $\mathbf{v}$ is in V, then we have $\mathbf{v} = a_1\mathbf{u}_1 + a_2\mathbf{u}_2 + \cdots + a_k\mathbf{u}_k$ for some scalars $a_1, a_2, \ldots, a_k$. For

$$\mathbf{a} = \begin{bmatrix} a_1 \\ a_2 \\ \vdots \\ a_k \end{bmatrix},$$

in $\mathcal{R}^k$, we have $T(\mathbf{a}) = \mathbf{v}$. Hence every vector in V is the image of a vector in $\mathcal{R}^k$. Thus the range of T is V.

85. (a) Let B be a 4×4 matrix such that $AB = O$. Then

$$O = AB = A[\mathbf{b}_1 \ \mathbf{b}_2 \ \mathbf{b}_3 \ \mathbf{b}_4]$$
$$= [A\mathbf{b}_1 \ A\mathbf{b}_2 \ A\mathbf{b}_3 \ A\mathbf{b}_4].$$

So each column of B is a solution of $A\mathbf{x} = \mathbf{0}$. The reduced row echelon form of A is

$$\begin{bmatrix} 1 & 0 & -1 & -2 \\ 0 & 1 & 1 & -1 \\ 0 & 0 & 0 & 0 \\ 0 & 0 & 0 & 0 \end{bmatrix},$$

and so the vector form of the general solution of $A\mathbf{x} = \mathbf{0}$ is

$$\begin{bmatrix} x_1 \\ x_2 \\ x_3 \\ x_4 \end{bmatrix} = x_3 \begin{bmatrix} 1 \\ -1 \\ 1 \\ 0 \end{bmatrix} + x_4 \begin{bmatrix} 2 \\ 1 \\ 0 \\ 1 \end{bmatrix}.$$

Hence

$$B = \begin{bmatrix} 1 & 2 & 0 & 0 \\ -1 & 1 & 0 & 0 \\ 1 & 0 & 0 & 0 \\ 0 & 1 & 0 & 0 \end{bmatrix}$$

is a 4×4 matrix with rank 2 such that $AB = O$.

(b) If C is a 4×4 matrix such that $AC = O$, then the preceding argument shows that each column of C is a vector in Null A, a 2-dimensional subspace. Hence C can have at most two linearly independent columns; so rank $C \leq 2$.

4.4 COORDINATE SYSTEMS

1. Because $[\mathbf{v}]_\mathcal{B} = \begin{bmatrix} 4 \\ 3 \end{bmatrix}$, we see that

$$\mathbf{v} = 4 \begin{bmatrix} 1 \\ -1 \end{bmatrix} + 3 \begin{bmatrix} -1 \\ 2 \end{bmatrix} = \begin{bmatrix} 1 \\ 2 \end{bmatrix}.$$

Equivalently, if B denotes the matrix whose columns are the vectors in $\mathcal{B}$, then

$$\mathbf{v} = B[\mathbf{v}]_\mathcal{B} = \begin{bmatrix} 1 & -1 \\ -1 & 2 \end{bmatrix} \begin{bmatrix} 4 \\ 3 \end{bmatrix} = \begin{bmatrix} 1 \\ 2 \end{bmatrix}.$$

5. As in Exercise 1, we have

$$\mathbf{v} = 2 \begin{bmatrix} 1 \\ -1 \end{bmatrix} + 5 \begin{bmatrix} -1 \\ 2 \end{bmatrix} = \begin{bmatrix} -3 \\ 8 \end{bmatrix}.$$

9. As in Exercise 1, we have

$$\mathbf{v} = (-1) \begin{bmatrix} 0 \\ 1 \\ 1 \end{bmatrix} + 5 \begin{bmatrix} -1 \\ 0 \\ 1 \end{bmatrix} + (-2) \begin{bmatrix} 1 \\ 1 \\ 1 \end{bmatrix}$$
$$= \begin{bmatrix} -7 \\ -3 \\ 2 \end{bmatrix}.$$

13. (a) Let B be the matrix whose columns are the vectors in $\mathcal{B}$. Since the reduced row echelon form of B is I_3, $\mathcal{B}$ is linearly independent. So $\mathcal{B}$ is a linearly independent set of 3 vectors from $\mathcal{R}^3$, and hence $\mathcal{B}$ is a basis for $\mathcal{R}^3$ by Theorem 4.7.

(b) The components of $[\mathbf{v}]_\mathcal{B}$ are the coefficients that express $\mathbf{v}$ as a linear combination of the vectors in $\mathcal{B}$. Thus

$$[\mathbf{v}]_\mathcal{B} = \begin{bmatrix} 3 \\ 0 \\ -1 \end{bmatrix}.$$

17. By Theorem 4.11,

$$[\mathbf{v}]_\mathcal{B} = \begin{bmatrix} 1 & -1 \\ -1 & 2 \end{bmatrix}^{-1} \begin{bmatrix} 5 \\ -3 \end{bmatrix} = \begin{bmatrix} 7 \\ 2 \end{bmatrix}.$$

21. By Theorem 4.11,

$$[\mathbf{v}]_\mathcal{B} = \begin{bmatrix} 0 & -1 & 1 \\ 1 & 0 & 1 \\ 1 & 1 & 1 \end{bmatrix}^{-1} \begin{bmatrix} 1 \\ -3 \\ -2 \end{bmatrix} = \begin{bmatrix} -5 \\ 1 \\ 2 \end{bmatrix}.$$

25. The unique representation of $\mathbf{u}$ as a linear combination of $\mathbf{b}_1$ and $\mathbf{b}_2$ is given by the coordinate vector of $\mathbf{u}$ relative to $\{\mathbf{b}_1, \mathbf{b}_2\}$, which is

$$[\mathbf{b}_1 \ \mathbf{b}_2]^{-1} \mathbf{u} = \begin{bmatrix} -2 & 3 \\ 3 & -5 \end{bmatrix}^{-1} \begin{bmatrix} a \\ b \end{bmatrix}$$
$$= \begin{bmatrix} -5a - 3b \\ -3a - 2b \end{bmatrix}.$$

Thus

$$\mathbf{u} = (-5a - 3b)\mathbf{b}_1 + (-3a - 2b)\mathbf{b}_2.$$

29. Proceeding as in Exercise 25, we have

$$[\mathbf{b}_1 \ \mathbf{b}_2 \ \mathbf{b}_3]^{-1} \mathbf{u}$$
$$= \begin{bmatrix} 1 & -1 & -2 \\ 0 & 1 & 0 \\ 1 & 0 & -1 \end{bmatrix}^{-1} \begin{bmatrix} a \\ b \\ c \end{bmatrix}$$
$$= \begin{bmatrix} -a - b + 2c \\ b \\ -a - b + c \end{bmatrix}.$$

Thus
$$\mathbf{u} = (-a - b + 2c)\mathbf{b}_1 + b\mathbf{b}_2 + (-a - b + c)\mathbf{b}_3.$$

31. False, every vector in V can be *uniquely* represented as a linear combination of the vectors in $\mathcal{S}$ if and only if $\mathcal{S}$ is a basis for V.

32. True   33. True   34. True

35. True   36. True   37. True

38. True   39. True   40. True

41. True   42. True   43. True

44. False, $\begin{bmatrix} x' \\ y' \end{bmatrix} = A_\theta^T \begin{bmatrix} x \\ y \end{bmatrix}.$

45. True   46. True   47. True

48. False, the graph of such an equation is a *hyperbola*.

49. True   50. True

51. (a) Since the reduced row echelon form of $\begin{bmatrix} 1 & 2 \\ 2 & 3 \end{bmatrix}$ is $\begin{bmatrix} 1 & 0 \\ 0 & 1 \end{bmatrix}$, $\mathcal{B}$ is a linearly independent subset of $\mathcal{R}^2$ containing 2 vectors. Hence $\mathcal{B}$ is a basis for $\mathcal{R}^2$ by Theorem 4.7.

(b) Let $B = [\mathbf{b}_1\ \mathbf{b}_2]$. Then
$$[\mathbf{e}_1]_\mathcal{B} = B^{-1}\mathbf{e}_1 = \begin{bmatrix} -3 \\ 2 \end{bmatrix}$$
and
$$[\mathbf{e}_2]_\mathcal{B} = B^{-1}\mathbf{e}_2 = \begin{bmatrix} 2 \\ -1 \end{bmatrix}.$$
Hence
$$A = \begin{bmatrix} -3 & 2 \\ 2 & -1 \end{bmatrix}.$$

(c) From (b), we see that
$$A = [B^{-1}\mathbf{e}_1\ B^{-1}\mathbf{e}_2]$$
$$= B^{-1}[\mathbf{e}_1\ \mathbf{e}_2] = B^{-1}I_2 = B^{-1}.$$
So A and B are inverses of each other.

55. Let $\mathbf{v} = \begin{bmatrix} x \\ y \end{bmatrix}$ and $[\mathbf{v}]_\mathcal{B} = \begin{bmatrix} x' \\ y' \end{bmatrix}$, where $\mathcal{B}$ is the basis obtained by rotating the vectors in the standard basis by $30°$. Then
$$\begin{bmatrix} x' \\ y' \end{bmatrix} = [\mathbf{v}]_\mathcal{B} = (A_{30°})^{-1}\mathbf{v}$$
$$= A_{30°}^T \mathbf{v} = \begin{bmatrix} \frac{\sqrt{3}}{2} & \frac{1}{2} \\ -\frac{1}{2} & \frac{\sqrt{3}}{2} \end{bmatrix} \begin{bmatrix} x \\ y \end{bmatrix}.$$
Hence
$$x' = \tfrac{\sqrt{3}}{2}x + \tfrac{1}{2}y$$
$$y' = -\tfrac{1}{2}x + \tfrac{\sqrt{3}}{2}y.$$

59. Let B be the matrix whose columns are the vectors in $\mathcal{B}$. Then
$$\begin{bmatrix} x' \\ y' \end{bmatrix} = B^{-1}\begin{bmatrix} x \\ y \end{bmatrix} = \begin{bmatrix} -5 & -3 \\ -2 & -1 \end{bmatrix} \begin{bmatrix} x \\ y \end{bmatrix}$$
$$= \begin{bmatrix} -5x - 3y \\ -2x - y \end{bmatrix}.$$
Hence
$$x' = -5x - 3y$$
$$y' = -2x - y.$$

63. Let
$$B = \begin{bmatrix} 1 & 1 & 0 \\ 0 & 1 & -2 \\ 1 & 0 & 1 \end{bmatrix}.$$
Then, as in Exercise 55,
$$\begin{bmatrix} x' \\ y' \\ z' \end{bmatrix} = [\mathbf{v}]_\mathcal{B} = B^{-1}\mathbf{v}$$

$$= \begin{bmatrix} -1 & 1 & 2 \\ 2 & -1 & -2 \\ 1 & -1 & -1 \end{bmatrix} \begin{bmatrix} x \\ y \\ z \end{bmatrix}.$$

Hence
$$x' = -x + y + 2z$$
$$y' = 2x - y - 2z$$
$$z' = x - y - z.$$

67. Let $\mathbf{v} = \begin{bmatrix} x \\ y \end{bmatrix}$ and $[\mathbf{v}]_{\mathcal{B}} = \begin{bmatrix} x' \\ y' \end{bmatrix}$, where $\mathcal{B}$ is the basis obtained by rotating the vectors in the standard basis by $60°$. As in Example 4,

$$\begin{bmatrix} x \\ y \end{bmatrix} = \mathbf{v} = A_{60°}[\mathbf{v}]_{\mathcal{B}}$$
$$= \begin{bmatrix} \frac{1}{2} & -\frac{\sqrt{3}}{2} \\ \frac{\sqrt{3}}{2} & \frac{1}{2} \end{bmatrix} \begin{bmatrix} x' \\ y' \end{bmatrix}.$$

Hence
$$x = \tfrac{1}{2}x' - \tfrac{\sqrt{3}}{2}y'$$
$$y = \tfrac{\sqrt{3}}{2}x' + \tfrac{1}{2}y'.$$

71. Let B be the matrix whose columns are the vectors in $\mathcal{B}$. Then
$$\begin{bmatrix} x \\ y \end{bmatrix} = B \begin{bmatrix} x' \\ y' \end{bmatrix} = \begin{bmatrix} x' + 3y' \\ 2x' + 4y' \end{bmatrix}.$$

Hence
$$x = x' + 3y'$$
$$y = 2x' + 4y'.$$

75. Let
$$\begin{bmatrix} 1 & -1 & 0 \\ 3 & 1 & -1 \\ 0 & 1 & 1 \end{bmatrix}.$$

As in Exercise 67, we have
$$\begin{bmatrix} x \\ y \\ z \end{bmatrix} = \mathbf{v} = B[\mathbf{v}]_{\mathcal{B}}$$

$$= \begin{bmatrix} 1 & -1 & 0 \\ 3 & 1 & -1 \\ 0 & 1 & 1 \end{bmatrix} \begin{bmatrix} x' \\ y' \\ z' \end{bmatrix}.$$

Thus
$$x = x' - y'$$
$$y = 3x' + y' - z'$$
$$z = y' + z'.$$

79. As in Exercise 55, we have
$$\begin{bmatrix} x' \\ y' \end{bmatrix} = A_{60°}^T \begin{bmatrix} x \\ y \end{bmatrix} = \begin{bmatrix} \frac{1}{2} & \frac{\sqrt{3}}{2} \\ -\frac{\sqrt{3}}{2} & \frac{1}{2} \end{bmatrix} \begin{bmatrix} x \\ y \end{bmatrix}.$$

Thus
$$x' = \tfrac{1}{2}x + \tfrac{\sqrt{3}}{2}y$$
$$y' = -\tfrac{\sqrt{3}}{2}x + \tfrac{1}{2}y.$$

Rewrite the given equation in the form
$$25(x')^2 + 16(y')^2 = 400.$$

Then substitute the expressions for x' and y' into this equation to obtain
$$\frac{73}{4}x^2 + \frac{9\sqrt{3}}{2}xy + \frac{91}{4}y^2 = 400,$$

that is,
$$73x^2 + 18\sqrt{3}xy + 91y^2 = 1600.$$

83. As in Exercise 79, we have
$$\begin{bmatrix} x' \\ y' \end{bmatrix} = A_{120°}^T \begin{bmatrix} x \\ y \end{bmatrix} = \begin{bmatrix} -\frac{1}{2} & \frac{\sqrt{3}}{2} \\ -\frac{\sqrt{3}}{2} & -\frac{1}{2} \end{bmatrix} \begin{bmatrix} x \\ y \end{bmatrix}.$$

Thus
$$x' = -\tfrac{1}{2}x + \tfrac{\sqrt{3}}{2}y$$
$$y' = -\tfrac{\sqrt{3}}{2}x - \tfrac{1}{2}y.$$

Rewrite the given equation in the form $4(x')^2 - 9(y')^2 = 36$. and substitute the preceding expressions for x' and y' to obtain
$$-23x^2 - 26\sqrt{3}xy + 3y^2 = 144.$$

87. As in Exercise 67, we have
$$\begin{bmatrix} x \\ y \end{bmatrix} = A_{45°} \begin{bmatrix} x' \\ y' \end{bmatrix} = \begin{bmatrix} \frac{\sqrt{2}}{2} & -\frac{\sqrt{2}}{2} \\ \frac{\sqrt{2}}{2} & \frac{\sqrt{2}}{2} \end{bmatrix} \begin{bmatrix} x' \\ y' \end{bmatrix}.$$
Thus
$$x = \frac{\sqrt{2}}{2}x' - \frac{\sqrt{2}}{2}y'$$
$$y = \frac{\sqrt{2}}{2}x' + \frac{\sqrt{2}}{2}y'.$$
Substituting these expressions for x and y into the given equation produces
$$4(x')^2 - 10(y')^2 = 20,$$
that is,
$$2(x')^2 - 5(y')^2 = 10.$$

91. As in Exercise 87, we have
$$\begin{bmatrix} x \\ y \end{bmatrix} = A_{30°} \begin{bmatrix} x' \\ y' \end{bmatrix} = \begin{bmatrix} \frac{\sqrt{3}}{2} & -\frac{1}{2} \\ \frac{1}{2} & \frac{\sqrt{3}}{2} \end{bmatrix} \begin{bmatrix} x' \\ y' \end{bmatrix}.$$
Thus
$$x = \frac{\sqrt{3}}{2}x' - \frac{1}{2}y'$$
$$y = \frac{1}{2}x' + \frac{\sqrt{3}}{2}y'.$$
Substituting these expressions for x and y into the given equation produces
$$16(x')^2 - 12(y')^2 = 240,$$
that is,
$$4(x')^2 - 3(y')^2 = 60.$$

95. By the definition of $[\mathbf{v}]_\mathcal{A}$, we have
$$\mathbf{v} = a_1 \mathbf{u}_1 + \cdots + a_n \mathbf{u}_n$$
$$= \frac{a_1}{c_1}(c_1 \mathbf{u}_1) + \cdots + \frac{a_n}{c_n}(c_n \mathbf{u}_n).$$
Hence
$$[\mathbf{v}]_\mathcal{B} = \begin{bmatrix} \frac{a_1}{c_1} \\ \vdots \\ \frac{a_n}{c_n} \end{bmatrix}.$$

99. Consider
$$\mathcal{A} = \{\mathbf{e}_1, \mathbf{e}_2\} \text{ and } \mathcal{B} = \{\mathbf{e}_1, 2\mathbf{e}_2\}.$$
Then $[\mathbf{e}_1]_\mathcal{A} = \mathbf{e}_1$ and $[\mathbf{e}_1]_\mathcal{B} = \mathbf{e}_1$, but $\mathcal{A} \neq \mathcal{B}$.

103. (a) Let B be the matrix whose columns are the vectors in $\mathcal{B}$. By Theorem 4.11,
$$T(\mathbf{v}) = [\mathbf{v}]_\mathcal{B} = B^{-1}\mathbf{v}$$
for every vector $\mathbf{v}$ in $\mathcal{R}^n$. Hence T is the matrix transformation induced by B^{-1}, and so T is a linear transformation.

(b) Because the standard matrix of T is B^{-1}, an invertible matrix, the columns of the standard matrix of T are linearly independent and form a generating set for $\mathcal{R}^n$. Hence T is one-to-one and onto by Theorems 2.11 and 2.10.

107. Suppose that $\mathcal{A} = \{\mathbf{u}_1, \ldots, \mathbf{u}_k\}$ is a linearly independent subset of $\mathcal{R}^n$, and let $c_1, \ldots, c_k$ be scalars such that
$$c_1[\mathbf{u}_1]_\mathcal{B} + \cdots + c_k[\mathbf{u}_k]_\mathcal{B} = \mathbf{0}.$$
Define $T: \mathcal{R}^n \to \mathcal{R}^n$ by $T(\mathbf{v}) = [\mathbf{v}]_\mathcal{B}$ for all $\mathbf{v}$ in $\mathcal{R}^n$. Then T is a linear transformation by Exercise 103(a), and so
$$c_1 T(\mathbf{u}_1) + \cdots + c_k T(\mathbf{u}_k) = \mathbf{0}$$
$$T(c_1 \mathbf{u}_1 + \cdots + c_k \mathbf{u}_k) = \mathbf{0}.$$
Therefore $c_1 \mathbf{u}_1 + \cdots + c_k \mathbf{u}_k$ is in the null space of T. Since T is one-to-one by Exercise 103(b), it follows that $c_1 \mathbf{u}_1 + \cdots + c_k \mathbf{u}_k = \mathbf{0}$. Hence the linear independence of $\{\mathbf{u}_1, \ldots, \mathbf{u}_k\}$ yields $c_1 = \cdots = c_k = 0$. It follows that $\{[\mathbf{u}_1]_\mathcal{B}, [\mathbf{u}_2]_\mathcal{B}, \ldots, [\mathbf{u}_k]_\mathcal{B}\}$ is linearly independent.

The proof of the converse is similar.

111. Let B be the matrix whose columns are the vectors in $\mathcal{B}$. Since $[\mathbf{v}]_\mathcal{B} = B^{-1}\mathbf{v}$, we must find a nonzero vector $\mathbf{v}$ in $\mathcal{R}^5$ such that
$$B^{-1}\mathbf{v} = .5\mathbf{v}$$
$$B^{-1}\mathbf{v} - .5\mathbf{v} = \mathbf{0}$$
$$(B^{-1} - .5I_5)\mathbf{v} = \mathbf{0}.$$

Because the reduced row echelon form of $B^{-1} - .5I_5$ is
$$\begin{bmatrix} 1 & 0 & 0 & 0 & 0 \\ 0 & 1 & 0 & 0 & -2 \\ 0 & 0 & 1 & 0 & 2 \\ 0 & 0 & 0 & 1 & -2 \\ 0 & 0 & 0 & 0 & 0 \end{bmatrix},$$

the vector form of the general solution of $(B^{-1} - .5I_5)\mathbf{x} = \mathbf{0}$ is
$$\begin{bmatrix} x_1 \\ x_2 \\ x_3 \\ x_4 \\ x_5 \end{bmatrix} = x_5 \begin{bmatrix} 0 \\ 2 \\ -2 \\ 2 \\ 1 \end{bmatrix}.$$

So by taking
$$\mathbf{v} = \begin{bmatrix} 0 \\ 2 \\ -2 \\ 2 \\ 1 \end{bmatrix},$$

we have $[\mathbf{v}]_\mathcal{B} = .5\mathbf{v}$.

4.5 MATRIX REPRESENTATIONS OF LINEAR OPERATORS

3. The standard matrix of T is
$$A = \begin{bmatrix} 1 & 2 \\ 1 & 1 \end{bmatrix}.$$

If B is the matrix whose columns are the vectors in $\mathcal{B}$, then, by Theorem 4.12, we have
$$[T]_\mathcal{B} = B^{-1}AB = \begin{bmatrix} 1 & 2 \\ 1 & 1 \end{bmatrix}.$$

7. The standard matrix of T is
$$A = \begin{bmatrix} 0 & 4 & 0 \\ 1 & 0 & 2 \\ 0 & -2 & 3 \end{bmatrix}.$$

If B is the matrix whose columns are the vectors in $\mathcal{B}$, then, by Theorem 4.12, we have
$$[T]_\mathcal{B} = B^{-1}AB = \begin{bmatrix} 0 & -19 & 28 \\ 3 & 34 & -47 \\ 3 & 23 & -31 \end{bmatrix}.$$

11. Let A be the standard matrix of T and B be the matrix whose columns are the vectors in $\mathcal{B}$. Then, by Theorem 4.12, we have
$$A = B[T]_\mathcal{B}B^{-1} = \begin{bmatrix} 10 & -19 \\ 3 & -4 \end{bmatrix}.$$

15. As in Exercise 11, if B is the matrix whose columns are the vectors in $\mathcal{B}$, then the standard matrix of T is
$$B[T]_\mathcal{B}B^{-1} = \begin{bmatrix} 2 & 5 & 10 \\ -6 & 1 & -7 \\ 2 & -2 & 0 \end{bmatrix}.$$

19. False, a linear operator on $\mathcal{R}^n$ is a linear transformation whose domain and codomain both equal $\mathcal{R}^n$.

20. True   21. True

22. False, the matrix representation of T with respect to $\mathcal{B}$ is
$$[[T(\mathbf{b}_1)]_\mathcal{B} \ [T(\mathbf{b}_2)]_\mathcal{B} \ \cdots \ [T(\mathbf{b}_n)]_\mathcal{B}].$$

23. True

24. False, $[T]_B = B^{-1}AB$.

25. False, $[T]_B = B^{-1}AB$.

26. True

27. False, $T(\mathbf{v}) = \mathbf{v}$ for every vector $\mathbf{v}$ on L.

28. False, $T(\mathbf{v}) = \mathbf{v}$ for every vector $\mathbf{v}$ on L.

29. False, there exists a basis $\mathcal{B}$ for $\mathcal{R}^n$ such that $[T]_\mathcal{B} = \begin{bmatrix} 1 & 0 \\ 0 & -1 \end{bmatrix}$.

30. True

31. False, $\mathcal{B}$ consists of one vector on L and one vector perpendicular to L.

32. False, an $n \times n$ matrix A is said to be similar to an $n \times n$ matrix B if $B = P^{-1}AP$ for some invertible matrix P.

33. True   34. True   35. True

36. True

37. False, $[T]_\mathcal{B}[\mathbf{v}]_\mathcal{B} = [T(\mathbf{v})]_\mathcal{B}$.

38. True

39. Because $T(\mathbf{b}_1) = \mathbf{b}_1 + 4\mathbf{b}_2$, the coordinate vector of $T(\mathbf{b}_1)$ with respect to $\mathcal{B}$ is $\begin{bmatrix} 1 \\ 4 \end{bmatrix}$. Similarly, the coordinate vector of $T(\mathbf{b}_2)$ with respect to $\mathcal{B}$ is $\begin{bmatrix} -3 \\ 0 \end{bmatrix}$. Hence

$$[T(\mathbf{b}_1)]_\mathcal{B} = [[T(\mathbf{b}_1)]_\mathcal{B} \ [T(\mathbf{b}_2)]_\mathcal{B}]$$
$$= \begin{bmatrix} 1 & -3 \\ 4 & 0 \end{bmatrix}.$$

43. Since $T(\mathbf{b}_1) = 0\mathbf{b}_1 - 5\mathbf{b}_2 + 4\mathbf{b}_3$, we have

$$[T(\mathbf{b}_1)]_\mathcal{B} = \begin{bmatrix} 0 \\ -5 \\ 4 \end{bmatrix}.$$

Likewise

$$[T(\mathbf{b}_2)]_\mathcal{B} = \begin{bmatrix} 2 \\ 0 \\ -7 \end{bmatrix}$$

and

$$[T(\mathbf{b}_2)]_\mathcal{B} = \begin{bmatrix} 3 \\ 0 \\ 1 \end{bmatrix}.$$

Hence

$$[T]_\mathcal{B} = \begin{bmatrix} 0 & 2 & 3 \\ -5 & 0 & 0 \\ 4 & -7 & 1 \end{bmatrix}.$$

47. (a) Let $\mathbf{b}_1 = \begin{bmatrix} 1 \\ 1 \end{bmatrix}$ and $\mathbf{b}_2 = \begin{bmatrix} 1 \\ 2 \end{bmatrix}$. Because $T(\mathbf{b}_1) = 0\mathbf{b}_1 + 1\mathbf{b}_2$ and $T(\mathbf{b}_2) = 3\mathbf{b}_1 + 0\mathbf{b}_2$, we have

$$[T(\mathbf{b}_1)]_\mathcal{B} = [[T(\mathbf{b}_1)]_\mathcal{B} \ [T(\mathbf{b}_2)]_\mathcal{B}]$$
$$= \begin{bmatrix} 0 & 3 \\ 1 & 0 \end{bmatrix}.$$

(b) The standard matrix A of T satisfies $A = B[T]_\mathcal{B} B^{-1}$ by Theorem 4.12. (Here $B = [\mathbf{b}_1 \ \mathbf{b}_2]$.) Hence

$$A = \begin{bmatrix} 1 & 1 \\ 1 & 2 \end{bmatrix} \begin{bmatrix} 0 & 3 \\ 1 & 0 \end{bmatrix} \begin{bmatrix} 1 & 1 \\ 1 & 2 \end{bmatrix}^{-1}$$
$$= \begin{bmatrix} -1 & 2 \\ 1 & 1 \end{bmatrix}.$$

(c) Because A is the standard matrix of T, we have

$$T(\mathbf{x}) = A \begin{bmatrix} x_1 \\ x_2 \end{bmatrix} = \begin{bmatrix} -1 & 2 \\ 1 & 1 \end{bmatrix} \begin{bmatrix} x_1 \\ x_2 \end{bmatrix}$$
$$= \begin{bmatrix} -x_1 + 2x_2 \\ x_1 + x_2 \end{bmatrix}.$$

51. Let

$$\mathbf{b}_1 = \begin{bmatrix} 1 \\ 0 \\ 1 \end{bmatrix}, \mathbf{b}_2 = \begin{bmatrix} 0 \\ 1 \\ 0 \end{bmatrix}, \text{ and } \mathbf{b}_3 = \begin{bmatrix} 1 \\ 1 \\ 0 \end{bmatrix}.$$

(a) Since $T(\mathbf{b}_1) = 0\mathbf{b}_1 - \mathbf{b}_2 + 0\mathbf{b}_3$, we have

$$[T(\mathbf{b}_1)]_\mathcal{B} = \begin{bmatrix} 0 \\ -1 \\ 0 \end{bmatrix}.$$

Likewise

$$[T(\mathbf{b}_2)]_\mathcal{B} = \begin{bmatrix} 0 \\ 0 \\ 2 \end{bmatrix}$$

and

$$[T(\mathbf{b}_3)]_\mathcal{B} = \begin{bmatrix} 1 \\ 2 \\ 0 \end{bmatrix}.$$

Hence

$$[T]_\mathcal{B} = \begin{bmatrix} 0 & 0 & 1 \\ -1 & 0 & 2 \\ 0 & 2 & 0 \end{bmatrix}.$$

(b) The standard matrix A of T is given by

$$A = B[T]_\mathcal{B} B^{-1}$$

$$= \begin{bmatrix} -1 & 2 & 1 \\ 0 & 2 & -1 \\ 1 & 0 & -1 \end{bmatrix},$$

where $B = [\mathbf{b}_1 \ \mathbf{b}_2 \ \mathbf{b}_3]$.

(c) For any vector $\mathbf{x}$ in $\mathcal{R}^3$, we have

$$T(\mathbf{x}) = A\mathbf{x}$$

$$= \begin{bmatrix} -1 & 2 & 1 \\ 0 & 2 & -1 \\ 1 & 0 & -1 \end{bmatrix} \begin{bmatrix} x_1 \\ x_2 \\ x_3 \end{bmatrix}$$

$$= \begin{bmatrix} -x_1 + 2x_2 + x_3 \\ 2x_2 - x_3 \\ x_1 - x_3 \end{bmatrix}.$$

55. From Exercise 39, we have

$$[T]_\mathcal{B} = \begin{bmatrix} 1 & -3 \\ 4 & 0 \end{bmatrix}.$$

Hence, by the comment preceding Example 1, we have

$$[T(3\mathbf{b}_1 - 2\mathbf{b}_2)]_\mathcal{B} = [T]_\mathcal{B}[3\mathbf{b}_1 - 2\mathbf{b}_2]_\mathcal{B}$$

$$= [T]_\mathcal{B} \begin{bmatrix} 3 \\ -2 \end{bmatrix} = \begin{bmatrix} 9 \\ 12 \end{bmatrix}.$$

Therefore $T(3\mathbf{b}_1 - 2\mathbf{b}_2) = 9\mathbf{b}_1 + 12\mathbf{b}_2$.

Equivalently, using the linear transformation properties of T, we have

$$T(3\mathbf{b}_1 - 2\mathbf{b}_2) = 3T(\mathbf{b}_1) - 2T(\mathbf{b}_2)$$
$$= 3(\mathbf{b}_1 + 4\mathbf{b}_2) - 2(-3\mathbf{b}_1)$$
$$= 9\mathbf{b}_1 + 12\mathbf{b}_2.$$

59. Proceeding as in Exercise 55 and using the answer to Exercise 43, we have

$$[T(2\mathbf{b}_1 - \mathbf{b}_2)]_\mathcal{B} = [T]_\mathcal{B}[2\mathbf{b}_1 - \mathbf{b}_2]_\mathcal{B}$$

$$= \begin{bmatrix} 0 & 2 & 3 \\ -5 & 0 & 0 \\ 4 & -7 & 1 \end{bmatrix} \begin{bmatrix} 2 \\ -1 \\ 0 \end{bmatrix}$$

$$= \begin{bmatrix} -2 \\ -10 \\ 15 \end{bmatrix}.$$

Therefore

$$T(2\mathbf{b}_1 - \mathbf{b}_2) = -2\mathbf{b}_1 - 10\mathbf{b}_2 + 15\mathbf{b}_3.$$

63. For any $\mathbf{v}$ in $\mathcal{R}^n$, we have $I(\mathbf{v}) = \mathbf{v}$. Hence if $\mathcal{B} = \{\mathbf{b}_1, \mathbf{b}_2, \ldots, \mathbf{b}_n\}$, then

$$[I]_\mathcal{B} = [[I(\mathbf{b}_1)]_\mathcal{B} \ \cdots \ [I(\mathbf{b}_n)]_\mathcal{B}]$$
$$= [[\mathbf{b}_1]_\mathcal{B} \ \cdots \ [\mathbf{b}_n]_\mathcal{B}]$$
$$= [\mathbf{e}_1 \ \cdots \ \mathbf{e}_n] = I_n.$$

67. Take
$$\mathbf{b}_1 = \begin{bmatrix} 1 \\ -2 \end{bmatrix} \quad \text{and} \quad \mathbf{b}_2 = \begin{bmatrix} 2 \\ 1 \end{bmatrix}.$$

Then $\mathbf{b}_1$ lies on the line with equation $y = -2x$, and $\mathbf{b}_2$ is perpendicular to this line. Hence if $\mathcal{B} = \{\mathbf{b}_1, \mathbf{b}_2\}$, then
$$[T]_\mathcal{B} = \begin{bmatrix} 1 & 0 \\ 0 & -1 \end{bmatrix}.$$

So the standard matrix of T is
$$B[T]_\mathcal{B} B^{-1} = \begin{bmatrix} -.6 & -.8 \\ -.8 & .6 \end{bmatrix},$$

where $B = [\mathbf{b}_1 \ \mathbf{b}_2]$. Thus
$$T\left(\begin{bmatrix} x_1 \\ x_2 \end{bmatrix}\right) = \begin{bmatrix} -.6x_1 - .8x_2 \\ -.8x_1 + .6x_2 \end{bmatrix}.$$

71. Take
$$\mathbf{b}_1 = \begin{bmatrix} 1 \\ -3 \end{bmatrix} \quad \text{and} \quad \mathbf{b}_2 = \begin{bmatrix} 3 \\ 1 \end{bmatrix}.$$

Then $\mathbf{b}_1$ lies on the line with equation $y = -3x$, and $\mathbf{b}_2$ is perpendicular to this line. Hence $U(\mathbf{b}_1) = \mathbf{b}_1$ and $U(\mathbf{b}_2) = \mathbf{0}$, so that for $\mathcal{B} = \{\mathbf{b}_1, \mathbf{b}_2\}$, we have
$$[U]_\mathcal{B} = \begin{bmatrix} 1 & 0 \\ 0 & 0 \end{bmatrix}.$$

It follows that the standard matrix of U is
$$B[U]_\mathcal{B} B^{-1} = \begin{bmatrix} .1 & -.3 \\ -.3 & .9 \end{bmatrix},$$

where $B = [\mathbf{b}_1 \ \mathbf{b}_2]$. Therefore
$$U\left(\begin{bmatrix} x_1 \\ x_2 \end{bmatrix}\right) = \begin{bmatrix} .1x_1 - .3x_2 \\ -.3x_1 + .9x_2 \end{bmatrix}.$$

75. We must determine a basis for $\mathcal{R}^3$ consisting of two vectors in W and one vector perpendicular to W. Solving the equation defining W, we obtain
$$\begin{aligned} x &= 4y - 3z \\ y &\quad \text{free} \\ z &\quad \text{free.} \end{aligned}$$

Hence the vector form of this general solution is
$$\begin{bmatrix} x \\ y \\ z \end{bmatrix} = y \begin{bmatrix} 4 \\ 1 \\ 0 \end{bmatrix} + z \begin{bmatrix} -3 \\ 0 \\ 1 \end{bmatrix}.$$

Thus
$$\left\{ \begin{bmatrix} 4 \\ 1 \\ 0 \end{bmatrix}, \begin{bmatrix} -3 \\ 0 \\ 1 \end{bmatrix} \right\}$$

is a basis for W. Also, the vector $\begin{bmatrix} 1 \\ -4 \\ 3 \end{bmatrix}$, whose components are the coefficients in the equation defining W, is perpendicular to W. Therefore
$$\mathcal{B} = \left\{ \begin{bmatrix} 4 \\ 1 \\ 0 \end{bmatrix}, \begin{bmatrix} -3 \\ 0 \\ 1 \end{bmatrix}, \begin{bmatrix} 1 \\ -4 \\ 3 \end{bmatrix} \right\}$$

is a basis for $\mathcal{R}^3$ consisting of two vectors in W and one vector perpendicular to W. For every vector $\mathbf{u}$ in W, $T_W(\mathbf{u}) = \mathbf{u}$, and for every vector $\mathbf{v}$ perpendicular to W, $T_W(\mathbf{v}) = -\mathbf{v}$. Hence
$$[T_W]_\mathcal{B} = \begin{bmatrix} 1 & 0 & 0 \\ 0 & 1 & 0 \\ 0 & 0 & -1 \end{bmatrix}.$$

Thus, if B is the matrix whose columns are the vectors in $\mathcal{B}$, then by Theorem 4.10, the standard matrix of T_W is
$$A = B[T_W]_\mathcal{B} B^{-1}$$
$$= \frac{1}{13} \begin{bmatrix} 12 & 4 & -3 \\ 4 & -3 & 12 \\ -3 & 12 & 4 \end{bmatrix}.$$

It follows that

$$T_W\left(\begin{bmatrix}x_1\\x_2\\x_3\end{bmatrix}\right) = A\begin{bmatrix}x_1\\x_2\\x_3\end{bmatrix}$$

$$= \frac{1}{13}\begin{bmatrix}12x_1 + 4x_2 - 3x_3\\ 4x_1 - 3x_2 + 12x_3\\ -3x_1 + 12x_2 + 4x_3\end{bmatrix}.$$

79. We proceed as in Exercise 75, using the basis

$$\mathcal{B} = \left\{\begin{bmatrix}2\\1\\0\end{bmatrix}, \begin{bmatrix}4\\0\\1\end{bmatrix}, \begin{bmatrix}1\\-2\\-4\end{bmatrix}\right\}.$$

Then

$$T_W\left(\begin{bmatrix}x_1\\x_2\\x_3\end{bmatrix}\right)$$

$$= \frac{1}{21}\begin{bmatrix}19x_1 + 4x_2 + 8x_3\\ 4x_1 + 13x_2 - 16x_3\\ 8x_1 - 16x_2 - 11x_3\end{bmatrix}.$$

81. Let

$$\mathbf{b}_1 = \begin{bmatrix}-2\\1\\0\end{bmatrix}, \quad \mathbf{b}_2 = \begin{bmatrix}3\\0\\1\end{bmatrix},$$

and

$$\mathbf{b}_3 = \begin{bmatrix}1\\2\\-3\end{bmatrix}.$$

(a) Since $\mathbf{b}_1$ and $\mathbf{b}_2$ lie in W, we have $U_W(\mathbf{b}_1) = \mathbf{b}_1$ and $U_W(\mathbf{b}_2) = \mathbf{b}_2$. Moreover, since $\mathbf{b}_3$ is perpendicular to W, $U_W(\mathbf{b}_3) = \mathbf{0}$.

(b) By definition, the columns of $[U_W]_\mathcal{B}$ are

$$[U_W(\mathbf{b}_1)]_\mathcal{B} = [\mathbf{b}_1]_\mathcal{B} = \mathbf{e}_1,$$
$$[U_W(\mathbf{b}_2)]_\mathcal{B} = [\mathbf{b}_2]_\mathcal{B} = \mathbf{e}_2,$$

and

$$[U_W(\mathbf{b}_3)]_\mathcal{B} = [\mathbf{0}]_\mathcal{B} = \mathbf{0}.$$

Therefore

$$[U_W]_\mathcal{B} = [\mathbf{e}_1 \ \mathbf{e}_2 \ \mathbf{0}]$$

$$= \begin{bmatrix}1 & 0 & 0\\ 0 & 1 & 0\\ 0 & 0 & 0\end{bmatrix}.$$

(c) Let $B = [\mathbf{b}_1 \ \mathbf{b}_2 \ \mathbf{b}_3]$. Then by Theorem 4.12, the standard matrix of U_W is

$$B[U_W]_\mathcal{B}B^{-1}$$

$$= \frac{1}{14}\begin{bmatrix}13 & -2 & 3\\ -2 & 10 & 6\\ 3 & 6 & 5\end{bmatrix}.$$

(d) Using the preceding standard matrix of U_W, we have

$$U_W\left(\begin{bmatrix}x_1\\x_2\\x_3\end{bmatrix}\right)$$

$$= \frac{1}{14}\begin{bmatrix}13x_1 - 2x_2 + 3x_3\\ -2x_1 + 10x_2 + 6x_3\\ 3x_1 + 6x_2 + 5x_3\end{bmatrix}.$$

85. We proceed as in Exercise 81, using the basis

$$\mathcal{B} = \left\{\begin{bmatrix}3\\1\\0\end{bmatrix}, \begin{bmatrix}5\\0\\1\end{bmatrix}, \begin{bmatrix}1\\-3\\-5\end{bmatrix}\right\}.$$

Then

$$U_W\left(\begin{bmatrix}x_1\\x_2\\x_3\end{bmatrix}\right)$$

$$= \frac{1}{35}\begin{bmatrix}34x_1 + 3x_2 + 5x_3\\ 3x_1 + 26x_2 - 15x_3\\ 5x_1 - 15x_2 + 10x_3\end{bmatrix}.$$

89. Let T be a linear operator on $\mathcal{R}^n$, A be the standard matrix of T, $\mathcal{B}$ be a basis for $\mathcal{R}^n$, and B be the matrix whose columns are the vectors in $\mathcal{B}$. Because $\mathcal{B}$ is linearly independent, B is invertible by the Invertible Matrix Theorem.

    By Theorem 4.12, $A = B[T]_\mathcal{B} B^{-1}$. Hence if $[T]_\mathcal{B}$ is invertible, then A is a product of invertible matrices, and so A is invertible by Theorem 2.2(b). Thus T is invertible by Theorem 2.13.

    Conversely, if T is invertible, then A is invertible by Theorem 2.13. But, by Theorem 4.12, $[T]_\mathcal{B} = B^{-1}AB$, and so $[T]_\mathcal{B}$ is a product of invertible matrices. Thus $[T]_\mathcal{B}$ is invertible by Theorem 2.2(b).

91. Let A be the standard matrix of T and B be the matrix whose columns are the vectors in $\mathcal{B}$. Then
$$[T]_\mathcal{B} = B^{-1}AB$$
by Theorem 4.12. Since B and B^{-1} are invertible, rank $[T]_\mathcal{B}$ = rank A by Exercises 68 and 70 of Section 2.4. Because the range of T equals the column space of A, the dimension of the range of T equals the dimension of the column space of A, which is rank A. Thus the dimension of the range of T equals rank $[T]_\mathcal{B}$.

95. Let A and B be the matrices whose columns are the vectors in $\mathcal{A}$ and $\mathcal{B}$, respectively, and let C be the standard matrix of T. Then, by Theorem 4.12,
$$[T]_\mathcal{A} = A^{-1}CA$$
and
$$[T]_\mathcal{B} = B^{-1}CB.$$

Solving the second equation for C, we obtain $C = B[T]_\mathcal{B} B^{-1}$. Hence
$$[T]_\mathcal{A} = A^{-1}CA = A^{-1}(B[T]_\mathcal{B} B^{-1})A$$
$$= (A^{-1}B)[T]_\mathcal{B}(B^{-1}A)$$
$$= (B^{-1}A)^{-1}[T]_\mathcal{B}(B^{-1}A)$$
by Theorem 2.2(b). Thus $[T]_\mathcal{A}$ and $[T]_\mathcal{B}$ are similar.

99. Consider column j of $[T]_\mathcal{B}$. Let
$$[T(\mathbf{b}_j)]_\mathcal{B} = \begin{bmatrix} c_1 \\ c_2 \\ \vdots \\ c_n \end{bmatrix}.$$

Because the jth column of $[T]_\mathcal{B}$ is $[T(\mathbf{b}_j)]_\mathcal{B}$, if $[T]_\mathcal{B}$ is an upper triangular matrix, then $c_i = 0$ for $i > j$. Thus we have
$$T(\mathbf{b}_j) = c_1\mathbf{b}_1 + c_2\mathbf{b}_2 + \cdots + c_n\mathbf{b}_n$$
$$= c_1\mathbf{b}_1 + c_2\mathbf{b}_2 + \cdots + c_j\mathbf{b}_j,$$
which is a linear combination of $\mathbf{b}_1, \mathbf{b}_2, \ldots, \mathbf{b}_j$. Conversely, if $T(\mathbf{b}_j)$ is a linear combination of $\mathbf{b}_1, \mathbf{b}_2, \ldots, \mathbf{b}_j$ for each j, then the (i,j)-entry of the matrix $[T]_\mathcal{B}$ equals 0 for $i > j$. Hence $[T]_\mathcal{B}$ is an upper triangular matrix.

103. (a) Let $B = [\mathbf{b}_1\ \mathbf{b}_2\ \mathbf{b}_3\ \mathbf{b}_4]$. The standard matrices of T and U are
$$A = \begin{bmatrix} 1 & -2 & 0 & 0 \\ 0 & 0 & 1 & 0 \\ -1 & 0 & 3 & 0 \\ 0 & 2 & 0 & -1 \end{bmatrix}$$
and
$$C = \begin{bmatrix} 0 & 1 & -1 & 2 \\ -2 & 0 & 0 & 3 \\ 0 & 2 & -1 & 0 \\ 3 & 0 & 0 & 1 \end{bmatrix},$$

respectively. So, by Theorem 2.12, the standard matrix of UT is

$$CA = \begin{bmatrix} 1 & 4 & -2 & -2 \\ -2 & 10 & 0 & -3 \\ 1 & 0 & -1 & 0 \\ 3 & -4 & 0 & -1 \end{bmatrix}.$$

Thus, by Theorem 4.12, we have

$$[T]_\mathcal{B} = B^{-1}AB$$
$$= \begin{bmatrix} 11 & 5 & 13 & 1 \\ -2 & 0 & -5 & -3 \\ -8 & -3 & -9 & 0 \\ 6 & 1 & 8 & 1 \end{bmatrix},$$

$$[U]_\mathcal{B} = B^{-1}CB$$
$$= \begin{bmatrix} -5 & 10 & -38 & -31 \\ 2 & -3 & 9 & 6 \\ 6 & -10 & 27 & 17 \\ -4 & 7 & -25 & -19 \end{bmatrix},$$

and

$$[UT]_\mathcal{B} = B^{-1}(CA)B$$
$$= \begin{bmatrix} 43 & 58 & -21 & -66 \\ -8 & -11 & 8 & 17 \\ -28 & -34 & 21 & 53 \\ 28 & 36 & -14 & -44 \end{bmatrix}.$$

(b) From (a), we see that

$$[U]_\mathcal{B}[T]_\mathcal{B} = (B^{-1}CB)(B^{-1}AB)$$
$$= B^{-1}CI_4AB$$
$$= B^{-1}(CA)B$$
$$= [UT]_\mathcal{B}.$$

107. We will show that $[T^{-1}]_\mathcal{B} = ([T]_\mathcal{B})^{-1}$. By Theorem 4.12,

$$[T]_\mathcal{B} = B^{-1}AB,$$

where A is the standard matrix of T and B is the matrix whose columns are the vectors in $\mathcal{B}$. Recall from Theorem 2.13 that A is invertible and that the standard matrix of T^{-1} is A^{-1}. Hence, by Theorems 4.12 and 2.2(b), we have

$$[T^{-1}]_\mathcal{B} = B^{-1}A^{-1}B$$
$$= (B^{-1}AB)^{-1} = [T]_\mathcal{B}^{-1}.$$

CHAPTER 4 REVIEW

1. True   2. True

3. False, the null space of an $m \times n$ matrix is contained in $\mathcal{R}^n$.

4. False, the column space of an $m \times n$ matrix is contained in $\mathcal{R}^m$.

5. False, the row space of an $m \times n$ matrix is contained in $\mathcal{R}^n$.

6. True   7. True

8. False, the range of every linear transformation equals the *column space* of its standard matrix.

9. False, a nonzero subspace of $\mathcal{R}^n$ has infinitely many bases.

10. False, every basis for a particular subspace contains the same number of vectors.

11. True   12. True   13. True

14. True   15. True   16. True

17. False, the dimension of the null space of a matrix equals the *nullity* of the matrix.

18. True

19. False, the dimension of the row space of a matrix equals the *rank* of the matrix.

20. False, consider $\begin{bmatrix} 1 & 2 \\ 1 & 2 \end{bmatrix}$.

21. True   **22.** True

23. False, $[T]_\mathcal{B} = B^{-1}AB$.

24. True   **25.** True

27. (a) There are at most k vectors in a linearly independent subset of V.
(b) No conclusions can be drawn about the values of k and m in this case.
(c) There are at least k vectors in a generating set for V.

31. The reduced row echelon form of the given matrix A is

$$R = \begin{bmatrix} 1 & 0 & 3 \\ 0 & 1 & -2 \\ 0 & 0 & 0 \\ 0 & 0 & 0 \end{bmatrix}.$$

(a) The null space of A consists of the solutions of $A\mathbf{x} = \mathbf{0}$. The general solution of this system is

$$\begin{aligned} x_1 &= -3x_3 \\ x_2 &= 2x_3 \\ x_2 &\quad \text{free.} \end{aligned}$$

Hence the vector form of the general solution of $A\mathbf{x} = \mathbf{0}$ is

$$\begin{bmatrix} x_1 \\ x_2 \\ x_3 \end{bmatrix} = x_1 \begin{bmatrix} -3 \\ 2 \\ 1 \end{bmatrix}.$$

Thus

$$\left\{ \begin{bmatrix} -3 \\ 2 \\ 1 \end{bmatrix} \right\}$$

is a basis for the null space of A.

(b) The pivot columns of the given matrix form a basis for its column space. From R above, we see that the pivot columns of the given matrix are columns 1 and 2. Hence

$$\left\{ \begin{bmatrix} 1 \\ -1 \\ 2 \\ 1 \end{bmatrix}, \begin{bmatrix} 2 \\ -1 \\ 1 \\ 4 \end{bmatrix} \right\}$$

is a basis for the column space of the given matrix.

(c) A basis for the row space of the given matrix consists of the nonzero rows in its reduced row echelon form. From R above, we see that this basis is

$$\left\{ \begin{bmatrix} 1 \\ 0 \\ 3 \end{bmatrix}, \begin{bmatrix} 0 \\ 1 \\ -2 \end{bmatrix} \right\}.$$

33. The standard matrix of T is

$$A = \begin{bmatrix} 0 & 1 & -2 \\ -1 & 3 & 1 \\ 1 & -4 & 1 \\ 2 & -1 & 3 \end{bmatrix},$$

and its reduced row echelon form is $[\mathbf{e}_1 \ \mathbf{e}_2 \ \mathbf{e}_3]$.

(a) The set

$$\left\{ \begin{bmatrix} 0 \\ -1 \\ 1 \\ 2 \end{bmatrix}, \begin{bmatrix} 1 \\ 3 \\ -4 \\ -1 \end{bmatrix}, \begin{bmatrix} -2 \\ 1 \\ 1 \\ 3 \end{bmatrix} \right\}$$

of pivot columns of A is a basis for the range of T.

(b) The only solution of $A\mathbf{x} = \mathbf{0}$ is $\mathbf{x} = \mathbf{0}$; so the null space of T is the zero subspace.

37. Let B be the matrix whose columns are the vectors in $\mathcal{B}$.

(a) Since the reduced row echelon form of B is I_3, $\mathcal{B}$ is a linearly independent subset of $\mathcal{R}^3$. Since $\mathcal{B}$ contains exactly 3 vectors, $\mathcal{B}$ is a basis for $\mathcal{R}^3$ by Theorem 4.7.

(b) We have
$$\mathbf{v} = 4\begin{bmatrix}0\\-1\\1\end{bmatrix} - 3\begin{bmatrix}1\\0\\-1\end{bmatrix} - 2\begin{bmatrix}-1\\-1\\1\end{bmatrix}$$
$$= \begin{bmatrix}-1\\-2\\5\end{bmatrix}.$$

(c) By Theorem 4.11, we have
$$[\mathbf{w}]_{\mathcal{B}} = B^{-1}\mathbf{w} = \begin{bmatrix}1\\-8\\-6\end{bmatrix}.$$

39. (a) Let B be the matrix whose columns are the vectors in $\mathcal{B}$. Then
$$[T(\mathbf{b}_1)]_{\mathcal{B}} = B^{-1}(T(\mathbf{b}_1))$$
$$= B^{-1}\begin{bmatrix}3\\4\end{bmatrix} = \begin{bmatrix}-17\\-10\end{bmatrix}$$
and
$$[T(\mathbf{b}_2)]_{\mathcal{B}} = B^{-1}(T(\mathbf{b}_2))$$
$$= B^{-1}\begin{bmatrix}-1\\1\end{bmatrix} = \begin{bmatrix}1\\1\end{bmatrix}.$$
Therefore
$$[T]_{\mathcal{B}} = [[T(\mathbf{b}_1)]_{\mathcal{B}} \ [T(\mathbf{b}_2)]_{\mathcal{B}}]$$
$$= \begin{bmatrix}-17 & 1\\-10 & 1\end{bmatrix}.$$

(b) Let A denote the standard matrix of T. By Theorem 4.12, we have
$$A = B[T]_{\mathcal{B}}B^{-1}$$
$$= \begin{bmatrix}-7 & -5\\-14 & -9\end{bmatrix}.$$

(c) Using the result of (b), we have
$$T\left(\begin{bmatrix}x_1\\x_2\end{bmatrix}\right) = A\begin{bmatrix}x_1\\x_2\end{bmatrix}$$
$$= \begin{bmatrix}-7x_1 - 5x_2\\-14x_1 - 9x_2\end{bmatrix}.$$

43. Let $\mathbf{v} = \begin{bmatrix}x\\y\end{bmatrix}$ and $[\mathbf{v}]_{\mathcal{B}} = \begin{bmatrix}x'\\y'\end{bmatrix}$, where $\mathcal{B}$ is the basis obtained by rotating the vectors in the standard basis by $120°$. As in Section 4.4, we have
$$\begin{bmatrix}x'\\y'\end{bmatrix} = [\mathbf{v}]_{\mathcal{B}} = (A_{120°})^{-1}\mathbf{v}$$
$$= (A_{120°})^T\mathbf{v} = \begin{bmatrix}-\frac{1}{2} & \frac{\sqrt{3}}{2}\\-\frac{\sqrt{3}}{2} & -\frac{1}{2}\end{bmatrix}\begin{bmatrix}x\\y\end{bmatrix}.$$
Hence
$$x' = -\tfrac{1}{2}x + \tfrac{\sqrt{3}}{2}y$$
$$y' = -\tfrac{\sqrt{3}}{2}x - \tfrac{1}{2}y.$$
Rewrite the given equation in the form
$$9(x')^2 + 4(y')^2 = 36,$$
and substitute the expressions above for x' and y'. The resulting equation is
$$\frac{21}{4}x^2 - \frac{5\sqrt{3}}{2}xy + \frac{31}{4}y^2 = 36,$$
that is,
$$21x^2 - 10\sqrt{3}xy + 31y^2 = 144.$$

47. Take
$$\mathbf{b}_1 = \begin{bmatrix}2\\-3\end{bmatrix} \text{ and } \mathbf{b}_2 = \begin{bmatrix}3\\2\end{bmatrix}.$$
Then $\mathbf{b}_1$ lies on the line with equation $y = -\tfrac{3}{2}x$, and $\mathbf{b}_2$ is perpendicular to this line. Hence if $\mathcal{B} = \{\mathbf{b}_1, \mathbf{b}_2\}$, then
$$[T]_{\mathcal{B}} = \begin{bmatrix}1 & 0\\0 & -1\end{bmatrix}.$$

So the standard matrix of T is

$$A = B[T]_B B^{-1} = \frac{1}{13}\begin{bmatrix} -5 & -12 \\ -12 & 5 \end{bmatrix},$$

where $B = [\mathbf{b}_1 \; \mathbf{b}_2]$. Thus

$$T\left(\begin{bmatrix} x_1 \\ x_2 \end{bmatrix}\right) = A\begin{bmatrix} x_1 \\ x_2 \end{bmatrix}$$
$$= \frac{1}{13}\begin{bmatrix} -5x_1 - 12x_2 \\ -12x_1 + 5x_2 \end{bmatrix}.$$

51. See the solution to Exercise 107 in Section 4.5.

CHAPTER 4 MATLAB EXERCISES

1. (a) For the given vector $\mathbf{v}$, the equation $A\mathbf{x} = \mathbf{v}$ is consistent. Thus $\mathbf{v}$ is in the column space of A.

   (b) For the given vector $\mathbf{v}$, the equation $A\mathbf{x} = \mathbf{v}$ is not consistent. Thus $\mathbf{v}$ is not in the column space of A.

   (c) For the given vector $\mathbf{v}$, the equation $A\mathbf{x} = \mathbf{v}$ is not consistent. Thus $\mathbf{v}$ is not in the column space of A.

   (d) For the given vector $\mathbf{v}$, the equation $A\mathbf{x} = \mathbf{v}$ is consistent. Thus $\mathbf{v}$ is in the column space of A.

5. Let $\mathbf{b}_i$ ($1 \leq i \leq 6$) denote the ith vector in $\mathcal{B}$.

   (a) $\mathcal{B}$ is a linearly independent set of 6 vectors from $\mathcal{R}^6$.

   (b) For each vector $\mathbf{v}$, the coefficients that express $\mathbf{v}$ as a linear combination of the vectors in $\mathcal{B}$ are the components of $[\mathbf{v}]_\mathcal{B} = B^{-1}\mathbf{v}$, where B is the matrix whose columns are the vectors in $\mathcal{B}$. In this manner, we find that

   (i) $2\mathbf{b}_1 - \mathbf{b}_2 - 3\mathbf{b}_3 + 2\mathbf{b}_5 - \mathbf{b}_6$
   (ii) $\mathbf{b}_1 - \mathbf{b}_2 + \mathbf{b}_3 + 2\mathbf{b}_4 - 3\mathbf{b}_5 + \mathbf{b}_6$
   (iii) $-3\mathbf{b}_2 + \mathbf{b}_3 + 2\mathbf{b}_4 - 4\mathbf{b}_5$

   (c) From (b), we obtain the following coordinate vectors.

   (i) $\begin{bmatrix} 2 \\ -1 \\ -3 \\ 0 \\ 2 \\ -1 \end{bmatrix}$ (ii) $\begin{bmatrix} 1 \\ -1 \\ 1 \\ 2 \\ -3 \\ 1 \end{bmatrix}$ (iii) $\begin{bmatrix} 0 \\ -3 \\ 1 \\ 2 \\ -4 \\ 0 \end{bmatrix}$

9. Let

$$\mathbf{b}_1 = \begin{bmatrix} 1 \\ 3 \\ -1 \\ 0 \\ 2 \end{bmatrix}, \quad \mathbf{b}_2 = \begin{bmatrix} -1 \\ 0 \\ 1 \\ 2 \\ 1 \end{bmatrix},$$

and

$$\mathbf{b}_3 = \begin{bmatrix} 0 \\ 2 \\ 0 \\ 2 \\ 3 \end{bmatrix}.$$

Form the matrix

$$A = [\mathbf{b}_1 \; \mathbf{b}_2 \; \mathbf{b}_3 \; \mathbf{e}_1 \; \mathbf{e}_2 \; \mathbf{e}_3 \; \mathbf{e}_4 \; \mathbf{e}_5],$$

where $\mathbf{e}_i$ denotes the ith standard vector in $\mathcal{R}^5$. From the reduced row echelon form of A, we see that the pivot columns of A are columns 1, 2, 3, 4, and 6. Thus $\mathcal{B} = \{\mathbf{b}_1, \mathbf{b}_2, \mathbf{b}_3, \mathbf{e}_1, \mathbf{e}_3\}$ is a linearly independent set of 5 vectors from $\mathcal{R}^5$, so that $\mathcal{B}$ is a basis for $\mathcal{R}^5$. Let B be the matrix whose columns are the vectors in $\mathcal{B}$ and

$$C = [\mathbf{0} \; \mathbf{0} \; \mathbf{0} \; \mathbf{e}_1 \; \mathbf{e}_3].$$

Then, as explained in Exercise 98 of Section 4.5 and Exercise 8 in the Chapter 4 MATLAB exercises, the matrix

$$A = CB^{-1}$$
$$= \begin{bmatrix} 1 & 0 & 0 & 0.75 & -0.50 \\ 0 & 0 & 0 & 0.00 & 0.00 \\ 0 & 0 & 1 & -0.75 & 0.50 \\ 0 & 0 & 0 & 0.00 & 0.00 \\ 0 & 0 & 0 & 0.00 & 0.00 \end{bmatrix}$$

has the property that $W = \text{Null } A$.

Chapter 5

Eigenvalues, Eigenvectors, and Diagonalization

5.1 EIGENVALUES AND EIGENVECTORS

1. The eigenvalue is 6 because
$$\begin{bmatrix} -10 & -8 \\ 24 & 18 \end{bmatrix} \begin{bmatrix} 1 \\ -2 \end{bmatrix} = \begin{bmatrix} 6 \\ -12 \end{bmatrix} = 6 \begin{bmatrix} 1 \\ -2 \end{bmatrix}.$$

5. The eigenvalue is -2 because
$$\begin{bmatrix} 19 & -7 \\ 42 & -16 \end{bmatrix} \begin{bmatrix} 1 \\ 3 \end{bmatrix} = \begin{bmatrix} -2 \\ -6 \end{bmatrix} = -2 \begin{bmatrix} 1 \\ 3 \end{bmatrix}.$$

9. The eigenvalue is -4 because
$$\begin{bmatrix} 2 & -6 & 6 \\ 1 & 9 & -6 \\ -2 & 16 & -13 \end{bmatrix} \begin{bmatrix} -1 \\ 1 \\ 2 \end{bmatrix} = \begin{bmatrix} 4 \\ -4 \\ -8 \end{bmatrix}$$
$$= (-4) \begin{bmatrix} -1 \\ 1 \\ 2 \end{bmatrix}.$$

13. Let A denote the given matrix. The reduced row echelon form of $A - 3I_3$ is
$$\begin{bmatrix} 1 & 1 \\ 0 & 0 \end{bmatrix},$$

and so
$$\left\{ \begin{bmatrix} -1 \\ 1 \end{bmatrix} \right\}$$
is a basis for the eigenspace of A corresponding to eigenvalue 3.

17. Let A denote the given matrix. The reduced row echelon form of $A - 3I_3$ is
$$\begin{bmatrix} 1 & 1 & 0 \\ 0 & 0 & 1 \\ 0 & 0 & 0 \end{bmatrix},$$

and so
$$\left\{ \begin{bmatrix} -1 \\ 1 \\ 0 \end{bmatrix} \right\}$$
is a basis for the eigenspace of A corresponding to eigenvalue 3.

21. Let A denote the given matrix. The reduced row echelon form of $A - (-1)I_3 = A + I_3$ is
$$\begin{bmatrix} 1 & \frac{1}{3} & -\frac{2}{3} \\ 0 & 0 & 0 \\ 0 & 0 & 0 \end{bmatrix},$$

and so
$$\left\{\begin{bmatrix}-1\\3\\0\end{bmatrix}, \begin{bmatrix}2\\0\\3\end{bmatrix}\right\}$$
is a basis for the eigenspace corresponding to eigenvalue -1.

25. The eigenvalue is 6 because
$$T\left(\begin{bmatrix}-2\\3\end{bmatrix}\right) = \begin{bmatrix}-12\\18\end{bmatrix} = 6\begin{bmatrix}-2\\3\end{bmatrix}.$$

29. The eigenvalue is -3 because
$$T\left(\begin{bmatrix}3\\2\\1\end{bmatrix}\right) = \begin{bmatrix}-9\\-6\\-3\end{bmatrix} = -3\begin{bmatrix}3\\2\\1\end{bmatrix}.$$

33. The standard matrix of T is
$$A = \begin{bmatrix}1 & -2\\6 & -6\end{bmatrix},$$
and the reduced row echelon form of $A\,(-2)I_2 = A + 2I_2$ is
$$\begin{bmatrix}1 & -\frac{2}{3}\\0 & 0\end{bmatrix}.$$

Thus the eigenvectors corresponding to eigenvalue -2 are the nonzero solutions of $x_1 - \frac{2}{3}x_2 = 0$. So a basis for the eigenspace of T corresponding to eigenvalue -2 is
$$\left\{\begin{bmatrix}2\\3\end{bmatrix}\right\}.$$

37. The standard matrix of T is
$$A = \begin{bmatrix}1 & -1 & -3\\-3 & -1 & -9\\1 & 1 & 5\end{bmatrix},$$
and the reduced row echelon form of $A - 2I_3$ is
$$\begin{bmatrix}1 & 1 & 3\\0 & 0 & 0\\0 & 0 & 0\end{bmatrix}.$$

Hence
$$\left\{\begin{bmatrix}-1\\1\\0\end{bmatrix}, \begin{bmatrix}-3\\0\\1\end{bmatrix}\right\}$$
is a basis for the eigenspace corresponding to eigenvalue 2.

41. False, if $A\mathbf{v} = \lambda\mathbf{v}$ for some *nonzero* vector $\mathbf{v}$, then λ is an eigenvalue of A.

42. False, if $A\mathbf{v} = \lambda\mathbf{v}$ for some *nonzero* vector $\mathbf{v}$, then $\mathbf{v}$ is an eigenvector of A.

43. True   **44.** True   **45.** True

46. False, the eigenspace of A corresponding to eigenvalue λ is the null space of $A - \lambda I_n$.

47. True   **48.** True

49. False, the linear operator on $\mathcal{R}^2$ that rotates a vector by $90°$ has no real eigenvalues. (See pages 298–299.)

50. True

51. False, the exception is the zero vector.

52. True   **53.** True   **54.** True

55. False, the exception is $c = 0$.

56. True

57. False, if $A = B = I_n$, then $\lambda = 1$ is an eigenvalue of I_n, but not of $A + B = 2I_n$.

58. True

59. False, if $A = B = 2I_n$, then $\lambda = 2$ is an eigenvalue of $2I_n$, but not of $AB = 4I_n$.

60. True

63. Let λ be the eigenvalue of A corresponding to $\mathbf{v}$. Then $\mathbf{v} \neq \mathbf{0}$, and
$$A(c\mathbf{v}) = c(A\mathbf{v}) = c(\lambda\mathbf{v}) = \lambda(c\mathbf{v}).$$

65. The eigenspace of A corresponding to 0 is the set of vectors $\mathbf{v}$ such that $A\mathbf{v} = 0\mathbf{v}$, that is, such that $A\mathbf{v} = \mathbf{0}$. So the eigenspace corresponding to 0 is the null space of A.

69. If $\mathbf{v}$ is an eigenvector of A with λ as the corresponding eigenvalue, then $A\mathbf{v} = \lambda \mathbf{v}$. So

$$A^2\mathbf{v} = A(A\mathbf{v}) = A(\lambda\mathbf{v}) = \lambda(A\mathbf{v})$$
$$= \lambda(\lambda\mathbf{v}) = \lambda^2\mathbf{v}.$$

Hence λ^2 is an eigenvalue of A^2.

73. Suppose that $c_1\mathbf{v}_1 + c_2\mathbf{v}_2 = \mathbf{0}$ for some scalars c_1 and c_2. Then

$$\mathbf{0} = T(\mathbf{0}) = T(c_1\mathbf{v}_1 + c_2\mathbf{v}_2)$$
$$= c_1\lambda_1\mathbf{v}_1 + c_2\lambda_2\mathbf{v}_2$$
$$= \lambda_1(-c_2\mathbf{v}_2) + c_2\lambda_2\mathbf{v}_2$$
$$= (\lambda_2 - \lambda_1)c_2\mathbf{v}_2.$$

Since $\lambda_1 \neq \lambda_2$ and $\mathbf{v}_2 \neq \mathbf{0}$, we have $c_2 = 0$. Thus we also have $c_1 = 0$, and so $\{\mathbf{v}_1, \mathbf{v}_2\}$ is linearly independent.

77. The eigenvalues of A are -2.7, 2.3, and -1.1 (with multiplicity 2), but the eigenvalues of $3A$ are -8.1, 6.9, and -3.3 (with multiplicity 2).

81. Yes, the eigenvalues of A^T are the same as those of A. Eigenvectors of A^T are found by solving $(A^T - \lambda I_4)\mathbf{x} = \mathbf{0}$ for each eigenvalue λ. Four eigenvectors of A^T are

$$\begin{bmatrix}-1\\1\\-2\\1\end{bmatrix}, \begin{bmatrix}2\\0\\3\\3\end{bmatrix}, \begin{bmatrix}1\\-1\\2\\0\end{bmatrix}, \text{ and } \begin{bmatrix}0\\-1\\0\\1\end{bmatrix}.$$

5.2 THE CHARACTERISTIC POLYNOMIAL

1. The eigenvalues of the given matrix A are the roots of its characteristic polynomial, which are 5 and 6. The reduced row echelon form of $A - 5I_2$ is

$$\begin{bmatrix}1 & 1.5\\0 & 0\end{bmatrix},$$

and so the vector form of the general solution of $(A - 5I_2)\mathbf{x} = \mathbf{0}$ is

$$\begin{bmatrix}x_1\\x_2\end{bmatrix} = x_2\begin{bmatrix}-1.5\\1\end{bmatrix}.$$

So

$$\left\{\begin{bmatrix}-1.5\\1\end{bmatrix}\right\} \text{ or } \left\{\begin{bmatrix}-3\\2\end{bmatrix}\right\}$$

is a basis for the eigenspace of A corresponding to eigenvalue 5. Also, the reduced row echelon form of $A - 6I_2$ is

$$\begin{bmatrix}1 & 1\\0 & 0\end{bmatrix},$$

and so the vector form of the general solution of $(A - 6I_2)\mathbf{x} = \mathbf{0}$ is

$$\begin{bmatrix}x_1\\x_2\end{bmatrix} = x_2\begin{bmatrix}-1\\1\end{bmatrix}.$$

Thus

$$\left\{\begin{bmatrix}-1\\1\end{bmatrix}\right\}$$

is a basis for the eigenspace of A corresponding to eigenvalue 6.

5. The eigenvalues of the given matrix A are the roots of its characteristic polynomial, which are -3 and 2. Since the reduced row echelon form of the matrix $A - (-3)I_3 = A + 3I_3$ is

$$\begin{bmatrix} 1 & 0 & -1 \\ 0 & 1 & -1 \\ 0 & 0 & 0 \end{bmatrix},$$

we see as in Exercise 1 that a basis for the eigenspace corresponding to eigenvalue -3 is

$$\left\{ \begin{bmatrix} 1 \\ 1 \\ 1 \end{bmatrix} \right\}.$$

Likewise, the reduced row echelon form of $A - 2I_3$ is

$$\begin{bmatrix} 1 & 0 & -1 \\ 0 & 1 & 0 \\ 0 & 0 & 0 \end{bmatrix},$$

and so

$$\left\{ \begin{bmatrix} 1 \\ 0 \\ 1 \end{bmatrix} \right\}$$

is a basis for the eigenspace corresponding to eigenvalue 2.

9. Proceeding as in Exercises 1 and 5, we see that the eigenvalues of the given matrix A are -3, -2, and 1. Bases for the respective eigenspaces are $\left\{ \begin{bmatrix} -1 \\ 1 \\ 1 \end{bmatrix} \right\}$,

$$\left\{ \begin{bmatrix} -1 \\ 1 \\ 0 \end{bmatrix} \right\}, \text{ and } \left\{ \begin{bmatrix} 1 \\ 0 \\ 1 \end{bmatrix} \right\}.$$

13. The characteristic polynomial of the given matrix A is

$$\det(A - tI_2) = \det \begin{bmatrix} 1-t & 3 \\ 0 & -4-t \end{bmatrix}$$
$$= (t-1)(t+4).$$

So the eigenvalues of A are 1 and -4. Proceeding as in Exercises 1 and 5, we see that bases for the respective eigenspaces are

$$\left\{ \begin{bmatrix} 1 \\ 0 \end{bmatrix} \right\} \text{ and } \left\{ \begin{bmatrix} -3 \\ 5 \end{bmatrix} \right\}.$$

17. The characteristic polynomial of the given matrix A is

$$\det(A - tI_3)$$
$$= \det \begin{bmatrix} -7-t & 5 & 4 \\ 0 & -3-t & 0 \\ -8 & 9 & 5-t \end{bmatrix}$$
$$= -(t-1)(t+3)^2.$$

Proceeding as in Exercises 1 and 5, we see that bases for the eigenspaces corresponding to eigenvalues 1 and -3 are

$$\left\{ \begin{bmatrix} 1 \\ 0 \\ 2 \end{bmatrix} \right\} \text{ and } \left\{ \begin{bmatrix} 1 \\ 0 \\ 1 \end{bmatrix} \right\}.$$

21. The characteristic polynomial of the given matrix A is

$$\det(A - tI_3)$$
$$= \det \begin{bmatrix} -4-t & 0 & 2 \\ 2 & 4-t & -8 \\ 2 & 0 & -4-t \end{bmatrix}$$
$$= -(t-4)(t+2)(t+6).$$

Proceeding as in Exercises 1 and 5, we see that bases for the eigenspaces corresponding to eigenvalues 4, -2, and -6 are

$$\left\{ \begin{bmatrix} 0 \\ 1 \\ 0 \end{bmatrix} \right\}, \left\{ \begin{bmatrix} 1 \\ 1 \\ 1 \end{bmatrix} \right\}, \text{ and } \left\{ \begin{bmatrix} -1 \\ 1 \\ 1 \end{bmatrix} \right\}.$$

25. The standard matrix of T is
$$A = \begin{bmatrix} -1 & 6 \\ -8 & 13 \end{bmatrix}.$$

The eigenvalues of T are the roots of its characteristic polynomial, which are 5 and 7. The eigenvectors of T are the same as the eigenvectors of A; so we must find bases for the null spaces of $A - 5I_2$ and $A - 7I_2$. As in Exercise 1, we obtain the following bases for the eigenspaces corresponding to 5 and 7:

$$\left\{ \begin{bmatrix} 1 \\ 1 \end{bmatrix} \right\} \text{ and } \left\{ \begin{bmatrix} 3 \\ 4 \end{bmatrix} \right\}.$$

29. The standard matrix of T is
$$A = \begin{bmatrix} 0 & -2 & 4 \\ -3 & 1 & 3 \\ -1 & 1 & 5 \end{bmatrix},$$

and its characteristic polynomial is
$$\det(A - tI_3) = -(t-4)^2(t+2).$$

Bases for the eigenspaces corresponding to 4 and -2 are

$$\left\{ \begin{bmatrix} 1 \\ 0 \\ 1 \end{bmatrix} \right\} \text{ and } \left\{ \begin{bmatrix} 1 \\ 1 \\ 0 \end{bmatrix} \right\}.$$

33. The standard matrix of T is
$$A = \begin{bmatrix} -4 & 1 \\ -2 & -1 \end{bmatrix},$$

and its characteristic polynomial is
$$\det(A - tI_2) = (t+2)(t+3).$$

Bases for the eigenspaces corresponding to -2 and -3 are

$$\left\{ \begin{bmatrix} 1 \\ 2 \end{bmatrix} \right\} \text{ and } \left\{ \begin{bmatrix} 1 \\ 1 \end{bmatrix} \right\}.$$

37. The standard matrix of T is
$$\begin{bmatrix} 7 & -10 & 0 \\ 5 & -8 & 0 \\ -1 & 1 & 2 \end{bmatrix},$$

and its characteristic polynomial is
$$-t^3 + t^2 + 8t - 12 = -(t+3)(t-2)^2.$$

So the eigenvalues of T are -3 and 2. As in Exercise 25, we find the following bases for the eigenspaces:

$$\left\{ \begin{bmatrix} 1 \\ 1 \\ 0 \end{bmatrix} \right\} \text{ and } \left\{ \begin{bmatrix} 0 \\ 0 \\ 1 \end{bmatrix} \right\}.$$

41. The characteristic polynomial of the given matrix is $t^2 - 3t + 10$, which has no (real) roots.

45. Let A denote this matrix. The characteristic polynomial of A is $t^2 - 2t + 5$. So the eigenvalues of A, which are the roots of this polynomial, are $1 - 2i$ and $1 + 2i$. The vector form of the general solution of $(A - (1 - 2i)I_2)\mathbf{x} = \mathbf{0}$ is

$$\begin{bmatrix} x_1 \\ x_2 \end{bmatrix} = x_2 \begin{bmatrix} -\frac{1}{2} \\ 1 \end{bmatrix}$$

So

$$\left\{ \begin{bmatrix} -\frac{1}{2} \\ 1 \end{bmatrix} \right\} \text{ or } \left\{ \begin{bmatrix} 1 \\ -2 \end{bmatrix} \right\}$$

is a basis for the eigenspace of A corresponding to the eigenvalue $1 - 2i$.

Likewise, the vector form of the general solution of $(A - (1 + 2i)I_2)\mathbf{x} = \mathbf{0}$ is

$$\begin{bmatrix} x_1 \\ x_2 \end{bmatrix} = x_2 \begin{bmatrix} -\frac{1}{3} \\ 1 \end{bmatrix}$$

So

$$\left\{ \begin{bmatrix} -\frac{1}{3} \\ 1 \end{bmatrix} \right\} \text{ or } \left\{ \begin{bmatrix} -1 \\ 3 \end{bmatrix} \right\}$$

is a basis for the eigenspace of A corresponding to the eigenvalue $1 + 2i$.

49. The given matrix is upper triangular; so its eigenvalues are its diagonal entries $2i$, 4, and 1. Since the reduced row echelon form of $A - (2i)I_3$ is

$$\begin{bmatrix} 0 & 1 & 0 \\ 0 & 0 & 1 \\ 0 & 0 & 0 \end{bmatrix},$$

a basis for the eigenspace corresponding to the eigenvalue $2i$ is

$$\left\{ \begin{bmatrix} 1 \\ 0 \\ 0 \end{bmatrix} \right\}.$$

Since the reduced row echelon form of $A - 4I_3$ is

$$\begin{bmatrix} 1 & -\frac{i}{2} & 0 \\ 0 & 0 & 1 \\ 0 & 0 & 0 \end{bmatrix},$$

a basis for the eigenspace corresponding to the eigenvalue 4 is

$$\left\{ \begin{bmatrix} i \\ 2 \\ 0 \end{bmatrix} \right\}.$$

Since the reduced row echelon form of $A - I_3$ is

$$\begin{bmatrix} 1 & 0 & 2i \\ 0 & 1 & i \\ 0 & 0 & 0 \end{bmatrix},$$

a basis for the eigenspace corresponding to the eigenvalue 1 is

$$\left\{ \begin{bmatrix} 2 \\ 1 \\ i \end{bmatrix} \right\}.$$

53. False, consider the matrix A in Example 1 and $B = \begin{bmatrix} -3 & 0 \\ 0 & 5 \end{bmatrix}$, which both have $(t+3)(t-5)$ as their characteristic polynomial.

54. True   55. True

56. False, see page 303.

57. False, see page 303.

58. False, consider I_n.

59. False, the rotation matrix $A_{90°}$ has no eigenvectors in $\mathcal{R}^2$.

60. True

61. False, $\begin{bmatrix} 0 & -1 \\ 1 & 0 \end{bmatrix}$ has a characteristic polynomial of $t^2 + 1$.

62. True

63. False, consider $4I_3$; here 4 is an eigenvalue of multiplicity 3.

64. False, see Example 4.

65. True

66. False, consider the matrix given in Exercise 49.

67. True

68. False, see Example 3 of a matrix with no (real) eigenvalues.

69. True   70. True

71. False, it has the eigenvalue 0.

72. True

73. If the reduced row echelon form of $A - cI_n$ is I_n, then $(A - cI_n)\mathbf{x} = \mathbf{0}$ has no solutions except $\mathbf{0}$. Thus there can be no eigenvector corresponding to c, and so c is not an eigenvalue of A.

77. (a) By Theorem 5.1, the eigenvalue 5 must have a multiplicity of 3 or more. In addition, the eigenvalue -9 must have a multiplicity of 1 or more. Since A is a 4×4 matrix, the sum of the multiplicities of its two eigenvalues must be 4. Hence eigenvalue 5 must have multiplicity 3, and eigenvalue -9 must have multiplicity 1. Thus the characteristic polynomial of A must be $(t-5)^3(t+9)$.

(b) By Theorem 5.1, the eigenvalue -9 must have a multiplicity of 1 or more. As in (a), the sum of the multiplicities of the two eigenvalues of A must be 4. Since eigenvalue 5 must have a multiplicity of at least one, there are three possibilities:

(i) Eigenvalue 5 has multiplicity 1, and eigenvalue -9 has multiplicity 3, in which case the characteristic polynomial of A is $(t-5)(t+9)^3$.

(ii) Eigenvalue 5 has multiplicity 2, and eigenvalue -9 has multiplicity 2, in which case the characteristic polynomial of A is $(t-5)^2(t+9)^2$.

(iii) Eigenvalue 5 has multiplicity 3, and eigenvalue -9 has multiplicity 1, in which case the characteristic polynomial of A is $(t-5)^3(t+9)$.

(c) If $\dim W_1 = 2$, then eigenvalue 5 must have a multiplicity of 2 or more. This leads to the two cases described in (ii) and (iii) of (b).

81. (a) Matrix A has eigenvalues of 1 and 2, and

$$\left\{\begin{bmatrix}-1\\1\end{bmatrix}\right\} \text{ and } \left\{\begin{bmatrix}-2\\1\end{bmatrix}\right\}$$

are bases for the corresponding eigenspaces.

(b) Matrix $3A$ has eigenvalues of 3 and 6, and

$$\left\{\begin{bmatrix}-1\\1\end{bmatrix}\right\} \text{ and } \left\{\begin{bmatrix}-2\\1\end{bmatrix}\right\}$$

are bases for the corresponding eigenspaces.

(c) Matrix $5A$ has eigenvalues of 5 and 10, and

$$\left\{\begin{bmatrix}-1\\1\end{bmatrix}\right\} \text{ and } \left\{\begin{bmatrix}-2\\1\end{bmatrix}\right\}$$

are bases for the corresponding eigenspaces.

(d) If c is a nonzero scalar, then $\mathbf{v}$ is an eigenvector of B if and only if $\mathbf{v}$ is an eigenvector of cB because

$$(cB)\mathbf{v} = c(B\mathbf{v}) = c(\lambda \mathbf{v}) = (c\lambda)\mathbf{v}.$$

(e) If c is a nonzero scalar, then λ is an eigenvalue of B if and only if $c\lambda$ is an eigenvalue of cB because

$$(cB)\mathbf{v} = c(B\mathbf{v}) = c(\lambda \mathbf{v}) = (c\lambda)\mathbf{v}.$$

85. If

$$A = \begin{bmatrix} a & b \\ b & c \end{bmatrix},$$

then the characteristic polynomial of A is

$$t^2 - (a+c)t + (ac - b^2).$$

Since the discriminant of this quadratic polynomial is

$$(a-c)^2 + 4b^2 \geq 0,$$

A has real eigenvalues.

89. From the result of Exercise 88, we expect that the matrix

$$\begin{bmatrix} 0 & 0 & 0 & 5 \\ 1 & 0 & 0 & -7 \\ 0 & 1 & 0 & -23 \\ 0 & 0 & 1 & 11 \end{bmatrix}$$

might have the desired characteristic polynomial, and it does!

5.3 DIAGONALIZATION OF MATRICES

1. The eigenvalues of A are 4 and 5. Eigenvectors corresponding to eigenvalue 4 are solutions of $(A - 4I_2)\mathbf{x} = \mathbf{0}$. Since the reduced row echelon form of $A - 4I_2$ is

$$\begin{bmatrix} 1 & 2 \\ 0 & 0 \end{bmatrix},$$

these solutions have the form

$$\begin{bmatrix} x_1 \\ x_2 \end{bmatrix} = x_2 \begin{bmatrix} -2 \\ 1 \end{bmatrix}.$$

Hence

$$\left\{ \begin{bmatrix} -2 \\ 1 \end{bmatrix} \right\}$$

is a basis for the eigenspace of A corresponding to eigenvalue 4. Likewise, the reduced row echelon form of $A - 5I_2$ is

$$\begin{bmatrix} 1 & 3 \\ 0 & 0 \end{bmatrix},$$

and so

$$\left\{ \begin{bmatrix} -3 \\ 1 \end{bmatrix} \right\}$$

is a basis for the eigenspace of A corresponding to eigenvalue 5. Let P be the matrix whose columns are the vectors in the bases for the eigenspaces, and let D be the diagonal matrix whose diagonal entries are the corresponding eigenvalues:

$$P = \begin{bmatrix} -2 & -3 \\ 1 & 1 \end{bmatrix} \text{ and } D = \begin{bmatrix} 4 & 0 \\ 0 & 5 \end{bmatrix}.$$

Then $A = PDP^{-1}$.

5. The eigenvalues of A are -5, 2, and 3. Eigenvectors corresponding to eigenvalue -5 are solutions of $(A+5I_3)\mathbf{x} = \mathbf{0}$, and so a basis for this eigenspace is

$$\left\{ \begin{bmatrix} 0 \\ 1 \\ 1 \end{bmatrix} \right\}.$$

Eigenvectors corresponding to eigenvalue 2 are solutions of $(A - 2I_3)\mathbf{x} = \mathbf{0}$, and so a basis for this eigenspace is

$$\left\{ \begin{bmatrix} -1 \\ \frac{3}{2} \\ 1 \end{bmatrix} \right\} \text{ or } \left\{ \begin{bmatrix} -2 \\ 3 \\ 2 \end{bmatrix} \right\}.$$

Eigenvectors corresponding to eigenvalue 3 are solutions of $(A - 3I_3)\mathbf{x} = \mathbf{0}$, and so a basis for this eigenspace is

$$\left\{ \begin{bmatrix} -1 \\ 1 \\ 1 \end{bmatrix} \right\}.$$

Thus if we take

$$P = \begin{bmatrix} 0 & -2 & -1 \\ 1 & 3 & 1 \\ 1 & 2 & 1 \end{bmatrix}$$

and

$$D = \begin{bmatrix} -5 & 0 & 0 \\ 0 & 2 & 0 \\ 0 & 0 & 3 \end{bmatrix},$$

then $A = PDP^{-1}$.

9. The eigenvalues of A are 5 (with multiplicity 1) and 3 (with multiplicity

2). Eigenvectors corresponding to eigenvalue 5 are solutions of $(A - 5I_3)\mathbf{x} = \mathbf{0}$, and so a basis for this eigenspace is

$$\left\{\begin{bmatrix} -\frac{1}{2} \\ 2 \\ 1 \end{bmatrix}\right\} \quad \text{or} \quad \left\{\begin{bmatrix} -1 \\ 4 \\ 2 \end{bmatrix}\right\}.$$

Eigenvectors corresponding to eigenvalue 3 are solutions of $(A - 3I_3)\mathbf{x} = \mathbf{0}$, and so a basis for this eigenspace is

$$\left\{\begin{bmatrix} -1 \\ 1 \\ 0 \end{bmatrix}, \begin{bmatrix} 1 \\ 0 \\ 1 \end{bmatrix}\right\}.$$

Thus if we take

$$P = \begin{bmatrix} -1 & -1 & 1 \\ 4 & 1 & 0 \\ 2 & 0 & 1 \end{bmatrix}$$

and

$$D = \begin{bmatrix} 5 & 0 & 0 \\ 0 & 3 & 0 \\ 0 & 0 & 3 \end{bmatrix},$$

then $A = PDP^{-1}$.

13. The characteristic polynomial of A is

$$t^2 - 2t + 1 = (t - 1)^2.$$

Since the rank of $A - I_2$ is 1, the eigenspace of A corresponding to eigenvalue 1 is 1-dimensional. Hence A is not diagonalizable because the eigenvalue 1 has multiplicity 2 and its eigenspace is 1-dimensional.

17. Since the given matrix is upper triangular, its eigenvalues are its diagonal entries, which are -1, -3, and 2. Bases for the corresponding eigenspaces are

$$\left\{\begin{bmatrix} 1 \\ 0 \\ 0 \end{bmatrix}\right\}, \left\{\begin{bmatrix} -1 \\ 1 \\ 0 \end{bmatrix}\right\}, \text{ and } \left\{\begin{bmatrix} -1 \\ 1 \\ 5 \end{bmatrix}\right\},$$

respectively. Hence we may take

$$P = \begin{bmatrix} 1 & -1 & -1 \\ 0 & 1 & 1 \\ 0 & 0 & 5 \end{bmatrix}$$

and

$$D = \begin{bmatrix} -1 & 0 & 0 \\ 0 & -3 & 0 \\ 0 & 0 & 2 \end{bmatrix}.$$

21. The characteristic polynomial of A is $t^2 - 4t + 5$, and hence the eigenvalues of A are $2 - i$ and $2 + i$. Bases for the corresponding eigenspaces are

$$\left\{\begin{bmatrix} -i \\ 1 \end{bmatrix}\right\} \quad \text{and} \quad \left\{\begin{bmatrix} i \\ 1 \end{bmatrix}\right\},$$

respectively. Hence we may take

$$P = \begin{bmatrix} -i & i \\ 1 & 1 \end{bmatrix}$$

and

$$D = \begin{bmatrix} 2-i & 0 \\ 0 & 2+i \end{bmatrix}.$$

25. The characteristic polynomial of A is

$$-(t^2 - 2t + 2)(t - 2)$$
$$= -(t - 1 - i)(t - 1 + i)(t - 2),$$

and so the eigenvalues of A are $1 + i$, $1 - i$, and 2. Bases for the corresponding eigenspaces are

$$\left\{\begin{bmatrix} -1 \\ 2+i \\ 1 \end{bmatrix}\right\}, \left\{\begin{bmatrix} -1 \\ 2-i \\ 1 \end{bmatrix}\right\},$$

and

$$\left\{\begin{bmatrix} 1 \\ -1 \\ 1 \end{bmatrix}\right\},$$

respectively. Hence we may take

$$P = \begin{bmatrix} -1 & -1 & 1 \\ 2+i & 2-i & -1 \\ 1 & 1 & 1 \end{bmatrix}$$

and

$$D = \begin{bmatrix} 1+i & 0 & 0 \\ 0 & 1-i & 0 \\ 0 & 0 & 2 \end{bmatrix}.$$

29. False, see Example 1.

30. True   31. True   32. True

33. False, the eigenvalues of A may occur in any sequence as the diagonal entries of D.

34. False, if an $n \times n$ matrix has n *linearly independent* eigenvectors, then it is diagonalizable.

35. False, I_n is diagonalizable and has only one eigenvalue.

36. True

37. False, see Example 1.

38. False, for A to be diagonalizable, its characteristic polynomial must also factor as a product of linear factors.

39. True

40. False, the dimension of the eigenspace corresponding to λ is the *nullity* of $A - \lambda I_n$.

41. False, for example, I_n has only one eigenvalue, namely 1.

42. True   43. True

44. False, $P^{-1}AP$ is a diagonal matrix.

45. False, for example, any nonzero multiple of an eigenvector is an eigenvector.

46. True

47. False, this is true only if the multiplicity of each eigenvalue is equal to the dimension of the corresponding eigenspace.

48. False, this is true only if the sum of the multiplicities of the eigenvalues is equal to the number of columns in A.

49. The first boxed statement on page 318 implies that the matrix is diagonalizable.

53. The matrix is diagonalizable if and only if the eigenspace corresponding to the eigenvalue -3 is 4-dimensional.

57. We have $A = PDP^{-1}$, where

$$P = \begin{bmatrix} 1 & 2 \\ 1 & 1 \end{bmatrix} \quad \text{and} \quad D = \begin{bmatrix} 4 & 0 \\ 0 & 3 \end{bmatrix}.$$

Thus, as in the example on page 314, we have

$$A^k = PD^kP^{-1}$$
$$= \begin{bmatrix} 1 & 2 \\ 1 & 1 \end{bmatrix} \begin{bmatrix} 4^k & 0 \\ 0 & 3^k \end{bmatrix} \begin{bmatrix} -1 & 2 \\ 1 & -1 \end{bmatrix}$$
$$= \begin{bmatrix} 4^k & 2 \cdot 3^k \\ 4^k & 3^k \end{bmatrix} \begin{bmatrix} -1 & 2 \\ 1 & -1 \end{bmatrix}$$
$$= \begin{bmatrix} 2 \cdot 3^k - 4^k & 2 \cdot 4^k - 2 \cdot 3^k \\ 3^k - 4^k & 2 \cdot 4^k - 3^k \end{bmatrix}.$$

61. We have $A = PDP^{-1}$, where

$$P = \begin{bmatrix} -1 & 0 & -2 \\ 1 & 0 & 1 \\ 0 & 1 & 0 \end{bmatrix}$$

and

$$D = \begin{bmatrix} 5 & 0 & 0 \\ 0 & 5 & 0 \\ 0 & 0 & 1 \end{bmatrix}.$$

Thus, as in Exercise 57, we have

$$A^k = PD^kP^{-1}$$
$$= \begin{bmatrix} -5^k+2 & -2\cdot 5^k+2 & 0 \\ 5^k-1 & 2\cdot 5^k-1 & 0 \\ 0 & 0 & 5^k \end{bmatrix}.$$

65. Since there is only one eigenvalue (with multiplicity 1), this matrix is not diagonalizable for any scalar c.

69. It follows from the first boxed statement on page 318 that the given matrix is diagonalizable if $c \neq -2$ and $c \neq -1$. Thus we must check only the values of -2 and -1. For $c = -2$, we see that -2 is an eigenvalue of multiplicity 2, but the reduced row echelon form of $A+2I_3$, which is

$$\begin{bmatrix} 1 & 0 & 0 \\ 0 & 0 & 1 \\ 0 & 0 & 0 \end{bmatrix},$$

has rank 2. Hence A is not diagonalizable if $c = -2$. Likewise, for $c = -1$, the eigenvalue -1 has multiplicity 2, but the reduced row echelon form of $A+I_3$, which is

$$\begin{bmatrix} 1 & 0 & 0 \\ 0 & 0 & 1 \\ 0 & 0 & 0 \end{bmatrix},$$

has rank 2. Thus A is also not diagonalizable if $c = -1$.

73. The desired matrix A satisfies $A = PDP^{-1}$, where

$$P = \begin{bmatrix} 1 & 1 \\ 1 & 3 \end{bmatrix} \text{ and } D = \begin{bmatrix} -3 & 0 \\ 0 & 5 \end{bmatrix}.$$

(Here the columns of P are the given eigenvectors of A, and the diagonal entries of D are the corresponding eigenvalues.) Thus

$$A = \begin{bmatrix} -7 & 4 \\ -12 & 9 \end{bmatrix}.$$

77. The matrices

$$A = \begin{bmatrix} 0 & 0 \\ 0 & 1 \end{bmatrix} \text{ and } B = \begin{bmatrix} 0 & 1 \\ 0 & -1 \end{bmatrix}$$

are diagonalizable by the first boxed result on page 318, but their sum is the matrix in Example 1 that is not diagonalizable.

81. Let $A = PDP^{-1}$, where D is a diagonal matrix and P is an invertible matrix. Then for $Q = (P^T)^{-1}$, we have

$$A^T = (PDP^{-1})^T = (P^{-1})^T D^T P^T$$
$$= (P^T)^{-1} DP^T = QDQ^{-1},$$

and so A^T is also diagonalizable.

85. (a) Suppose that A is diagonalizable. Then $A = QDQ^{-1}$ for some diagonal matrix D and invertible matrix Q. Since

$$B = PAP^{-1} = P(QDQ^{-1})P^{-1}$$
$$= (PQ)D(PQ)^{-1},$$

B is also diagonalizable. The proof of the converse is similar.

(b) The eigenvalues of A and B are equal. See the box on page 307.

(c) We claim that $\mathbf{v}$ is an eigenvector of A if and only if $P\mathbf{v}$ is an eigenvector of B. For if $A\mathbf{v} = \lambda\mathbf{v}$, then

$$B(P\mathbf{v}) = (PAP^{-1})(P\mathbf{v})$$
$$= PA\mathbf{v} = P(\lambda\mathbf{v}) = \lambda(P\mathbf{v}).$$

Conversely, if $B(P\mathbf{v}) = \lambda(P\mathbf{v})$, then

$$A\mathbf{v} = (P^{-1}BP)\mathbf{v}$$
$$= P^{-1}(\lambda P\mathbf{v}) = \lambda\mathbf{v}.$$

89. (a) Let $A = PDP^{-1}$, where D is a diagonal matrix and P is an invertible matrix. By the hint, the trace of $A = PDP^{-1}$ equals the trace of $PP^{-1}D = D$, which is the sum of the eigenvalues of A.

(b) In $p(t)$, the characteristic polynomial of A, the coefficient of t^{n-1} is
$$(-1)^n(-\lambda_1 - \lambda_2 - \cdots - \lambda_n)$$
$$= (-1)^{n+1}(\lambda_1 + \lambda_2 + \cdots + \lambda_n),$$
which by (a) equals $(-1)^{n-1}$ times the trace of A.

(c) The constant term of $p(t)$ is $(-1)^n(-\lambda_1)(-\lambda_2)\cdots(-\lambda_n)$, that is, $\lambda_1 \lambda_2 \cdots \lambda_n = \det D = \det A$.

93. The characteristic polynomial of the given matrix is $-(t-2)^2(t-1)^3$, and thus the eigenvalue 1 has multiplicity 3. The rank of the matrix $A - 1I_5$ is 3, however, and so the eigenspace corresponding to this eigenvalue has dimension 2. Therefore the matrix is not diagonalizable by the test on page 319.

5.4 DIAGONALIZATION OF LINEAR OPERATORS

3. The standard matrix of T is
$$A = \begin{bmatrix} 0 & -1 & -2 \\ 0 & 2 & 0 \\ 1 & 1 & 3 \end{bmatrix}.$$

If B is the matrix whose columns are the vectors in $\mathcal{B}$, then
$$[T]_\mathcal{B} = B^{-1}AB = \begin{bmatrix} 2 & 1 & 0 \\ 0 & 2 & 0 \\ 0 & 0 & 1 \end{bmatrix}.$$

Since $[T]_\mathcal{B}$ is not a diagonal matrix, the basis $\mathcal{B}$ does not consist of eigenvectors of T.

7. The standard matrix of T is
$$A = \begin{bmatrix} -3 & 5 & -5 \\ 2 & -3 & 2 \\ 2 & -5 & 4 \end{bmatrix}.$$

If B is the matrix whose columns are the vectors in $\mathcal{B}$, then
$$[T]_\mathcal{B} = B^{-1}AB = \begin{bmatrix} 2 & 0 & 0 \\ 0 & -1 & 0 \\ 0 & 0 & -3 \end{bmatrix}.$$

Since $[T]_\mathcal{B}$ is a diagonal matrix, the basis $\mathcal{B}$ consists of eigenvectors of T.

11. The standard matrix of T is
$$A = \begin{bmatrix} 7 & -5 \\ 10 & -8 \end{bmatrix}.$$

A basis for the eigenspace of T corresponding to the eigenvalue -3 can be obtained by solving $(A + 3I_2)\mathbf{x} = \mathbf{0}$, and a basis for the eigenspace of T corresponding to eigenvalue 2 can be obtained by solving $(A - 2I_2)\mathbf{x} = \mathbf{0}$. The resulting bases are
$$\mathcal{B}_1 = \left\{ \begin{bmatrix} 1 \\ 2 \end{bmatrix} \right\} \quad \text{and} \quad \mathcal{B}_2 = \left\{ \begin{bmatrix} 1 \\ 1 \end{bmatrix} \right\}.$$

Combining these two sets produces a basis for $\mathcal{R}^2$ consisting of eigenvectors of T.

15. The standard matrix of T is
$$A = \begin{bmatrix} -1 & -1 & 0 \\ 0 & -1 & 0 \\ 1 & 1 & 0 \end{bmatrix}.$$

Since the reduced row echelon form of $A + I_3$ is
$$\begin{bmatrix} 1 & 0 & 1 \\ 0 & 1 & 0 \\ 0 & 0 & 0 \end{bmatrix},$$

the dimension of the eigenspace of T corresponding to eigenvalue -1 is

$$3 - \text{rank}\,(A + I_3) = 1.$$

But the multiplicity of the eigenvalue -1 is 2, so that T is not diagonalizable. That is, there is no basis for $\mathcal{R}^3$ consisting of eigenvectors of T.

19. The eigenvalues of T are -3 (with multiplicity 1) and 1 (with multiplicity 3). By solving $(A + 3I_4)\mathbf{x} = \mathbf{0}$, we obtain the basis

$$\left\{ \begin{bmatrix} 1 \\ 0 \\ 1 \\ 0 \end{bmatrix} \right\}$$

for the eigenspace corresponding to eigenvalue -3. Similarly, by solving $(A - 1I_4)\mathbf{x} = \mathbf{0}$, we obtain the basis

$$\left\{ \begin{bmatrix} -1 \\ 2 \\ 0 \\ 0 \end{bmatrix}, \begin{bmatrix} 1 \\ 0 \\ 2 \\ 0 \end{bmatrix}, \begin{bmatrix} -1 \\ 0 \\ 0 \\ 2 \end{bmatrix} \right\}$$

for the eigenspace corresponding to eigenvalue 1. Combining the bases for these two eigenspaces produces a basis

$$\left\{ \begin{bmatrix} 1 \\ 0 \\ 1 \\ 0 \end{bmatrix}, \begin{bmatrix} -1 \\ 2 \\ 0 \\ 0 \end{bmatrix}, \begin{bmatrix} 1 \\ 0 \\ 2 \\ 0 \end{bmatrix}, \begin{bmatrix} -1 \\ 0 \\ 0 \\ 2 \end{bmatrix} \right\}$$

for $\mathcal{R}^4$ consisting of eigenvectors of T.

23. The standard matrix of T is

$$\begin{bmatrix} -2 & 3 \\ 4 & -3 \end{bmatrix},$$

and its characteristic polynomial is $t^2 + 5t - 6 = (t - 1)(t + 6)$. As in Exercise 11, we find that

$$\mathcal{B}_1 = \left\{ \begin{bmatrix} 1 \\ 1 \end{bmatrix} \right\} \quad \text{and} \quad \mathcal{B}_2 = \left\{ \begin{bmatrix} -3 \\ 4 \end{bmatrix} \right\}$$

are bases for the eigenspaces of T corresponding to the eigenvalues -6 and 1, respectively. Combining these two sets produces a basis $\mathcal{B}$ for $\mathcal{R}^2$ consisting of eigenvectors of T, and so

$$[T]_\mathcal{B} = \begin{bmatrix} 1 & 0 \\ 0 & -6 \end{bmatrix}$$

is a diagonal matrix.

27. The standard matrix of T is

$$\begin{bmatrix} 1 & 0 & 0 \\ -1 & 1 & -1 \\ 0 & 0 & 1 \end{bmatrix},$$

and its characteristic polynomial is

$$-t^3 + 3t^2 - 3t + 1 = -(t-1)^3.$$

Since the reduced row echelon form of $A - I_3$ is

$$\begin{bmatrix} 1 & 0 & 1 \\ 0 & 0 & 0 \\ 0 & 0 & 0 \end{bmatrix},$$

the dimension of the eigenspace of T corresponding to eigenvalue 1 is

$$3 - \text{rank}\,(A - I_3) = 2.$$

Because this dimension does not equal the multiplicity of the eigenvalue 1, T is not diagonalizable. That is, there is no basis $\mathcal{B}$ for $\mathcal{R}^3$ such that $[T]_\mathcal{B}$ is a diagonal matrix.

29. False, its standard matrix is diagonalizable, that is, *similar* to a diagonal matrix.

30. False, the linear operator on $\mathcal{R}^2$ that rotates a vector by $90°$ is not diagonalizable.

31. True

32. False, $\mathcal{B}$ can be any basis for $\mathcal{R}^n$ consisting of eigenvectors of T.

33. False, the eigenvalues of T may occur in any sequence as the diagonal entries of D.

34. True   35. True   36. True

37. True

38. False, in addition, the multiplicity of each eigenvaue must equal the dimension of the corresponding eigenspace.

39. True

40. False, in addition, the sum of the multiplicities of the eigenvalues must equal n.

41. False, it is an eigenvector corresponding to the eigenvalue 1.

42. False, it is an eigenvector corresponding to the eigenvalue -1.

43. True

44. False, a linear operator on $\mathcal{R}^n$ may have no eigenvalues.

45. True   46. True   47. True

48. False, this statement is true only when T is diagonalizable.

51. The standard matrix of T is
$$A = \begin{bmatrix} c & 0 & 0 \\ -1 & -3 & -1 \\ -8 & 1 & -5 \end{bmatrix},$$
and so
$$A + 4I_3 = \begin{bmatrix} c+4 & 0 & 0 \\ -1 & 1 & -1 \\ -8 & 1 & -1 \end{bmatrix}.$$

Because the last two rows of $A+4I_3$ are linearly independent, the rank of $A+4I_3$ is at least 2. Hence the dimension of the eigenspace of T corresponding to eigenvalue -4 is 1. Since this dimension does not equal the multiplicity of the eigenvalue -4, T is not diagonalizable for any scalar c.

55. The only real eigenvalue of T is c, and its mutiplicity is 1. Thus T is not diagonalizable for any scalar c.

59. The vector form of the general solution of the equation $x + y + z = 0$ is
$$\begin{bmatrix} x \\ y \\ z \end{bmatrix} = y \begin{bmatrix} -1 \\ 1 \\ 0 \end{bmatrix} + z \begin{bmatrix} -1 \\ 0 \\ 1 \end{bmatrix}.$$

Hence
$$\left\{ \begin{bmatrix} -1 \\ 1 \\ 0 \end{bmatrix}, \begin{bmatrix} -1 \\ 0 \\ 1 \end{bmatrix} \right\}$$
is a basis for W, the eigenspace of T_W corresponding to eigenvalue 1. As on page 329, the vector
$$\begin{bmatrix} 1 \\ 1 \\ 1 \end{bmatrix}$$
whose components are the coefficients of the equation $x + y + z = 0$, is normal to W, and so is an eigenvector of T_W corresponding to eigenvalue -1. Thus
$$\mathcal{B} = \left\{ \begin{bmatrix} -1 \\ 1 \\ 0 \end{bmatrix}, \begin{bmatrix} -1 \\ 0 \\ 1 \end{bmatrix}, \begin{bmatrix} 1 \\ 1 \\ 1 \end{bmatrix} \right\}$$
is a basis for $\mathcal{R}^3$ consisting of eigenvectors of T_W. Hence
$$[T_W]_\mathcal{B} = \begin{bmatrix} 1 & 0 & 0 \\ 0 & 1 & 0 \\ 0 & 0 & -1 \end{bmatrix}.$$

Let B be the matrix whose columns are the vectors in $\mathcal{B}$. Then the standard matrix of T_W is

$$A = B[T_W]_\mathcal{B} B^{-1}$$

$$= \frac{1}{3}\begin{bmatrix} 1 & -2 & -2 \\ -2 & 1 & -2 \\ -2 & -2 & 1 \end{bmatrix}.$$

Therefore

$$T_W\left(\begin{bmatrix} x_1 \\ x_2 \\ x_3 \end{bmatrix}\right) = A \begin{bmatrix} x_1 \\ x_2 \\ x_3 \end{bmatrix}$$

$$= \frac{1}{3} \begin{bmatrix} x_1 - 2x_2 - 2x_3 \\ -2x_1 + x_2 - 2x_3 \\ -2x_1 - 2x_2 + x_3 \end{bmatrix}.$$

63. As in the solution to Exercise 59, we choose a basis

$$\mathcal{B} = \left\{ \begin{bmatrix} -8 \\ 1 \\ 0 \end{bmatrix}, \begin{bmatrix} 5 \\ 0 \\ 1 \end{bmatrix}, \begin{bmatrix} 1 \\ 8 \\ -5 \end{bmatrix} \right\}$$

for $\mathcal{R}^3$ consisting of eigenvectors of T_W. Here the first two vectors lie in the plane with equation $x + 8y - 5z = 0$ and the third vector is perpendicular to this plane. Let B be the matrix whose columns are the vectors in $\mathcal{B}$. Then the standard matrix of T_W is

$$A = B[T_W]_\mathcal{B} B^{-1}$$

$$= \frac{1}{45} \begin{bmatrix} 44 & -8 & 5 \\ -8 & -19 & 40 \\ 5 & 40 & 20 \end{bmatrix},$$

where

$$[T_W]_\mathcal{B} = \begin{bmatrix} 1 & 0 & 0 \\ 0 & 1 & 0 \\ 0 & 0 & -1 \end{bmatrix}$$

is as in Exercise 59. Thus

$$T_W\left(\begin{bmatrix} x_1 \\ x_2 \\ x_3 \end{bmatrix}\right) =$$

$$\frac{1}{45} \begin{bmatrix} 44x_1 - 8x_2 + 5x_3 \\ -8x_1 - 19x_2 + 40x_3 \\ 5x_1 + 40x_2 + 20x_3 \end{bmatrix}.$$

67. (a) Let $\{\mathbf{v}_1, \mathbf{v}_2\}$ be a basis for W and $\mathbf{v}_3$ be a nonzero vector perpendicular to W. Then $\mathcal{B} = \{\mathbf{v}_1, \mathbf{v}_2, \mathbf{v}_3\}$ is a basis for $\mathcal{R}^3$. Furthermore, $U_W(\mathbf{v}_1) = \mathbf{v}_1$, $U_W(\mathbf{v}_2) = \mathbf{v}_2$, and $U_W(\mathbf{v}_3) = \mathbf{0}$. So

$$[U_W]_\mathcal{B} = \begin{bmatrix} 1 & 0 & 0 \\ 0 & 1 & 0 \\ 0 & 0 & 0 \end{bmatrix}.$$

(b) Since $T_W(\mathbf{v}_1) = \mathbf{v}_1$, $T_W(\mathbf{v}_2) = \mathbf{v}_2$, and $T_W(\mathbf{v}_3) = -\mathbf{v}_3$, we have

$$[T_W]_\mathcal{B} = \begin{bmatrix} 1 & 0 & 0 \\ 0 & 1 & 0 \\ 0 & 0 & -1 \end{bmatrix}.$$

(c) $\frac{1}{2}([T_W]_\mathcal{B} + I_3)$

$$= \frac{1}{2}\left(\begin{bmatrix} 1 & 0 & 0 \\ 0 & 1 & 0 \\ 0 & 0 & -1 \end{bmatrix} + \begin{bmatrix} 1 & 0 & 0 \\ 0 & 1 & 0 \\ 0 & 0 & 1 \end{bmatrix} \right)$$

$$= \begin{bmatrix} 1 & 0 & 0 \\ 0 & 1 & 0 \\ 0 & 0 & 0 \end{bmatrix} = [U_W]_\mathcal{B}$$

(d) Let I denote the identity operator on $\mathcal{R}_3$. Then $[I]_\mathcal{B} = I_3$, and hence

$$[U_W]_\mathcal{B} = \frac{1}{2}([T_W]_\mathcal{B} + [I]_\mathcal{B})$$

$$= \frac{1}{2}([T_W + I]_\mathcal{B}).$$

So $U_W = \frac{1}{2}(T_W + I)$. Therefore

$$U_W\left(\begin{bmatrix} x_1 \\ x_2 \\ x_3 \end{bmatrix}\right)$$

$$= \tfrac{1}{2}\left(\tfrac{1}{3}\begin{bmatrix} x_1 - 2x_2 - 2x_3 \\ -2x_1 + x_2 - 2x_3 \\ -2x_1 - 2x_2 + x_3 \end{bmatrix}\right)$$

$$+ \tfrac{1}{2}\begin{bmatrix} x_1 \\ x_2 \\ x_3 \end{bmatrix}$$

$$= \tfrac{1}{3}\begin{bmatrix} 2x_1 - x_2 - x_3 \\ -x_1 + 2x_2 - x_3 \\ -x_1 - x_2 + 2x_3 \end{bmatrix}.$$

71. By Exercise 67(c) and the answer to Exercise 63, we have

$$U_W(\mathbf{x}) = \tfrac{1}{2}(T_W + I)(\mathbf{x})$$

$$= \tfrac{1}{90}\begin{bmatrix} 89x_1 - 8x_2 + 5x_3 \\ -8x_1 + 26x_2 + 40x_3 \\ 5x_1 + 40x_2 + 65x_3 \end{bmatrix}.$$

75. We combine (a) and (b). Let $\mathcal{B} = \{\mathbf{u}, \mathbf{v}, \mathbf{w}\}$. Observe that $T(\mathbf{u}) = \mathbf{u}$, $T(\mathbf{v}) = \mathbf{v}$, and $T(\mathbf{w}) = \mathbf{0}$. Hence $\mathbf{u}$ and $\mathbf{v}$ are eigenvectors of T with corresponding eigenvalue 1, and $\mathbf{w}$ is an eigenvector of T with corresponding eigenvalue 0. Consequently, $\{\mathbf{u}, \mathbf{v}\}$ is a basis for the eigenspace of T corresponding to the eigenvalue 1, and $\{\mathbf{w}\}$ is a basis for the eigenspace of T corresponding to the eigenvalue 0. It also follows that T is diagonalizable because there is a basis for $\mathcal{R}^3$ consisting of eigenvectors of T.

79. Let c be any scalar. If $\mathbf{v}$ is an eigenvector of T corresponding to eigenvalue λ, then it is also an eigenvector of cT corresponding to eigenvalue λ, because

$$cT(\mathbf{v}) = c(T(\mathbf{v})) = c(\lambda \mathbf{v}) = (c\lambda)\mathbf{v}.$$

Thus if $\mathcal{B}$ is a basis for $\mathcal{R}^n$ consisting of eigenvectors of T, then $\mathcal{B}$ is also a basis for $\mathcal{R}^n$ consisting of eigenvectors of cT for any scalar c.

83. Let U be a diagonalizable linear operator on $\mathcal{R}^n$ having only nonnegative eigenvalues, and let $\mathcal{B}$ be a basis for $\mathcal{R}^n$ consisting of eigenvectors of U. If C is the standard matrix of U and B is the matrix whose columns are the vectors in $\mathcal{B}$, then $[U]_\mathcal{B} = B^{-1}CB$. Let A be the diagonal matrix whose entries are the square roots of the entries of the diagonal matrix $[U]_\mathcal{B}$. Then $A^2 = [U]_\mathcal{B}$, and so

$$C = B[U]_\mathcal{B} B^{-1} = BA^2B^{-1}$$
$$= (BAB^{-1})^2.$$

So if T is the matrix transformation induced by BAB^{-1}, then T is a square root of U.

5.5 APPLICATIONS OF EIGENVALUES

1. False, the *column* sums of the transition matrix of a Markov chain are all 1.

2. False, see the matrix A on page 334.

3. True

4. False, consider $A = \begin{bmatrix} 0 & 1 \\ 1 & 0 \end{bmatrix}$ and $\mathbf{p} = \begin{bmatrix} .8 \\ .2 \end{bmatrix}$.

5. True 6. True 7. True

8. False, the general solution of $y' = ky$ is $y = ce^{kt}$.

9. False, the change of variable $\mathbf{y} = P\mathbf{z}$ transforms $\mathbf{y}' = A\mathbf{y}$ into $\mathbf{z}' = D\mathbf{z}$.

10. True

11. False, the solution is $\mathbf{y} = P\mathbf{z}$.

12. True

13. No, for the given transition matrix A, the second column of A^k equals $\mathbf{e}_2$ for every positive integer k. Thus the $(1,2)$-entry of A^k is always zero.

17. No, the $(1,2)$-entry of A^k is zero for every k.

19. Yes, $A^2 = \begin{bmatrix} .36 & .05 & .08 & .06 \\ .08 & .25 & .10 & .18 \\ .24 & .45 & .72 & .04 \\ .32 & .25 & .10 & .72 \end{bmatrix}$ has no zero entries.

23. A steady-state vector is a probability vector that is also an eigenvector corresponding to eigenvalue 1. We begin by finding the eigenvectors corresponding to eigenvalue 1. Since the reduced row echelon form of $A - 1I_3$ is

$$\begin{bmatrix} 1 & 0 & -.5 \\ 0 & 1 & -.5 \\ 0 & 0 & 0 \end{bmatrix},$$

a basis for the eigenspace corresponding to eigenvalue 1 is

$$\left\{ \begin{bmatrix} 1 \\ 1 \\ 2 \end{bmatrix} \right\}.$$

Thus the eigenvectors corresponding to eigenvalue 1 have the form

$$c \begin{bmatrix} 1 \\ 1 \\ 2 \end{bmatrix} = \begin{bmatrix} c \\ c \\ 2c \end{bmatrix}.$$

We seek a vector of this form that is also a probability vector, that is, such that

$$c + c + 2c = 1.$$

So $c = .25$, and the steady-state vector is

$$\begin{bmatrix} .25 \\ .25 \\ .50 \end{bmatrix}.$$

25. As in Exercise 23, the eigenvectors corresponding to eigenvalue 1 are multiples of $\begin{bmatrix} 1 \\ 3 \\ 2 \end{bmatrix}$. The desired steady-state vector is the multiple whose components sum to 1, that is, the vector

$$\frac{1}{6} \begin{bmatrix} 1 \\ 3 \\ 2 \end{bmatrix}.$$

29. (a) The two states of this Markov chain are buying a root beer float (F) and buying a chocolate sundae (S). A transition matrix for this Markov chain is

$$\text{Next visit} \begin{array}{c} \\ \text{F} \\ \text{S} \end{array} \overset{\begin{array}{cc} \text{Last visit} \\ \text{F} \quad \text{S} \end{array}}{\begin{bmatrix} .25 & .5 \\ .75 & .5 \end{bmatrix}} = A.$$

Note that the $(1,2)$-entry and the $(2,1)$-entry of A can be determined from the condition that each column sum in A must be 1.

(b) If Alison bought a sundae on her next-to-last visit, we can take

$$\mathbf{p} = \begin{bmatrix} 0 \\ 1 \end{bmatrix}.$$

Then the probabilities of each purchase on her last visit are

$$A\mathbf{p} = \begin{bmatrix} .5 \\ .5 \end{bmatrix},$$

and the probabilities of each purchase on her next visit are

$$A^2\mathbf{p} = A(A\mathbf{p}) = \begin{bmatrix} .375 \\ .625 \end{bmatrix}.$$

Thus the probability that she will buy a float on her next visit is .375.

(c) Over the long run, the proportion of purchases of each kind is given by the steady-state vector for A. As in Exercise 23, we first find a basis for the eigenspace corresponding to eigenvalue 1, which is

$$\left\{ \begin{bmatrix} 2 \\ 3 \end{bmatrix} \right\}.$$

The vector in this eigenspace that is also a probability vector is

$$\frac{1}{5} \begin{bmatrix} 2 \\ 3 \end{bmatrix} = \begin{bmatrix} .4 \\ .6 \end{bmatrix}.$$

Hence, over the long run, Alison buys a sundae on 60% of her trips to the ice cream store.

33. For a fixed j, $1 \leq j \leq n$, the probability of moving from page j to page i is a_{ij} for $1 \leq i \leq n$. Since it is certain that the surfer will move from page j to some page, the sum of all these probabilities must be 1. That is, the sum of the entries of the jth column of A is 1.

37. (a) In general, the probability of moving from state j to state i in one time period is a_{ij}. So the probability of moving from state 1 to state 2 in one time period is $a_{21} = .05$, and the probability of moving from state 1 to state 3 in one time period is $a_{31} = .05$.

(b) The probability of moving from state 2 to state 1 in one time period is $a_{12} = .1$, and the probability of moving from state 2 to state 3 in one time period is $a_{32} = .1$.

(c) Similarly, the probability of moving from state 3 to each of the other states in one time period is .3.

(d) In general, suppose we have a transition matrix of the form

$$M = \begin{bmatrix} 1-2a & b & c \\ a & 1-2b & c \\ a & b & 1-2c \end{bmatrix},$$

where $0 < a, b, c < 1$. For example, in the given matrix A, we have $a = .05$, $b = .1$, and $c = .3$. Suppose also that $\mathbf{p}$ is the steady state vector for M. Then $(M - I_3)\mathbf{p} = \mathbf{0}$, and hence

$$\begin{bmatrix} -2a & b & c \\ a & -2b & c \\ a & b & -2c \end{bmatrix} \begin{bmatrix} p_1 \\ p_2 \\ p_3 \end{bmatrix}$$

$$= \begin{bmatrix} -2 & 1 & 1 \\ 1 & -2 & 1 \\ 1 & 1 & -2 \end{bmatrix} \begin{bmatrix} ap_1 \\ bp_2 \\ cp_3 \end{bmatrix}$$

$$= \begin{bmatrix} 0 \\ 0 \\ 0 \end{bmatrix}.$$

Since a basis for the null space of the matrix

$$\begin{bmatrix} -2 & 1 & 1 \\ 1 & -2 & 1 \\ 1 & 1 & -2 \end{bmatrix} \text{ is } \left\{ \begin{bmatrix} 1 \\ 1 \\ 1 \end{bmatrix} \right\},$$

it follows that $ap_1 = bp_2 = cp_3$. So

$$p_1 = \frac{cp_3}{a}, \quad p_2 = \frac{cp_3}{b},$$

and

$$p_3 = \frac{cp_3}{c}.$$

It follows that

$$\mathbf{p} = k \begin{bmatrix} \frac{1}{a} \\ \frac{1}{b} \\ \frac{1}{c} \end{bmatrix},$$

where $k = \frac{1}{a} + \frac{1}{b} + \frac{1}{c}$. So, for the given matrix A,

$$\mathbf{p} = k \begin{bmatrix} \frac{1}{.05} \\ \frac{1}{.1} \\ \frac{1}{.3} \end{bmatrix} = \begin{bmatrix} .6 \\ .3 \\ .1 \end{bmatrix}.$$

(e) For the vector $\mathbf{p}$ in (d), we have $A\mathbf{p} = \mathbf{p}$.

41. Let A be an $n \times n$ stochastic matrix and $\mathbf{p}$ be a probability vector in $\mathcal{R}^n$. Then each component of $A\mathbf{p}$ is nonnegative, and the sum of the components is

$$(a_{11}p_1 + a_{12}p_2 + \cdots + a_{1n}p_n) + \cdots$$
$$+ (a_{n1}p_1 + a_{n2}p_2 + \cdots + a_{nn}p_n)$$
$$= (a_{11} + \cdots + a_{n1})p_1 + \cdots$$
$$+ (a_{1n} + \cdots + a_{nn})p_n$$
$$= p_1 + \cdots + p_n$$
$$= 1.$$

43. (a) The absolute value of the ith component of $A^T\mathbf{v}$ is

$$|a_{1i}v_1 + \cdots + a_{ni}v_n|$$
$$\leq |a_{1i}||v_1| + \cdots + |a_{ni}||v_n|$$
$$\leq (|a_{1i}| + \cdots + |a_{ni}|)|v_k|$$
$$\leq |v_k|.$$

(b) Let $\mathbf{v}$ be an eigenvector of A^T corresponding to eigenvalue λ. Then $A^T\mathbf{v} = \lambda\mathbf{v}$. It follows from (a) that the absolute value of the kth component of $A^T\mathbf{v}$ is $|\lambda v_k| \leq |v_k|$. Hence $|\lambda| \cdot |v_k| \leq |v_k|$.

(c) Since $|v_k| \neq 0$, the preceding inequality implies that $|\lambda| \leq 1$.

47. The given system of differential equations can be written in the form $\mathbf{y}' = A\mathbf{y}$, where

$$A = \begin{bmatrix} 2 & 4 \\ -6 & -8 \end{bmatrix}.$$

Since the characteristic polynomial of A is

$$t^2 + 6t + 8 = (t+4)(t+2),$$

A has the eigenvalues -4 and -2. Bases for the corresponding eigen-spaces of A are

$$\left\{ \begin{bmatrix} -2 \\ 3 \end{bmatrix} \right\} \text{ and } \left\{ \begin{bmatrix} -1 \\ 1 \end{bmatrix} \right\}.$$

Hence $A = PDP^{-1}$, where

$$P = \begin{bmatrix} -2 & -1 \\ 3 & 1 \end{bmatrix} \text{ and } D = \begin{bmatrix} -4 & 0 \\ 0 & -2 \end{bmatrix}.$$

The solution of $\mathbf{z}' = D\mathbf{z}$ is

$$z_1 = ae^{-4t}$$
$$z_2 = be^{-2t}.$$

The algorithm on page 341 gives the solution of the original system to be

$$\begin{bmatrix} y_1 \\ y_2 \end{bmatrix} = \mathbf{y} = P\mathbf{z} = \begin{bmatrix} -2 & -1 \\ 3 & 1 \end{bmatrix} \begin{bmatrix} ae^{-4t} \\ be^{-2t} \end{bmatrix}$$
$$= \begin{bmatrix} -2ae^{-4t} - be^{-2t} \\ 3ae^{-4t} + be^{-2t} \end{bmatrix}.$$

49. The given system of differential equations can be written in the form $\mathbf{y}' = A\mathbf{y}$, where

$$A = \begin{bmatrix} 2 & 0 & 0 \\ 3 & 2 & 3 \\ -3 & 0 & -1 \end{bmatrix}.$$

Here A has the eigenvalues -1 and 2 (with multiplicity 2). Bases for the corresponding eigenspaces of A are

$$\left\{\begin{bmatrix} 0 \\ -1 \\ 1 \end{bmatrix}\right\} \quad \text{and} \quad \left\{\begin{bmatrix} -1 \\ 0 \\ 1 \end{bmatrix}, \begin{bmatrix} 0 \\ 1 \\ 0 \end{bmatrix}\right\}.$$

Hence $A = PDP^{-1}$, where

$$P = \begin{bmatrix} 0 & -1 & 0 \\ -1 & 0 & 1 \\ 1 & 1 & 0 \end{bmatrix}$$

and

$$D = \begin{bmatrix} -1 & 0 & 0 \\ 0 & 2 & 0 \\ 0 & 0 & 2 \end{bmatrix}.$$

The solution of $\mathbf{z}' = D\mathbf{z}$ is

$$z_1 = ae^{-t}$$
$$z_2 = be^{2t}$$
$$z_3 = ce^{2t}$$

The algorithm on page 341 gives the solution of the original system to be

$$\begin{bmatrix} y_1 \\ y_2 \end{bmatrix} = \mathbf{y} = P\mathbf{z}$$

$$= \begin{bmatrix} 0 & -1 & 0 \\ -1 & 0 & 1 \\ 1 & 1 & 0 \end{bmatrix} \begin{bmatrix} ae^{-t} \\ be^{2t} \\ ce^{2t} \end{bmatrix}$$

$$= \begin{bmatrix} -be^{2t} \\ -ae^{-t} + ce^{2t} \\ ae^{-t} + be^{2t} \end{bmatrix}.$$

53. The given system of differential equations can be written in the form $\mathbf{y}' = A\mathbf{y}$, where

$$A = \begin{bmatrix} 1 & 1 \\ 4 & 1 \end{bmatrix}.$$

Here A has eigenvalues of -1 and 3, and bases for the corresponding eigenspaces of A are

$$\left\{\begin{bmatrix} -1 \\ 2 \end{bmatrix}\right\} \quad \text{and} \quad \left\{\begin{bmatrix} 1 \\ 2 \end{bmatrix}\right\}.$$

Hence $A = PDP^{-1}$, where

$$P = \begin{bmatrix} -1 & 1 \\ 2 & 2 \end{bmatrix} \quad \text{and} \quad D = \begin{bmatrix} -1 & 0 \\ 0 & 3 \end{bmatrix}.$$

The solution of $\mathbf{z}' = D\mathbf{z}$ is

$$z_1 = ae^{-t}$$
$$z_2 = be^{3t}.$$

Thus the general solution of the original system is

$$\begin{bmatrix} y_1 \\ y_2 \end{bmatrix} = \mathbf{y} = P\mathbf{z} = \begin{bmatrix} -1 & 1 \\ 2 & 2 \end{bmatrix} \begin{bmatrix} ae^{-t} \\ be^{3t} \end{bmatrix}$$

$$= \begin{bmatrix} -ae^{-t} + be^{3t} \\ 2ae^{-t} + 2be^{3t} \end{bmatrix}.$$

Taking $t = 0$, we obtain

$$15 = y_1(0) = -a + b$$

and

$$-10 = y_2(0) = 2a + 2b.$$

Solving this system, we obtain $a = -10$ and $b = 5$. Thus the solution of the original system of differential equations and initial conditions is

$$y_1 = 10e^{-t} + 5e^{3t}$$
$$y_2 = -20e^{-t} + 10e^{3t}.$$

57. The given system of differential equations can be written in the form $\mathbf{y}' = A\mathbf{y}$, where

$$A = \begin{bmatrix} 6 & -5 & -7 \\ 1 & 0 & -1 \\ 3 & -3 & -4 \end{bmatrix}.$$

Here A has eigenvalues of -1, 1, and 2, and bases for the corresponding eigenspaces of A are

$$\left\{\begin{bmatrix}1\\0\\1\end{bmatrix}\right\},\ \left\{\begin{bmatrix}1\\1\\0\end{bmatrix}\right\},$$

and

$$\left\{\begin{bmatrix}3\\1\\1\end{bmatrix}\right\}.$$

Hence $A = PDP^{-1}$, where

$$P = \begin{bmatrix}1 & 1 & 3\\0 & 1 & 1\\1 & 0 & 1\end{bmatrix}$$

and

$$D = \begin{bmatrix}-1 & 0 & 0\\0 & 1 & 0\\0 & 0 & 2\end{bmatrix}.$$

The solution of $\mathbf{z}' = D\mathbf{z}$ is

$$z_1 = ae^{-t}$$
$$z_2 = be^t$$
$$z_3 = ce^{2t}.$$

Thus the general solution of the original system is

$$\begin{bmatrix}y_1\\y_2\\y_3\end{bmatrix} = \mathbf{y} = P\mathbf{z}$$

$$= \begin{bmatrix}1 & 1 & 3\\0 & 1 & 1\\1 & 0 & 1\end{bmatrix}\begin{bmatrix}ae^{-t}\\be^t\\ce^{2t}\end{bmatrix}$$

$$= \begin{bmatrix}ae^{-t} + be^t + 3ce^{2t}\\be^t + ce^{2t}\\ae^{-t} + ce^{2t}\end{bmatrix}.$$

Taking $t = 0$, we obtain

$$0 = y_1(0) = a + b + 3c$$

$$2 = y_2(0) = \quad b + c$$

and

$$1 = y_3(0) = a + \quad c.$$

Solving this system, we obtain $a = 4$, $b = 5$, and $c = -3$. Thus the solution of the original system of differential equations and initial conditions is

$$y_1 = 4e^{-t} + 5e^t - 9e^{2t}$$
$$y_2 = \qquad 5e^t - 3e^{2t}$$
$$y_3 = 4e^{-t} \qquad - 3e^{2t}.$$

61. Setting $y_1 = y$ and $y_2 = y_1' = y''$, we obtain the system

$$y_1' = \qquad y_2$$
$$y_2' = 3y_1 + 2y_2,$$

which in matrix form can be written $\mathbf{y}' = A\mathbf{y}$, where

$$A = \begin{bmatrix}0 & 1\\3 & 2\end{bmatrix}.$$

Observe that $A = PDP^{-1}$, for

$$P = \begin{bmatrix}1 & 1\\3 & -1\end{bmatrix} \text{ and } D = \begin{bmatrix}3 & 0\\0 & -1\end{bmatrix},$$

and the general solution of $\mathbf{z}' = D\mathbf{z}$ is $z_1 = ae^{3t}$, $z_2 = be^{-t}$. It follows that

$$\begin{bmatrix}y_1\\y_2\end{bmatrix} = \mathbf{y} = P\mathbf{z} = \begin{bmatrix}1 & 1\\3 & -1\end{bmatrix}\begin{bmatrix}z_1\\z_2\end{bmatrix}$$

$$= \begin{bmatrix}ae^{3t} + be^{-t}\\3ae^{3t} - be^{-t}\end{bmatrix},$$

and hence $y = y_1 = ae^{3t} + be^{-t}$.

65. Take $w = 10$ lbs, $b = 0.625$, and $k = 1.25$ lbs/foot in the equation

$$\frac{w}{g}y''(t) + by'(t) + ky(t) = 0.$$

Then the equation simplifies to the form
$$y'' + 2y' + 4y = 0.$$

We transform this differential equation into a system of differential equations by letting $y_1 = y$ and $y_2 = y'$. These substitutions produce
$$y_1' = y_2$$
$$y_2' = -4y_1 - 2y_2,$$

which can be written as $\mathbf{y}' = A\mathbf{y}$, where
$$A = \begin{bmatrix} 0 & 1 \\ -4 & -2 \end{bmatrix}.$$

The characteristic polynomial of this matrix is $t^2 + 2t + 4$, which has the roots $-1 + \sqrt{3}\,i$ and $-1 - \sqrt{3}\,i$. The general solution of the original differential equation can be written using Euler's formula as
$$y = ae^{(-1-\sqrt{3}\,i)t} + be^{(-1+\sqrt{3}\,i)t}$$
$$= ae^{-t}(\cos\sqrt{3}\,t + i\sin\sqrt{3}\,t)$$
$$ + be^{-t}(\cos\sqrt{3}\,t - i\sin\sqrt{3}\,t)$$
or, equivalently, as
$$y = ce^{-t}\cos\sqrt{3}\,t + de^{-t}\sin\sqrt{3}\,t.$$

69. The differential equation
$$y''' + ay'' + by' + cy = 0$$
can be written as $\mathbf{y}' = A\mathbf{y}$, where
$$A = \begin{bmatrix} 0 & 1 & 0 \\ 0 & 0 & 1 \\ -c & -b & -a \end{bmatrix}.$$

The characteristic polynomial of A is $-t^3 - at^2 - bt - c$; so $\lambda_i^3 = -a\lambda_i^2 - b\lambda_i - c$ for $i = 1, 2, 3$. Now
$$A\begin{bmatrix} 1 \\ \lambda_i \\ \lambda_i^2 \end{bmatrix} = \begin{bmatrix} \lambda_i \\ \lambda_i^2 \\ -c - b\lambda_i - a\lambda_i^2 \end{bmatrix}$$

$$= \begin{bmatrix} \lambda_i \\ \lambda_i^2 \\ \lambda_i^3 \end{bmatrix} = \lambda_i \begin{bmatrix} 1 \\ \lambda_i \\ \lambda_i^2 \end{bmatrix}.$$

Thus
$$\mathbf{v}_i = \begin{bmatrix} 1 \\ \lambda_i \\ \lambda_i^2 \end{bmatrix}$$

is an eigenvector of A with λ_i as its corresponding eigenvalue. So $\{\mathbf{v}_1, \mathbf{v}_2, \mathbf{v}_3\}$ is a basis for $\mathcal{R}^3$ consisting of eigenvectors of A. Thus the solution of the given equation is given by the boxed result of $\mathbf{y}' = A\mathbf{y}$ on page 341 with $P = [\mathbf{v}_1\ \mathbf{v}_2\ \mathbf{v}_3]$ and
$$D = \begin{bmatrix} \lambda_1 & 0 & 0 \\ 0 & \lambda_2 & 0 \\ 0 & 0 & \lambda_3 \end{bmatrix}.$$

73. The given difference equation can be written as $\mathbf{s}_n = A\mathbf{s}_{n-1}$, where
$$\mathbf{s}_n = \begin{bmatrix} r_n \\ r_{n+1} \end{bmatrix} \quad \text{and} \quad A = \begin{bmatrix} 0 & 1 \\ 4 & 3 \end{bmatrix}.$$

Taking
$$P = \begin{bmatrix} 1 & 1 \\ -1 & 4 \end{bmatrix} \quad \text{and} \quad D = \begin{bmatrix} -1 & 0 \\ 0 & 4 \end{bmatrix},$$

we have $A = PDP^{-1}$. Hence $A^n = PD^nP^{-1}$, and so
$$\mathbf{s}_n = A^n \mathbf{s}_0 = PD^n P^{-1} \mathbf{s}_0$$
$$= P \begin{bmatrix} (-1)^n & 0 \\ 0 & 4^n \end{bmatrix} \begin{bmatrix} .8 & -.2 \\ .2 & .2 \end{bmatrix} \begin{bmatrix} 1 \\ 1 \end{bmatrix}$$
$$= \begin{bmatrix} (-1)^n & 4^n \\ (-1)^{n+1} & 4^{n+1}n \end{bmatrix} \begin{bmatrix} .6 \\ .4 \end{bmatrix}$$
$$= \begin{bmatrix} .6(-1)^n + .4(4^n) \\ .6(-1)^{n+1} + .4(4^{n+1}) \end{bmatrix}.$$

Equating the first components of $\mathbf{s}_n$ and the preceding vector, we have

$$r_n = .6(-1)^n + .4(4^n) \text{ for } n \geq 0.$$

Hence

$$r_6 = .6(-1)^6 + .4(4^6) = 1639.$$

77. The given difference equation can be written as $\mathbf{s}_n = A\mathbf{s}_{n-1}$, where

$$\mathbf{s}_n = \begin{bmatrix} r_n \\ r_{n+1} \\ r_{n+2} \end{bmatrix}$$

and

$$A = \begin{bmatrix} 0 & 1 & 0 \\ 0 & 0 & 1 \\ 0 & 2 & 1 \end{bmatrix}.$$

Taking

$$P = \begin{bmatrix} 1 & 1 & 1 \\ 0 & -1 & 2 \\ 0 & 1 & 4 \end{bmatrix}$$

and

$$D = \begin{bmatrix} 0 & 0 & 0 \\ 0 & -1 & 0 \\ 0 & 0 & 2 \end{bmatrix},$$

we have $A = PDP^{-1}$. Hence $A^n = PD^nP^{-1}$, and so

$$\mathbf{s}_n = A^n \mathbf{s}_0 = PD^n P^{-1} \mathbf{s}_0$$

$$= P \begin{bmatrix} 0 & 0 & 0 \\ 0 & (-1)^n & 0 \\ 0 & 0 & 2^n \end{bmatrix} P^{-1} \begin{bmatrix} r_0 \\ r_1 \\ r_2 \end{bmatrix}$$

$$= P \begin{bmatrix} 0 & 0 & 0 \\ 0 & (-1)^n & 0 \\ 0 & 0 & 2^n \end{bmatrix} P^{-1} \begin{bmatrix} 9 \\ 0 \\ 18 \end{bmatrix}$$

$$= \begin{bmatrix} 6(-1)^n + 3(2^n) \\ -6(-1)^n - 6(2^n) \\ 6(-1)^n + 12(2^n) \end{bmatrix}.$$

Thus $r_n = 6(-1)^n + 3(2^n)$ for $n \geq 0$, and so $r_6 = 198$.

81. Since the given difference equation is of the third order, $\mathbf{s}_n$ is a vector in $\mathcal{R}^3$ and A is a 3×3 matrix. Taking

$$\mathbf{s}_n = \begin{bmatrix} r_n \\ r_{n+1} \\ r_{n+2} \end{bmatrix} \text{ and } A = \begin{bmatrix} 0 & 1 & 0 \\ 0 & 0 & 1 \\ 5 & -2 & 4 \end{bmatrix},$$

we see that the matrix form of the given equation is $\mathbf{s}_n = A\mathbf{s}_{n-1}$.

85. We have

$$A\mathbf{s}_0 = \begin{bmatrix} 1 & 0 \\ c & a \end{bmatrix} \begin{bmatrix} 1 \\ r_0 \end{bmatrix}$$

$$= \begin{bmatrix} 1 \\ c + ar_0 \end{bmatrix} = \begin{bmatrix} 1 \\ r_1 \end{bmatrix} = \mathbf{s}_1,$$

$$A\mathbf{s}_1 = \begin{bmatrix} 1 & 0 \\ c & a \end{bmatrix} \begin{bmatrix} 1 \\ r_1 \end{bmatrix}$$

$$= \begin{bmatrix} 1 \\ c + ar_1 \end{bmatrix} = \begin{bmatrix} 1 \\ r_2 \end{bmatrix} = \mathbf{s}_2,$$

and, in general,

$$A\mathbf{s}_{n-1} = \begin{bmatrix} 1 & 0 \\ c & a \end{bmatrix} \begin{bmatrix} 1 \\ r_{n-1} \end{bmatrix}$$

$$= \begin{bmatrix} 1 \\ c + ar_{n-1} \end{bmatrix} = \begin{bmatrix} 1 \\ r_n \end{bmatrix} = \mathbf{s}_n.$$

Hence

$$\mathbf{s}_n = A\mathbf{s}_{n-1} = A(A\mathbf{s}_{n-2}) = A^2 \mathbf{s}_{n-2}$$
$$= A^2(A\mathbf{s}_{n-3}) = A^3 \mathbf{s}_{n-3} = \cdots$$
$$= A^n \mathbf{s}_0.$$

(For those familiar with mathematical induction, this proof can be written using induction.)

89. The given system of differential equations has the form $\mathbf{y}' = A\mathbf{y}$, where

$$\mathbf{y} = \begin{bmatrix} y_1 \\ y_2 \\ y_3 \\ y_4 \end{bmatrix}$$

and

$$A = \begin{bmatrix} 3.2 & 4.1 & 7.7 & 3.7 \\ -0.3 & 1.2 & 0.2 & 0.5 \\ -1.8 & -1.8 & -4.4 & -1.8 \\ 1.7 & -0.7 & 2.9 & 0.4 \end{bmatrix}.$$

The characteristic polynomial of A is

$$t^4 - 0.4t^3 - 0.79t^2 + .166t + 0.24$$
$$= (t + 0.8)(t + 0.1)(t - 0.3)(t - 1).$$

Since A has four distinct eigenvalues, it is diagonalizable; in fact, $A = PDP^{-1}$, where

$$P = \begin{bmatrix} 1 & -1 & -1 & 2 \\ 0 & -1 & -2 & -1 \\ -1 & 0 & 0 & -1 \\ 1 & 2 & 3 & 2 \end{bmatrix}$$

and

$$D = \begin{bmatrix} -0.8 & 0.0 & 0.0 & 0 \\ 0.0 & -0.1 & 0.0 & 0 \\ 0.0 & 0.0 & 0.3 & 0 \\ 0.0 & 0.0 & 0.0 & 1 \end{bmatrix}.$$

The solution of $\mathbf{z}' = D\mathbf{z}$ is

$$\begin{bmatrix} z_1 \\ z_2 \\ z_3 \\ z_4 \end{bmatrix} = \begin{bmatrix} ae^{-0.8t} \\ be^{-0.1t} \\ ce^{0.3t} \\ de^t \end{bmatrix}.$$

Hence the general solution of the original equation is

$$\begin{bmatrix} y_1 \\ y_2 \\ y_3 \\ y_4 \end{bmatrix} =$$

$$\begin{bmatrix} ae^{-0.8t} - be^{-0.1t} - ce^{0.3t} + 2de^t \\ -be^{-0.1t} - 2ce^{0.3t} - de^t \\ -ae^{-0.8t} \qquad\qquad\qquad - de^t \\ ae^{-0.8t} + 2be^{-0.1t} + 3ce^{0.3t} + 2de^t \end{bmatrix}.$$

When $t = 0$, the preceding equation takes the form

$$\begin{bmatrix} 1 \\ -4 \\ 2 \\ 3 \end{bmatrix} = P \begin{bmatrix} a \\ b \\ c \\ d \end{bmatrix},$$

and so $a = -6$, $b = 2$, $c = -1$, and $d = 4$. Thus the particular solution of the original system that satisfies the given initial condition is

$$\begin{aligned} y_1 &= -6e^{-0.8t} - 2e^{-0.1t} + e^{0.3t} + 8e^t \\ y_2 &= \qquad\qquad -2e^{-0.1t} + 2e^{0.3t} - 4e^t \\ y_3 &= 6e^{-0.8t} \qquad\qquad\qquad - 4e^t \\ y_4 &= -6e^{-0.8t} + 4e^{-0.1t} - 3e^{0.3t} + 8e^t. \end{aligned}$$

CHAPTER 5 REVIEW

1. True

2. False, there are infinitely many eigenvectors that correspond to a particular eigenvalue.

3. True   4. True   5. True

6. True

7. False, the linear operator on $\mathcal{R}^2$ that rotates a vector by 90° has no real eigenvalues.

8. False, the rotation matrix $A_{90°}$ has no (real) eigenvalues.

9. False, I_n has only one eigenvalue, namely 1.

10. False, if two $n \times n$ matrices have the same characteristic polynomial, they have the same *eigenvalues*.

11. True   12. True

13. False, if $A = PDP^{-1}$, where P is an invertible matrix and D is a diagonal matrix, then the columns of P are a basis for $\mathcal{R}^n$ consisting of eigenvectors of A.

14. True   15. True   16. True

17. True

19. The characteristic polynomial of the given matrix A is

$$\det(A - tI_2) = \begin{bmatrix} 5-t & 6 \\ -2 & -2-t \end{bmatrix}$$

$$= (t-1)(t-2),$$

and so its eigenvalues are 1 and 2. The eigenspace of A corresponding to eigenvalue 1 is the null space of $A - 1I_2$. Since the reduced row echelon form of this matrix is

$$\begin{bmatrix} 1 & \frac{3}{2} \\ 0 & 0 \end{bmatrix},$$

a basis for this eigenspace is

$$\left\{ \begin{bmatrix} -3 \\ 2 \end{bmatrix} \right\}.$$

Similarly, the eigenspace of A corresponding to eigenvalue 2 is the null space of $A - 1I_2$, and the reduced row echelon form of this matrix is

$$\begin{bmatrix} 1 & 2 \\ 0 & 0 \end{bmatrix},$$

Thus a basis for this eigenspace is

$$\left\{ \begin{bmatrix} -2 \\ 1 \end{bmatrix} \right\}.$$

23. The characteristic polynomial of A is

$$\det(A - tI_2) = (t-2)(t-7),$$

and so its eigenvalues are 2 and 7. As in Exercise 19, the eigenspace of A corresponding to eigenvalue 2 has the basis

$$\left\{ \begin{bmatrix} 2 \\ 1 \end{bmatrix} \right\},$$

and the eigenspace of A corresponding to eigenvalue 7 has the basis

$$\left\{ \begin{bmatrix} 1 \\ 3 \end{bmatrix} \right\}.$$

Take

$$P = \begin{bmatrix} 2 & 1 \\ 1 & 3 \end{bmatrix},$$

the matrix whose columns are the vectors in these bases, and

$$D = \begin{bmatrix} 2 & 0 \\ 0 & 7 \end{bmatrix},$$

the diagonal matrix whose diagonal entries are the eigenvalues of A that correspond to the respective columns of P. Then $A = PDP^{-1}$.

27. The standard matrix of T is

$$A = \begin{bmatrix} 4 & 2 \\ -4 & -5 \end{bmatrix}.$$

A basis for $\mathcal{R}^2$ consisting of eigenvectors of A is also a basis for $\mathcal{R}^2$ consisting of eigenvectors of T. So we proceed as in Exercise 23. Here the eigenvalues of A are -4 and 3, and bases for the corresponding eigenspaces are

$$\left\{ \begin{bmatrix} -1 \\ 4 \end{bmatrix} \right\} \quad \text{and} \quad \left\{ \begin{bmatrix} -2 \\ 1 \end{bmatrix} \right\}.$$

By combining these bases, we obtain a basis for $\mathcal{R}^2$ consisting of eigenvectors of A or T, namely.

$$\left\{ \begin{bmatrix} -1 \\ 4 \end{bmatrix}, \begin{bmatrix} -2 \\ 1 \end{bmatrix} \right\}.$$

31. The given matrix A has eigenvalues of 2, 3, and c. By the first boxed result on page 318, A is diagonalizable if $c \neq 2$ and $c \neq 3$. We must check these two cases separately. When $c = 2$, the eigenvalue 2 has multiplicity 2. Since the rank of $A - 2I_3$ equals 1 when $c = 2$, the dimension of the eigenspace corresponding to 2 is

$$\text{nullity}\,(A - 2I_3) = 3 - 1 = 2,$$

which is the multiplicity of the eigenvalue. Since the only other eigenvalue has multiplicity 1, it follows from the test on page 319 that A is diagonalizable in this case. Similarly, A is diagonalizable when $c = 3$. Thus there are no values of c for which A is not diagonalizable.

35. The characteristic polynomial of A is

$$\det(A - tI_2) = t^2 - t - 2 = (t+1)(t-2).$$

So the eigenvalues of A are -1 and 2. Bases for the eigenspaces of A corresponding to the eigenvalues -1 and 2 are

$$\left\{ \begin{bmatrix} 1 \\ 1 \end{bmatrix} \right\} \text{ and } \left\{ \begin{bmatrix} 2 \\ 1 \end{bmatrix} \right\},$$

respectively. Hence $A = PDP^{-1}$, where

$$P = \begin{bmatrix} 1 & 2 \\ 1 & 1 \end{bmatrix} \text{ and } D = \begin{bmatrix} -1 & 0 \\ 0 & 2 \end{bmatrix}.$$

So, for any positive integer k,

$$A^k = PD^k P^{-1}$$

$$= \begin{bmatrix} 1 & 2 \\ 1 & 1 \end{bmatrix} \begin{bmatrix} (-1)^k & 0 \\ 0 & 2^k \end{bmatrix} \begin{bmatrix} -1 & 2 \\ 1 & -1 \end{bmatrix}$$

$$= \begin{bmatrix} 1 & 2 \\ 1 & 1 \end{bmatrix} \begin{bmatrix} -(-1)^k & 2(-1)^k \\ 2^k & -2^k \end{bmatrix}$$

$$= \begin{bmatrix} (-1)^{k+1} + 2^{k+1} & 2(-1)^k - 2^{k+1} \\ (-1)^{k+1} + 2^k & 2(-1)^k - 2^k \end{bmatrix}.$$

39. If $a = b$, then the eigenvalue a has multiplicity 3, but its eigenspace has dimension 2. If $a \neq b$, then the eigenvalue a has multiplicity 2, but its eigenspace has dimension 1. In either case, A is not diagonalizable.

43. If B is the matrix whose columns are the vectors in $\mathcal{B}$, then $[T]_\mathcal{B} = B^{-1}AB$. So the characteristic polynomial of $[T]_\mathcal{B}$ is

$$\det(B^{-1}AB - tI_n)$$
$$= \det(B^{-1}(A - tI_n)B)$$
$$= (\det B)^{-1}(\det(A - tI_n))(\det B)$$
$$= \det(A - tI_n),$$

which is the characteristic polynomial of A.

CHAPTER 5 MATLAB EXERCISES

1. (a) Proceeding as in Exercise 23 of the Chapter 5 Review Exercises, we see that $A = PDP^{-1}$ for $P =$

$$\begin{bmatrix} 1.0 & 0.8 & 0.75 & 1 & 1.0 \\ -0.5 & -0.4 & -0.50 & 1 & -1.0 \\ 0.0 & -0.2 & -0.25 & 0 & -0.5 \\ 0.5 & 0.4 & 0.50 & 0 & 0.0 \\ 1.0 & 1.0 & 1.00 & 1 & 1.0 \end{bmatrix},$$

and

$$D = \begin{bmatrix} 3 & 0 & 0 & 0 & 0 \\ 0 & 1 & 0 & 0 & 0 \\ 0 & 0 & 0 & 0 & 0 \\ 0 & 0 & 0 & -1 & 0 \\ 0 & 0 & 0 & 0 & 2 \end{bmatrix}.$$

(b) For this 4 × 4 matrix, the eigenvalue $\frac{1}{2}$ has multiplicity 2, but rank $\left(A - \frac{1}{2}I_4\right) = 3$. Thus the matrix is not diagonalizable by the test on page 319.

(c) As in (a), the matrix is diagonalizable with $P =$
$$\begin{bmatrix} -1.25 & -1.00 & -0.50 & -1 \\ -0.25 & -0.50 & 0.50 & 0 \\ 0.75 & 0.50 & 1.00 & 0 \\ 1.00 & 1.00 & 0.00 & 1 \end{bmatrix}$$
and
$$D = \begin{bmatrix} -1 & 0 & 0 & 0 \\ 0 & 2 & 0 & 0 \\ 0 & 0 & 1 & 0 \\ 0 & 0 & 0 & 1 \end{bmatrix}.$$

(d) In this matrix, the eigenvalue 0 has multiplicity 2, but rank $(A - 0I_5) = 4$. So the test on page 319 fails.

5. (a) A basis does not exist because the sum of the multiplicities of the eigenvalues of the standard matrix of T is not 4.

(b) The vectors
$$\begin{bmatrix} -1 \\ -1 \\ 0 \\ 1 \\ 0 \end{bmatrix}, \begin{bmatrix} 0 \\ -1 \\ -1 \\ 0 \\ 1 \end{bmatrix}, \begin{bmatrix} 11 \\ 10 \\ -3 \\ -13 \\ 3 \end{bmatrix},$$
$$\begin{bmatrix} 15 \\ 8 \\ -4 \\ -15 \\ 1 \end{bmatrix}, \text{ and } \begin{bmatrix} 5 \\ 10 \\ 0 \\ -7 \\ 1 \end{bmatrix}$$
form a basis for $\mathcal{R}^5$ consisting of eigenvectors of T.

7. Let B be the matrix whose columns are the vectors $\mathbf{v}_1$, $\mathbf{v}_2$, $\mathbf{v}_3$, and $\mathbf{v}_4$.

(a) By the definition of T, we have
$$[T]_\mathcal{B} = \begin{bmatrix} 0 & 2 & 0 & 0 \\ 1 & 0 & 0 & 0 \\ 0 & 0 & -1 & 0 \\ 0 & 0 & 0 & 2 \end{bmatrix}.$$

Thus, by Theorem 4.12, the standard matrix of T is $B[T]_\mathcal{B}B^{-1} =$
$$\begin{bmatrix} 11.5 & -13.7 & 3.4 & -4.5 \\ 5.5 & -5.9 & 1.8 & -2.5 \\ -6.0 & 10.8 & -1.6 & 0.0 \\ 5.0 & -5.6 & 1.2 & -3.0 \end{bmatrix}.$$

So the rule for T is as follows:
$$T\left(\begin{bmatrix} x_1 \\ x_2 \\ x_3 \\ x_4 \end{bmatrix}\right) =$$
$$\begin{bmatrix} 11.5x_1 - 13.7x_2 + 3.4x_3 - 4.5x_4 \\ 5.5x_1 - 5.9x_2 + 1.8x_3 - 2.5x_4 \\ -6.0x_1 + 10.8x_2 - 1.6x_3 \\ 5.0x_1 - 5.6x_2 + 1.2x_3 - 3.0x_4 \end{bmatrix}.$$

(b) The vectors listed below, obtained using the MATLAB eig function, form a basis of eigenvectors of T. (Answers are correct to 4 places after the decimal point.)
$$\begin{bmatrix} 0.7746 \\ 0.5164 \\ 0.2582 \\ 0.2582 \end{bmatrix}, \begin{bmatrix} 0.0922 \\ 0.3147 \\ 0.9440 \\ -0.0382 \end{bmatrix},$$
$$\begin{bmatrix} 0.6325 \\ 0.3162 \\ -0.6325 \\ 0.3162 \end{bmatrix}, \text{ and } \begin{bmatrix} 0.3122 \\ 0.1829 \\ 0.5486 \\ 0.7537 \end{bmatrix}.$$

Chapter 6

Orthogonality

6.1 THE GEOMETRY OF VECTORS

1. $\|\mathbf{u}\| = \sqrt{5^2 + (-3)^2} = \sqrt{34}$,
   $\|\mathbf{v}\| = \sqrt{2^2 + 4^2} = \sqrt{20}$,
   and
   $$d = \|\mathbf{u} - \mathbf{v}\|$$
   $$= \sqrt{(5-2)^2 + (-3-4)^2}$$
   $$= \sqrt{58}$$

5. $\|\mathbf{u}\| = \sqrt{1^2 + (-1)^2 + 3^2} = \sqrt{11}$,
   $\|\mathbf{v}\| = \sqrt{2^2 + 1^2 + 0^2} = \sqrt{5}$,
   and
   $$d = \|\mathbf{u} - \mathbf{v}\|$$
   $$= \sqrt{(1-2)^2 + (-1-1)^2 + (3-0)^2}$$
   $$= \sqrt{14}$$

9. Since $\mathbf{u} \cdot \mathbf{v} = 3(4) + (-2)(6) = 0$, $\mathbf{u}$ and $\mathbf{v}$ are orthogonal.

11. Since $\mathbf{u} \cdot \mathbf{v} = (1)(2) + (-1)(1) = 1$, $\mathbf{u}$ and $\mathbf{v}$ are not orthogonal because $\mathbf{u} \cdot \mathbf{v} \neq 0$.

15. The dot product of $\mathbf{u}$ and $\mathbf{v}$ equals
    $$(1)(2) + (-1)(3) + (-2)(1) + 1(1) = -2,$$
    and so $\mathbf{u}$ and $\mathbf{v}$ are not orthogonal because $\mathbf{u} \cdot \mathbf{v} \neq 0$.

19. We have
    $$\|\mathbf{u}\|^2 = 2^2 + 3^2 = 13,$$
    $$\|\mathbf{v}\|^2 = 0^2 + 0^2 = 0,$$
    and
    $$\|\mathbf{u} + \mathbf{v}\|^2 = (2+0)^2 + (3+0)^2 = 13.$$
    Since
    $$\|\mathbf{u}\|^2 + \|\mathbf{v}\|^2 = 13 + 0 = 13 = \|\mathbf{u} + \mathbf{v}\|^2,$$
    the Pythagorean theorem shows that $\mathbf{u}$ and $\mathbf{v}$ are orthogonal.

23. We have
    $$\|\mathbf{u}\|^2 = 1^2 + 2^2 + 3^2 = 14,$$
    $$\|\mathbf{v}\|^2 = (-11)^2 + 4^2 + 1^2 = 138,$$
    and
    $$\|\mathbf{u} + \mathbf{v}\|^2$$
    $$= (1-11)^2 + (2+4)^2 + (3+1)^2$$
    $$= 152.$$
    Since
    $$\|\mathbf{u}\|^2 + \|\mathbf{v}\|^2 = 14 + 138$$
    $$= \|\mathbf{u} + \mathbf{v}\|^2,$$
    the Pythagorean theorem shows that $\mathbf{u}$ and $\mathbf{v}$ are orthogonal.

126 Chapter 6 Orthogonality

27. We have
$$\|\mathbf{u}\| = \sqrt{4^2 + 2^2} = \sqrt{20},$$
$$\|\mathbf{v}\| = \sqrt{3^2 + (-1)^2} = \sqrt{10},$$
and
$$\|\mathbf{u} + \mathbf{v}\|$$
$$= \sqrt{(4+3)^2 + (2-1)^2} = \sqrt{50}.$$
So
$$\|\mathbf{u} + \mathbf{v}\| = \sqrt{50} \le \sqrt{20} + \sqrt{10}$$
$$= \|\mathbf{u}\| + \|\mathbf{v}\|.$$

31. We have
$$\|\mathbf{u}\| = \sqrt{2^2 + (-1)^2 + 3^2} = \sqrt{14},$$
$$\|\mathbf{v}\| = \sqrt{4^2 + 0^2 + 1^2} = \sqrt{17},$$
and
$$\|\mathbf{u} + \mathbf{v}\|$$
$$= \sqrt{(2+4)^2 + (-1+0)^2 + (3+1)^2}$$
$$= \sqrt{53}.$$
So
$$\|\mathbf{u} + \mathbf{v}\| = \sqrt{53} \le \sqrt{14} + \sqrt{17}$$
$$= \|\mathbf{u}\| + \|\mathbf{v}\|.$$

35. We have
$$\|\mathbf{u}\| = \sqrt{4^2 + 1^2} = \sqrt{17},$$
$$\|\mathbf{v}\| = \sqrt{0^2 + (-2)^2} = \sqrt{4} = 2,$$
and
$$\mathbf{u} \cdot \mathbf{v} = 4(0) + 1(-2) = -2.$$
This illustrates the Cauchy-Schwarz inequality because
$$|\mathbf{u} \cdot \mathbf{v}| = |-2| = 2$$
$$\le \sqrt{17} \cdot 2 = \|\mathbf{u}\|\|\mathbf{v}\|.$$

39. We have
$$\|\mathbf{u}\| = \sqrt{4^2 + 2^2 + 1^2} = \sqrt{21},$$
$$\|\mathbf{v}\| = \sqrt{2^2 + (-1)^2 + (-1)^2} = \sqrt{6},$$
and
$$\mathbf{u} \cdot \mathbf{v} = (4)(2) + 2(-1) + (1)(-1) = -5.$$
So $|\mathbf{u} \cdot \mathbf{v}| = 5 \le \sqrt{21}\sqrt{6} = \|\mathbf{u}\|\|\mathbf{v}\|$.

43. Let $\mathbf{v} = \begin{bmatrix} -1 \\ 1 \end{bmatrix}$, a nonzero vector that lies along the line $y = -x$. Then $\mathbf{w} = c\mathbf{v}$, where
$$c = \frac{\mathbf{u} \cdot \mathbf{v}}{\mathbf{v} \cdot \mathbf{v}} = \frac{1}{2},$$
and hence $\mathbf{w} = \frac{1}{2}\begin{bmatrix} -1 \\ 1 \end{bmatrix}$. Therefore
$$d = \|\mathbf{u} - \mathbf{w}\| = \left\|\begin{bmatrix} 3 \\ 4 \end{bmatrix} - \frac{1}{2}\begin{bmatrix} -1 \\ 1 \end{bmatrix}\right\|$$
$$= \left\|\frac{1}{2}\begin{bmatrix} 7 \\ 7 \end{bmatrix}\right\| = \left\|\frac{7}{2}\begin{bmatrix} 1 \\ 1 \end{bmatrix}\right\| = \frac{7\sqrt{2}}{2}.$$

47. As in Exercise 43, let $\mathbf{v} = \begin{bmatrix} -1 \\ 3 \end{bmatrix}$, a nonzero vector that lies along the line $y = -3x$. Then $\mathbf{w} = c\mathbf{v}$, where
$$c = \frac{\mathbf{u} \cdot \mathbf{v}}{\mathbf{v} \cdot \mathbf{v}} = \frac{13}{10} = 1.3,$$
and hence $\mathbf{w} = 1.3\begin{bmatrix} -1 \\ 3 \end{bmatrix}$. Therefore
$$d = \|\mathbf{u} - \mathbf{w}\| = \left\|\begin{bmatrix} 2 \\ 5 \end{bmatrix} - 1.3\begin{bmatrix} -1 \\ 3 \end{bmatrix}\right\|$$
$$= \left\|\begin{bmatrix} 3.3 \\ 1.1 \end{bmatrix}\right\| = \left\|1.1\begin{bmatrix} 3 \\ 1 \end{bmatrix}\right\| = 1.1\sqrt{10}.$$

51. Using Theorem 6.1, we obtain
$$\|\mathbf{u} + \mathbf{v}\|^2 = \|\mathbf{u}\|^2 + 2\mathbf{u} \cdot \mathbf{v} + \|\mathbf{v}\|^2$$
$$= 2^2 + 2(-1) + 3^2 = 11.$$

53. As in Exercise 51, we obtain
$$\|\mathbf{v} - 4\mathbf{w}\|^2 = \|\mathbf{v}\|^2 - 8\mathbf{v}\cdot\mathbf{w} + 16\|\mathbf{w}\|^2$$
$$= 3^2 - 8(-4) + 16(5^2)$$
$$= 441.$$

57. $\mathbf{v}\cdot\mathbf{u} = \mathbf{u}\cdot\mathbf{v} = 7$

61. True

62. False, the dot product of two vectors is a scalar.

63. False, the norm of a vector equals the square root of the dot product of the vector with itself.

64. False, the norm is the product of the *absolute value* of the multiple and the norm of the vector.

65. False, for example, if $\mathbf{v}$ is a nonzero vector, then
$$\|\mathbf{v} + (-\mathbf{v})\| = 0 \ne \|\mathbf{v}\| + \|-\mathbf{v}\|.$$

66. True   **67.** True   **68.** True

69. True

70. False, consider nonzero orthogonal vectors.

71. False, we need to replace $=$ by $\le$.

72. True   **73.** True   **74.** True

75. True

76. False, $A\mathbf{u}\cdot\mathbf{v} = \mathbf{u}\cdot A^T\mathbf{v}$.

77. True

78. False, we need to replace $=$ by $\le$.

79. True   **80.** True

81.
$$\mathbf{u}\cdot\mathbf{u} = u_1(u_1) + u_2(u_2) + \cdots + u_n(u_n)$$
$$= \left(\sqrt{u_1^2 + u_2^2 + \cdots + u_n^2}\right)^2 = \|\mathbf{u}\|^2$$

85. Suppose that $\mathbf{u}$ and $\mathbf{v}$ are in $\mathcal{R}^n$. Then
$$(c\mathbf{u})\cdot\mathbf{v} = (cu_1)v_1 + \cdots + (cu_n)v_n$$
$$= c(u_1v_1 + \cdots + u_nv_n)$$
$$= c(\mathbf{u}\cdot\mathbf{v}).$$

The proof that $(c\mathbf{u})\cdot\mathbf{v} = \mathbf{u}\cdot(c\mathbf{v})$ is similar.

89. Let $\mathbf{u}$ and $\mathbf{v}$ be vectors in $\mathcal{R}^n$. We show that
$$\|\mathbf{u} + \mathbf{v}\| = \|\mathbf{u}\| + \|\mathbf{v}\|$$
if and only if $\mathbf{u}$ is a nonnegative multiple of $\mathbf{v}$ or $\mathbf{v}$ is a nonnegative multiple of $\mathbf{u}$.

Suppose first that $\|\mathbf{u} + \mathbf{v}\| = \|\mathbf{u}\| + \|\mathbf{v}\|$. If $\mathbf{u} = \mathbf{0}$ or $\mathbf{v} = \mathbf{0}$, the result is immediate; so assume that the vectors are nonzero. Then
$$\|\mathbf{u}\|^2 + 2\mathbf{u}\cdot\mathbf{v} + \|\mathbf{v}\|^2$$
$$= \|\mathbf{u} + \mathbf{v}\|^2$$
$$= (\|\mathbf{u}\| + \|\mathbf{v}\|)^2$$
$$= \|\mathbf{u}\|^2 + 2\|\mathbf{u}\|\cdot\|\mathbf{v}\| + \|\mathbf{v}\|^2,$$

and therefore $\mathbf{u}\cdot\mathbf{v} = \|\mathbf{u}\|\cdot\|\mathbf{v}\|$. Thus $\mathbf{u}\cdot\mathbf{v}$ is nonnegative. By Exercise 88, it follows that $\mathbf{u}$ or $\mathbf{v}$ is a multiple of the other. Suppose $\mathbf{u} = k\mathbf{v}$ for some scalar k. Then
$$0 \le \mathbf{u}\cdot\mathbf{v} = k\mathbf{v}\cdot\mathbf{v} = k(\mathbf{v}\cdot\mathbf{v}) = k\|\mathbf{v}\|^2.$$

So $k \ge 0$. A similar argument may be used if $\mathbf{v} = k\mathbf{u}$.

Conversely, suppose one of the vectors is a nonnegative multiple of the other.

Since both of these vectors are nonzero, we may assume that $\mathbf{v} = c\mathbf{u}$, where c is a nonnegative scalar. Then

$$\|\mathbf{u} + \mathbf{v}\| = \|\mathbf{u} + c\mathbf{u}\| = \|(1+c)\mathbf{u}\|$$
$$= |1+c|\|\mathbf{u}\| = (1+c)\|\mathbf{u}\|,$$

and

$$\|\mathbf{u}\| + \|\mathbf{v}\| = \|\mathbf{u}\| + \|c\mathbf{u}\| = \|\mathbf{u}\| + |c|\|\mathbf{u}\|$$
$$= (1+|c|)\|\mathbf{u}\| = (1+c)\|\mathbf{u}\|.$$

Therefore $\|\mathbf{u} + \mathbf{v}\| = \|\mathbf{u}\| + \|\mathbf{v}\|$.

93. Since $\mathbf{0} \cdot \mathbf{z} = 0$ for all $\mathbf{z}$ in $\mathcal{S}$, we see that $\mathbf{0}$ is in W. Now suppose that $\mathbf{u}$ and $\mathbf{v}$ are in W. Then, for all $\mathbf{z}$ in $\mathcal{S}$,

$$(\mathbf{u} + \mathbf{v}) \cdot \mathbf{z} = \mathbf{u} \cdot \mathbf{z} + \mathbf{v} \cdot \mathbf{z} = 0 + 0 = 0,$$

and hence $\mathbf{u} + \mathbf{v}$ is in W. Finally, let c be any scalar. Then, for all $\mathbf{z}$ in $\mathcal{S}$,

$$(c\mathbf{u}) \cdot \mathbf{z} = c(\mathbf{u} \cdot \mathbf{z}) = c \cdot 0 = 0,$$

and hence $c\mathbf{u}$ is in W. We conclude that W is a subspace.

97. (a) Suppose that $\mathbf{v}$ is in Null A. Then

$$(A^T A)\mathbf{v} = A^T(A\mathbf{v}) = A^T \mathbf{0} = \mathbf{0},$$

and hence $\mathbf{v}$ is in Null $A^T A$. Thus Null A is contained in Null $A^T A$. Conversely, suppose that $\mathbf{v}$ is in Null $A^T A$. Then

$$0 = \mathbf{0} \cdot \mathbf{v}$$
$$= (A^T A)\mathbf{v} \cdot \mathbf{v}$$
$$= (A^T A\mathbf{v})^T \mathbf{v}$$
$$= \mathbf{v}^T A^T A\mathbf{v}$$
$$= (A\mathbf{v})^T (A\mathbf{v})$$
$$= (A\mathbf{v}) \cdot (A\mathbf{v})$$
$$= \|A\mathbf{v}\|^2,$$

and so $\|A\mathbf{v}\| = 0$. Thus $A\mathbf{v} = \mathbf{0}$, and it follows that Null $A^T A$ is contained in Null A. Therefore

$$\text{Null } A^T A = \text{Null } A.$$

(b) Since

$$\text{Null } A^T A = \text{Null } A,$$

we have that

$$\text{nullity } A^T A = \text{nullity } A.$$

Notice that $A^T A$ and A each have n columns, and hence

$$\text{rank } A^T A = n - \text{nullity } A^T A$$
$$= n - \text{nullity } A$$
$$= \text{rank } A.$$

101. Let θ denote the angle between $\mathbf{u}$ and $\mathbf{v}$. Since

$$\mathbf{u} \cdot \mathbf{v} = \|\mathbf{u}\|\|\mathbf{v}\| \cos \theta$$
$$(-2)(1) + 4(-2) = \sqrt{20}\sqrt{5} \cos \theta$$
$$-10 = 10 \cos \theta$$
$$-1 = \cos \theta,$$

we see that $\theta = 180°$.

105. Let θ denote the angle between $\mathbf{u}$ and $\mathbf{v}$. Since

$$\mathbf{u} \cdot \mathbf{v} = \|\mathbf{u}\|\|\mathbf{v}\| \cos \theta$$
$$1(-1) + (-2)(1) + 1(0) = \sqrt{6}\sqrt{2} \cos \theta$$
$$-3 = \sqrt{12} \cos \theta$$
$$-\frac{\sqrt{3}}{2} = \cos \theta,$$

we see that $\theta = 150°$.

109. For $\mathbf{u}$ in $\mathcal{R}^3$,

$$\mathbf{u} \times \mathbf{0} = \begin{bmatrix} u_2 0 - u_3 0 \\ u_3 0 - u_1 0 \\ u_1 0 - u_2 0 \end{bmatrix} = \begin{bmatrix} 0 \\ 0 \\ 0 \end{bmatrix} = \mathbf{0}.$$

By Exercise 108,
$$\mathbf{0} \times \mathbf{u} = -(\mathbf{u} \times \mathbf{0}) = -\mathbf{0} = \mathbf{0}.$$

113. Let $\mathbf{u}$, $\mathbf{v}$, and $\mathbf{w}$ be in $\mathcal{R}^3$. By Exercises 108 and 112,
$$\begin{aligned}(\mathbf{u} + \mathbf{v}) \times \mathbf{w} &= -(\mathbf{w} \times (\mathbf{u} + \mathbf{v})) \\ &= -(\mathbf{w} \times \mathbf{u} + \mathbf{w} \times \mathbf{v}) \\ &= -(\mathbf{w} \times \mathbf{u}) + -(\mathbf{w} \times \mathbf{v}) \\ &= \mathbf{u} \times \mathbf{w} + \mathbf{v} \times \mathbf{w}.\end{aligned}$$

117. Let $\mathbf{u}$, $\mathbf{v}$, and $\mathbf{w}$ be in $\mathcal{R}^3$. By Exercises 108 and 116, we have
$$\begin{aligned}(\mathbf{u} \times \mathbf{v}) \times \mathbf{w} &= -(\mathbf{w} \times (\mathbf{u} \times \mathbf{v})) \\ &= -((\mathbf{w} \cdot \mathbf{v})\mathbf{u} - (\mathbf{w} \cdot \mathbf{u})\mathbf{v}) \\ &= (\mathbf{w} \cdot \mathbf{u})\mathbf{v} - (\mathbf{w} \cdot \mathbf{v})\mathbf{u}.\end{aligned}$$

121. The supervisor polls all 20 students and finds that the students are divided among three sections. The first has 8 students, the second has 12 students, and the class has 6 students. She divides the total number of students by the number of sections and computes
$$\bar{v} = \frac{8 + 12 + 6}{3} = \frac{26}{3} = 8.6667.$$

When the investigator polls 8 students in Section 1, they all report that their class sizes are 8. Likewise, for the other two sections, 12 students report that their class sizes are 12, and 6 students report that their class sizes are 6. Adding these sizes and dividing by the total number of polled students, the investigator obtains
$$\begin{aligned}v^* &= \frac{8 \cdot 8 + 12 \cdot 12 + 6 \cdot 6}{8 + 12 + 6} = \frac{244}{26} \\ &= 9.3846.\end{aligned}$$

125. In (a), (b), and (c), we describe the use of MATLAB in the default (*short*) format.

(a) Entering `norm(u + v)` produces the output 13.964, and entering `norm(u) + norm(v)` yields 17.3516. We conclude that
$$\|\mathbf{u} + \mathbf{v}\| < \|\mathbf{u}\| + \|\mathbf{v}\|.$$

(b) As in (a), we obtain the outputs `norm(u + v`$_1$`)` = 16.449 and `norm(u) + norm(v`$_1$`)` = 16.449. However, these outputs are rounded to 4 places after the decimal, so we need an additional test for equality. Entering the difference
`norm(u) + norm(v`$_1$`)`
 `- norm(u + v`$_1$`)`
yields the output 1.0114×10^{-6}, which indicates that the two are unequal but the difference is small. Thus
$$\|\mathbf{u} + \mathbf{v}_1\| < \|\mathbf{u}\| + \|\mathbf{v}_1\|.$$

(c) As in (b), we obtain a strict inequality. In this case, the difference, given by the MATLAB output, is 5.0617×10^{-7}.

(d) Notice that in (b) and (c), $\mathbf{v}_1$ and $\mathbf{v}_2$ are "nearly" positive multiples of $\mathbf{u}$ and the triangle inequality is almost an equality. Thus we conjecture that $\|\mathbf{u}+\mathbf{v}\| = \|\mathbf{u}\| + \|\mathbf{v}\|$ if and only if $\mathbf{u}$ is a nonnegative multiple of $\mathbf{v}$ or $\mathbf{v}$ is a nonnegative multiple of $\mathbf{u}$.

(e) If two sides of a triangle are parallel, then the triangle is "degenerate," that is, the third side coincides with the union of the other two sides.

6.2 ORTHOGONAL VECTORS

3. Since
$$\begin{bmatrix} 1 \\ -1 \\ 1 \end{bmatrix} \cdot \begin{bmatrix} 2 \\ -1 \\ 0 \end{bmatrix}$$
$$= 1(2) + (-1)(-1) + 1(0)$$
$$= 3 \neq 0,$$

some pair of distinct vectors in the given set is not orthogonal. Thus the set itself is not orthogonal.

7. We have
$$\begin{bmatrix} 1 \\ 2 \\ 3 \\ -3 \end{bmatrix} \cdot \begin{bmatrix} 1 \\ 1 \\ -1 \\ 0 \end{bmatrix}$$
$$= 1(1) + 2(1) + 3(-1) + (-3)(0)$$
$$= 0,$$

$$\begin{bmatrix} 1 \\ 2 \\ 3 \\ -3 \end{bmatrix} \cdot \begin{bmatrix} 3 \\ -3 \\ 0 \\ -1 \end{bmatrix}$$
$$= 1(3) + 2(-3) + 3(0) + (-3)(-1)$$
$$= 0,$$

and
$$\begin{bmatrix} 1 \\ 1 \\ -1 \\ 0 \end{bmatrix} \cdot \begin{bmatrix} 3 \\ -3 \\ 0 \\ -1 \end{bmatrix}$$
$$= 1(3) + 1(-3) + (-1)(0) + 0(-1)$$
$$= 0.$$

Therefore the set is orthogonal.

11. (a) Let
$$\mathbf{u}_1 = \begin{bmatrix} 1 \\ -2 \\ -1 \end{bmatrix} \quad \text{and} \quad \mathbf{u}_2 = \begin{bmatrix} 7 \\ 7 \\ 5 \end{bmatrix}.$$
Then
$$\mathbf{v}_1 = \mathbf{u}_1 = \begin{bmatrix} 1 \\ -2 \\ -1 \end{bmatrix}$$
and
$$\mathbf{v}_2 = \mathbf{u}_2 - \frac{\mathbf{u}_2 \cdot \mathbf{v}_1}{\|\mathbf{v}_1\|^2} \mathbf{v}_1$$

$$= \begin{bmatrix} 7 \\ 7 \\ 5 \end{bmatrix} - \frac{\begin{bmatrix} 7 \\ 7 \\ 5 \end{bmatrix} \cdot \begin{bmatrix} 1 \\ -2 \\ -1 \end{bmatrix}}{\left\| \begin{bmatrix} 1 \\ -2 \\ -1 \end{bmatrix} \right\|^2} \begin{bmatrix} 1 \\ -2 \\ -1 \end{bmatrix}$$

$$= \begin{bmatrix} 7 \\ 7 \\ 5 \end{bmatrix} - \frac{(-12)}{6} \begin{bmatrix} 1 \\ -2 \\ -1 \end{bmatrix}$$

$$= \begin{bmatrix} 9 \\ 3 \\ 3 \end{bmatrix}.$$

So the desired orthogonal set is
$$\left\{ \begin{bmatrix} 1 \\ -2 \\ -1 \end{bmatrix}, \begin{bmatrix} 9 \\ 3 \\ 3 \end{bmatrix} \right\}.$$

(b) Normalizing the vectors above, we obtain the orthonormal set
$$\left\{ \frac{1}{\sqrt{6}} \begin{bmatrix} 1 \\ -2 \\ -1 \end{bmatrix}, \frac{1}{\sqrt{11}} \begin{bmatrix} 3 \\ 1 \\ 1 \end{bmatrix} \right\}.$$

13. (a) Let
$$\mathbf{u}_1 = \begin{bmatrix} 0 \\ 1 \\ 1 \\ 1 \end{bmatrix}, \quad \mathbf{u}_2 = \begin{bmatrix} 1 \\ 0 \\ 1 \\ 1 \end{bmatrix},$$

and
$$\mathbf{u}_3 = \begin{bmatrix} 1 \\ 1 \\ 0 \\ 1 \end{bmatrix}.$$

Set
$$\mathbf{v}_1 = \mathbf{u}_1,$$
$$\mathbf{v}_2 = \mathbf{u}_2 - \frac{\mathbf{u}_2 \cdot \mathbf{v}_1}{\|\mathbf{v}_1\|^2} \mathbf{v}_1$$
$$= \begin{bmatrix} 1 \\ 0 \\ 1 \\ 1 \end{bmatrix} - \frac{2}{3} \begin{bmatrix} 0 \\ 1 \\ 1 \\ 1 \end{bmatrix} = \frac{1}{3} \begin{bmatrix} 3 \\ -2 \\ 1 \\ 1 \end{bmatrix},$$

and
$$\mathbf{v}_3 = \mathbf{u}_3 - \frac{\mathbf{u}_3 \cdot \mathbf{v}_1}{\|\mathbf{v}_1\|^2} \mathbf{v}_1 - \frac{\mathbf{u}_3 \cdot \mathbf{v}_2}{\|\mathbf{v}_2\|^2} \mathbf{v}_2$$
$$= \begin{bmatrix} 1 \\ 1 \\ 0 \\ 1 \end{bmatrix} - \frac{2}{3} \begin{bmatrix} 0 \\ 1 \\ 1 \\ 1 \end{bmatrix} - \frac{2}{15} \begin{bmatrix} 3 \\ -2 \\ 1 \\ 1 \end{bmatrix}$$
$$= \frac{1}{5} \begin{bmatrix} 3 \\ 3 \\ -4 \\ 1 \end{bmatrix}.$$

Thus
$$\{\mathbf{v}_1, \mathbf{v}_2, \mathbf{v}_3\}$$
$$= \left\{ \begin{bmatrix} 0 \\ 1 \\ 1 \\ 1 \end{bmatrix}, \frac{1}{3}\begin{bmatrix} 3 \\ -2 \\ 1 \\ 1 \end{bmatrix}, \frac{1}{5}\begin{bmatrix} 3 \\ 3 \\ -4 \\ 1 \end{bmatrix} \right\}$$

is the corresponding orthogonal set.

(b) To obtain the vectors in the orthonormal set, we normalize each of the vectors $\mathbf{v}_1$, $\mathbf{v}_2$, $\mathbf{v}_3$ in (a). The resulting vectors are

$$\frac{1}{\|\mathbf{v}_1\|^2}\mathbf{v}_1 = \frac{1}{\sqrt{3}} \begin{bmatrix} 0 \\ 1 \\ 1 \\ 1 \end{bmatrix},$$

$$\frac{1}{\|\mathbf{v}_2\|^2}\mathbf{v}_2 = \frac{1}{\sqrt{15}} \begin{bmatrix} 3 \\ -2 \\ 1 \\ 1 \end{bmatrix},$$

and

$$\frac{1}{\|\mathbf{v}_3\|^2}\mathbf{v}_3 = \frac{1}{\sqrt{35}} \begin{bmatrix} 3 \\ 3 \\ -4 \\ 1 \end{bmatrix}.$$

17. Using the boxed result on page 376, we obtain

$$\mathbf{v} = \frac{\begin{bmatrix} 1 \\ 8 \end{bmatrix} \cdot \begin{bmatrix} 2 \\ 1 \end{bmatrix}}{\left\| \begin{bmatrix} 2 \\ 1 \end{bmatrix} \right\|^2} \begin{bmatrix} 2 \\ 1 \end{bmatrix}$$
$$+ \frac{\begin{bmatrix} 1 \\ 8 \end{bmatrix} \cdot \begin{bmatrix} -1 \\ 2 \end{bmatrix}}{\left\| \begin{bmatrix} -1 \\ 2 \end{bmatrix} \right\|^2} \begin{bmatrix} -1 \\ 2 \end{bmatrix}$$
$$= \frac{10}{5} \begin{bmatrix} 2 \\ 1 \end{bmatrix} + \frac{15}{5} \begin{bmatrix} -1 \\ 2 \end{bmatrix}$$
$$= 2 \begin{bmatrix} 2 \\ 1 \end{bmatrix} + 3 \begin{bmatrix} -1 \\ 2 \end{bmatrix}.$$

21. Suppose that c_1, c_2, and c_3 are the scalars such that $\mathbf{v} = c_1\mathbf{v}_1 + c_2\mathbf{v}_2 + c_3\mathbf{v}_3$, where

$$\mathbf{v}_1 = \begin{bmatrix} 1 \\ 0 \\ 1 \end{bmatrix}, \mathbf{v}_2 = \begin{bmatrix} 1 \\ 2 \\ -1 \end{bmatrix}, \mathbf{v}_3 = \begin{bmatrix} 1 \\ -1 \\ -1 \end{bmatrix}.$$

Then, by the boxed result on page 376, we have

$$c_1 = \frac{\mathbf{v} \cdot \mathbf{v}_1}{\|\mathbf{v}_1\|^2} = \frac{5}{2},$$

$$c_2 = \frac{\mathbf{v} \cdot \mathbf{v}_2}{\|\mathbf{v}_2\|^2} = \frac{3}{6} = \frac{1}{2},$$

and

$$c_3 = \frac{\mathbf{v} \cdot \mathbf{v}_3}{\|\mathbf{v}_3\|^2} = \frac{0}{3} = 0.$$

Thus $\mathbf{v} = \dfrac{5}{2}\begin{bmatrix}1\\0\\1\end{bmatrix} + \dfrac{1}{2}\begin{bmatrix}1\\2\\-1\end{bmatrix} + 0\begin{bmatrix}1\\-1\\-1\end{bmatrix}$.

25. Let $A = \begin{bmatrix} 1 & 5 \\ 1 & -1 \\ 1 & 2 \end{bmatrix}$. In Exercise 9, we obtained an orthonormal basis for Col A:

$$\left\{ \frac{1}{\sqrt{3}}\begin{bmatrix}1\\1\\1\end{bmatrix}, \frac{1}{\sqrt{2}}\begin{bmatrix}1\\-1\\0\end{bmatrix} \right\}.$$

These vectors are the columns of Q, so that

$$Q = \begin{bmatrix} \frac{1}{\sqrt{3}} & \frac{1}{\sqrt{2}} \\ \frac{1}{\sqrt{3}} & -\frac{1}{\sqrt{2}} \\ \frac{1}{\sqrt{3}} & 0 \end{bmatrix}.$$

The entries of R can be found as in Example 4:

$$r_{11} = \mathbf{a}_1 \cdot \mathbf{q}_1 = \frac{3}{\sqrt{3}} = \sqrt{3},$$

$$r_{12} = \mathbf{a}_2 \cdot \mathbf{q}_1 = \frac{6}{\sqrt{3}} = 2\sqrt{3}$$

$$r_{22} = \mathbf{a}_2 \cdot \mathbf{q}_2 = \frac{6}{\sqrt{2}} = 3\sqrt{2}.$$

Thus $R = \begin{bmatrix} \sqrt{3} & 2\sqrt{3} \\ 0 & 3\sqrt{2} \end{bmatrix}$.

29. Let $A = \begin{bmatrix} 0 & 1 & 1 \\ 1 & 0 & 1 \\ 1 & 1 & 0 \\ 1 & 1 & 1 \end{bmatrix}$. The vectors in the answer to Exercise 13(b) are the columns of Q, so that

$$Q = \begin{bmatrix} 0 & \frac{3}{\sqrt{15}} & \frac{3}{\sqrt{35}} \\ \frac{1}{\sqrt{3}} & -\frac{2}{\sqrt{15}} & \frac{3}{\sqrt{35}} \\ \frac{1}{\sqrt{3}} & \frac{1}{\sqrt{15}} & -\frac{4}{\sqrt{35}} \\ \frac{1}{\sqrt{3}} & \frac{1}{\sqrt{15}} & \frac{1}{\sqrt{35}} \end{bmatrix}.$$

For $i \leq j$, the entries of the 3×3 upper triangular matrix R can be computed using $r_{ij} = \mathbf{a}_j \cdot \mathbf{q}_i$, so that

$$R = \begin{bmatrix} \sqrt{3} & \frac{2}{\sqrt{3}} & \frac{2}{\sqrt{3}} \\ 0 & \frac{\sqrt{15}}{3} & \frac{2}{\sqrt{15}} \\ 0 & 0 & \frac{7}{\sqrt{35}} \end{bmatrix}.$$

33. We proceed as in Example 5 to solve $A\mathbf{x} = \mathbf{b}$, where

$$A = \begin{bmatrix} 1 & 5 \\ 1 & -1 \\ 1 & 2 \end{bmatrix} \text{ and } \mathbf{b} = \begin{bmatrix} -3 \\ 3 \\ 0 \end{bmatrix}.$$

We use the matrices Q and R from the QR factorization of A found in Exercise 25:

$$Q = \begin{bmatrix} \frac{1}{\sqrt{3}} & \frac{1}{\sqrt{2}} \\ \frac{1}{\sqrt{3}} & -\frac{1}{\sqrt{2}} \\ \frac{1}{\sqrt{3}} & 0 \end{bmatrix}$$

and

$$R = \begin{bmatrix} \sqrt{3} & 2\sqrt{3} \\ 0 & 3\sqrt{2} \end{bmatrix}.$$

6.2 Orthogonal Vectors

We must solve the equivalent system $R\mathbf{x} = Q^T\mathbf{b}$, which has the form

$$\sqrt{3}x_1 + 2\sqrt{3}x_2 = 0$$
$$3\sqrt{2}x_2 = -3\sqrt{2}$$

or

$$x_1 + 2x_2 = 0$$
$$x_2 = -1.$$

The latter system is easily solved by starting with the last equation: $x_2 = -1$ and $x_1 = -2(-1) = 2$. In vector form, the solution is $\mathbf{x} = \begin{bmatrix} 2 \\ -1 \end{bmatrix}$.

37. We proceed as in Example 5 to solve $A\mathbf{x} = \mathbf{b}$, where

$$A = \begin{bmatrix} 0 & 1 & 1 \\ 1 & 0 & 1 \\ 1 & 1 & 0 \\ 1 & 1 & 1 \end{bmatrix} \text{ and } \mathbf{b} = \begin{bmatrix} 4 \\ 1 \\ -1 \\ 2 \end{bmatrix}.$$

We use the matrices Q and R from the QR factorization of A found in Exercise 29. We must solve the equivalent system $R\mathbf{x} = Q^T\mathbf{b}$, which has the form

$$\sqrt{3}x_1 + \tfrac{2}{\sqrt{3}}x_2 + \tfrac{2}{\sqrt{3}}x_3 = \tfrac{2}{\sqrt{3}}$$
$$\tfrac{\sqrt{15}}{3}x_2 + \tfrac{2}{\sqrt{15}}x_3 = \tfrac{11}{\sqrt{15}}$$
$$\tfrac{7}{\sqrt{35}}x_3 = \tfrac{21}{\sqrt{35}}$$

or

$$3x_1 + 2x_2 + 2x_3 = 2$$
$$5x_2 + 2x_3 = 11$$
$$x_3 = 3.$$

The latter system is easily solved: $x_3 = 3$, $x_2 = 1$, and $x_1 = -2$. In vector form, the solution is $\mathbf{x} = \begin{bmatrix} -2 \\ 1 \\ 3 \end{bmatrix}$.

41. False, if $\mathbf{0}$ lies in the set, then the set is linearly dependent.

42. True   **43.** True   **44.** True

45. True   **46.** True   **47.** True

48. True

49. False, consider the sets $\{\mathbf{e}_1\}$ and $\{-\mathbf{e}_1\}$. The combined set is $\{\mathbf{e}_1, -\mathbf{e}_1\}$, which is not orthonormal.

50. False, consider $\mathbf{x} = \mathbf{e}_1$, $\mathbf{y} = \mathbf{0}$, and $\mathbf{z} = \mathbf{e}_1$.

51. True

52. False, in Example 4, Q is not upper triangular.

53. For any $i \neq j$,

$$(c_i\mathbf{v}_i) \cdot (c_j\mathbf{v}_j) = (c_ic_j)(\mathbf{v}_i \cdot \mathbf{v}_j)$$
$$= (c_ic_j) \cdot 0 = 0.$$

Hence $c_i\mathbf{v}_i$ and $c_j\mathbf{v}_j$ are orthogonal.

57. By Exercise 56, $\mathcal{S}$ can be extended to an orthonormal basis

$$\{\mathbf{v}_1, \mathbf{v}_2, \ldots, \mathbf{v}_k, \mathbf{v}_{k+1}, \ldots, \mathbf{v}_n\}$$

for $\mathcal{R}^n$. By Exercise 55(c),

$$\|\mathbf{u}\|^2 = (\mathbf{u} \cdot \mathbf{v}_1)^2 + \cdots + (\mathbf{u} \cdot \mathbf{v}_k)^2$$
$$+ (\mathbf{u} \cdot \mathbf{v}_{k+1})^2 + \cdots + (\mathbf{u} \cdot \mathbf{v}_n)^2.$$

(a) The desired inequality follows immediately from the equation above since

$$(\mathbf{u} \cdot \mathbf{v}_{k+1})^2 + \cdots + (\mathbf{u} \cdot \mathbf{v}_n)^2 \geq 0.$$

(b) The inequality in (a) is an equality if and only if

$$(\mathbf{u} \cdot \mathbf{v}_{k+1})^2 + \cdots + (\mathbf{u} \cdot \mathbf{v}_n)^2 = 0,$$

that is, if and only if $\mathbf{u} \cdot \mathbf{v}_i = 0$ for $i > k$. In this case,

$$\mathbf{u} = (\mathbf{u} \cdot \mathbf{v}_1)\mathbf{v}_1 + \cdots + (\mathbf{u} \cdot \mathbf{v}_k)\mathbf{v}_k,$$

which is true if and only if $\mathbf{u}$ is in Span $\mathcal{S}$.

61. From Exercise 60, we have $r_{ii} \neq 0$ for every i. If $r_{ii} < 0$, then replacing $\mathbf{q}_i$ by $-\mathbf{q}_i$ changes the corresponding entry of R to $-r_{ii}$, which is positive.

65. Suppose $QR = Q'R'$, where both R and R' have positive diagonal entries. Multiplying both sides on the left by Q^T and on the right by R'^{-1}, we obtain

$$Q^T Q R R'^{-1} = Q^T Q' R R'^{-1},$$

which, by Exercise 63, reduces to $RR'^{-1} = Q^T Q'$. By Exercises 42 and 43 of Section 2.6, RR'^{-1}, and hence $Q^T Q'$, is an upper triangular matrix with positive diagonal entries. By Exercise 64, the columns of $Q^T Q'$ form an orthonormal basis. Hence, by Exercise 58, $Q^T Q'$ is a diagonal matrix. But a diagonal matrix with positive diagonal entries whose columns are unit vectors must equal I_3. So

$$RR'^{-1} = Q^T Q' = I_3,$$

and therefore $Q = Q'$ and $R = R'$.

6.3 ORTHOGONAL PROJECTIONS

1. A vector $\mathbf{v}$ is in $\mathcal{S}^\perp$ if and only if

$$\mathbf{v} \cdot \begin{bmatrix} 1 \\ -1 \\ 2 \end{bmatrix} = v_1 - v_2 + 2v_3 = 0.$$

So a basis for $\mathcal{S}^\perp$ is a basis for the solution set of this system. One such basis is

$$\left\{ \begin{bmatrix} 1 \\ 1 \\ 0 \end{bmatrix}, \begin{bmatrix} -2 \\ 0 \\ 1 \end{bmatrix} \right\}.$$

5. As in Exercise 1, we must find a basis for the solution set of

$$x_1 - 2x_2 + x_3 + x_4 = 0$$
$$x_1 - x_2 + 3x_3 + 2x_4 = 0.$$

One such basis is

$$\left\{ \begin{bmatrix} -5 \\ -2 \\ 1 \\ 0 \end{bmatrix}, \begin{bmatrix} -3 \\ -1 \\ 0 \\ 1 \end{bmatrix} \right\}.$$

9. Let $\mathbf{v}_1 = \dfrac{1}{\sqrt{2}} \begin{bmatrix} 1 \\ -1 \end{bmatrix}$.

(a) As in Example 3, we have

$$\mathbf{w} = (\mathbf{u} \cdot \mathbf{v}_1)\mathbf{v}_1$$
$$= \frac{-2}{\sqrt{2}}\mathbf{v}_1 = \frac{-2}{\sqrt{2}} \frac{1}{\sqrt{2}} \begin{bmatrix} 1 \\ -1 \end{bmatrix}$$
$$= \begin{bmatrix} -1 \\ 1 \end{bmatrix},$$

and

$$\mathbf{z} = \mathbf{u} - \mathbf{w} = \begin{bmatrix} 1 \\ 3 \end{bmatrix} - \begin{bmatrix} -1 \\ 1 \end{bmatrix} = \begin{bmatrix} 2 \\ 2 \end{bmatrix}.$$

(b) The orthogonal projection of $\mathbf{u}$ on W is the vector $\mathbf{w} = \begin{bmatrix} -1 \\ 1 \end{bmatrix}$ in (a).

(c) The distance from $\mathbf{u}$ to W is $\|\mathbf{z}\| = \|\mathbf{u} - \mathbf{w}\| = \sqrt{8}$.

13. Let $\mathbf{v}_1$, $\mathbf{v}_2$, and $\mathbf{v}_3$ denote the vectors in $\mathcal{S}$, in the order listed.

(a) As in Example 3, we have

$$\mathbf{w} = (\mathbf{u} \cdot \mathbf{v}_1)\mathbf{v}_1 + (\mathbf{u} \cdot \mathbf{v}_2)\mathbf{v}_2$$

6.3 Orthogonal Projections

$$+ (\mathbf{u} \cdot \mathbf{v}_3)\mathbf{v}_3$$
$$= \frac{6}{\sqrt{3}}\mathbf{v}_1 + \frac{3}{\sqrt{3}}\mathbf{v}_2 + \frac{3}{\sqrt{3}}\mathbf{v}_3$$
$$= \begin{bmatrix} 2 \\ 2 \\ 3 \\ 1 \end{bmatrix},$$

and

$$\mathbf{z} = \mathbf{u} - \mathbf{w}$$
$$= \begin{bmatrix} 2 \\ 4 \\ 1 \\ 3 \end{bmatrix} - \begin{bmatrix} 2 \\ 2 \\ 3 \\ 1 \end{bmatrix} = \begin{bmatrix} 0 \\ 2 \\ -2 \\ 2 \end{bmatrix}.$$

(b) The orthogonal projection of $\mathbf{u}$ on W is the vector $\mathbf{w} = \begin{bmatrix} 2 \\ 2 \\ 3 \\ 1 \end{bmatrix}$ in (a).

(c) The distance from $\mathbf{u}$ to W is $\|\mathbf{z}\| = \|\mathbf{u} - \mathbf{w}\| = \sqrt{12}$.

17. (a) Clearly $\left\{ \begin{bmatrix} -3 \\ 4 \end{bmatrix} \right\}$ is a basis for W. Let $C = \begin{bmatrix} -3 \\ 4 \end{bmatrix}$, the matrix whose column is this basis vector. Then
$$P_W = C(C^T C)^{-1} C^T$$
$$= \frac{1}{25} \begin{bmatrix} 9 & -12 \\ -12 & 16 \end{bmatrix}.$$

(b) The orthogonal projection of $\mathbf{u}$ on W is the vector
$$\mathbf{w} = P_W \mathbf{u}$$
$$= \frac{1}{25} \begin{bmatrix} 9 & -12 \\ -12 & 16 \end{bmatrix} \begin{bmatrix} -10 \\ 5 \end{bmatrix}$$
$$= \begin{bmatrix} -6 \\ 8 \end{bmatrix},$$

and
$$\mathbf{z} = \mathbf{u} - \mathbf{w}$$
$$= \begin{bmatrix} -10 \\ 5 \end{bmatrix} - \begin{bmatrix} -6 \\ 8 \end{bmatrix} = \begin{bmatrix} -4 \\ -3 \end{bmatrix}.$$

(c) The distance from $\mathbf{u}$ to W is
$$\|\mathbf{z}\| = 5.$$

21. (a) Choose the pivot columns of
$$\begin{bmatrix} 1 & 1 & 5 \\ -1 & 2 & 1 \\ -1 & 1 & -1 \\ 2 & -1 & 4 \end{bmatrix}$$
to obtain a basis
$$\left\{ \begin{bmatrix} 1 \\ -1 \\ -1 \\ 2 \end{bmatrix}, \begin{bmatrix} 1 \\ 2 \\ 1 \\ -1 \end{bmatrix} \right\}$$
for W. Let C be the matrix whose columns are these basis vectors. Then
$$P_W = C(C^T C)^{-1} C^T$$
$$= \frac{1}{33} \begin{bmatrix} 22 & 11 & 0 & 11 \\ 11 & 19 & 9 & -8 \\ 0 & 9 & 6 & -9 \\ 11 & -8 & -9 & 19 \end{bmatrix}.$$

(b) The orthogonal projection of $\mathbf{u}$ on W is the vector
$$\mathbf{w} = P_W \mathbf{u} = \begin{bmatrix} 3 \\ 0 \\ -1 \\ 3 \end{bmatrix},$$

and
$$\mathbf{z} = \mathbf{u} - \mathbf{w} = \begin{bmatrix} -2 \\ 1 \\ 3 \\ 3 \end{bmatrix}.$$

(c) The distance from **u** to W is
$$\|\mathbf{z}\| = \sqrt{23}.$$

25. By solving the equation
$$x_1 + 2x_2 - x_3 = 0,$$
we obtain the basis
$$\left\{ \begin{bmatrix} -2 \\ 1 \\ 0 \end{bmatrix}, \begin{bmatrix} 1 \\ 0 \\ 1 \end{bmatrix} \right\}$$
for W. Proceeding as in the solution to Exercise 21, we obtain the following results.

(a) $P_W = \dfrac{1}{6}\begin{bmatrix} 5 & -2 & 1 \\ -2 & 2 & 2 \\ 1 & 2 & 5 \end{bmatrix}$

(b) $\mathbf{w} = \begin{bmatrix} 2 \\ -1 \\ 0 \end{bmatrix}$ and $\mathbf{z} = \begin{bmatrix} 1 \\ 2 \\ -1 \end{bmatrix}$

(c) The distance from **u** to W is $\sqrt{6}$.

29. As in Exercise 25, a basis for W is
$$\left\{ \begin{bmatrix} -1 \\ 2 \\ 1 \\ 0 \end{bmatrix}, \begin{bmatrix} -1 \\ 0 \\ 0 \\ 1 \end{bmatrix} \right\}.$$
Thus we take
$$C = \begin{bmatrix} -1 & -1 \\ 2 & 0 \\ 1 & 0 \\ 0 & 1 \end{bmatrix},$$
and proceed as in the solution to Exercise 21 to obtain the following results.

(a) $P_W = \dfrac{1}{11}\begin{bmatrix} 6 & -2 & -1 & -5 \\ -2 & 8 & 4 & -2 \\ -1 & 4 & 2 & -1 \\ -5 & -2 & -1 & 6 \end{bmatrix}$

(b) $\mathbf{w} = \begin{bmatrix} 0 \\ 4 \\ 2 \\ -2 \end{bmatrix}$ and $\mathbf{z} = \begin{bmatrix} 1 \\ 1 \\ -1 \\ 1 \end{bmatrix}$

(c) $\|\mathbf{z}\| = 2$

33. False, $(S^\perp)^\perp$ is a subspace for any set S. So, if S is not a subspace, then necessarily $S \neq (S^\perp)^\perp$.

34. False, in $\mathcal{R}^2$, let $F = \{\mathbf{e}_1, \mathbf{e}_2\}$ and $G = \{\mathbf{e}_1, 2\mathbf{e}_2\}$. Then $F^\perp = \{\mathbf{0}\} = G^\perp$, but $F \neq G$.

35. True

36. False, (Row A)$^\perp$ = Null A.

37. True   38. True   39. True

40. True   41. True

42. False, $\dim W = n - \dim W^\perp$.

43. False, we need the given basis to be orthonormal.

44. True   45. True

46. False, the only invertible orthogonal projection matrix is the identity matrix.

47. True

48. False, the columns of C can form *any* basis for W.

49. False, we need the columns of C to form a basis for W.

50. True

51. False, see Example 4.

52. True   53. True

54. False, the distance is $\|\mathbf{u} - P_W \mathbf{u}\|$.

55. True   56. True

57. Suppose that $\mathbf{v}$ is in $W^\perp$. Because every vector in $\mathcal{S}$ is also in W, it follows that $\mathbf{v}$ is orthogonal to each vector in $\mathcal{S}$. Therefore $W^\perp$ is contained in $\mathcal{S}^\perp$.

Conversely, let $\mathbf{u}$ be in $\mathcal{S}^\perp$ and $\mathbf{w}$ in W. There exist scalars $a_1, a_2, \ldots, a_k$ and vectors $\mathbf{v}_1, \mathbf{v}_2, \ldots, \mathbf{v}_k$ in $\mathcal{S}$ such that
$$\mathbf{w} = a_1 \mathbf{v}_1 + \cdots + a_k \mathbf{v}_k.$$
Thus
$$\begin{aligned}\mathbf{w} \cdot \mathbf{u} &= (a_1 \mathbf{v}_1 + \cdots + a_k \mathbf{v}_k) \cdot \mathbf{u} \\ &= a_1(\mathbf{v}_1 \cdot \mathbf{u}) + \cdots + a_k(\mathbf{v}_k \cdot \mathbf{u}) \\ &= a_1 \cdot 0 + \cdots + a_k \cdot 0 \\ &= 0,\end{aligned}$$
and therefore $\mathbf{u}$ is in $W^\perp$. Thus $\mathcal{S}^\perp$ is contained in $W^\perp$, and we conclude that $W^\perp = \mathcal{S}^\perp$.

61. (a) Suppose that $\mathbf{v}$ is in $(\text{Row } A)^\perp$. Then $\mathbf{v}$ is orthogonal to every row in A. But each component of $A\mathbf{v}$ is the dot product of a row of A with $\mathbf{v}$, and hence every component of $A\mathbf{v}$ is zero. So $A\mathbf{v} = \mathbf{0}$, and hence $\mathbf{v}$ is in Null A. Thus $(\text{Row } A)^\perp$ is contained in Null A.

Now suppose that $\mathbf{v}$ is in Null A. Then $A\mathbf{v} = \mathbf{0}$, and hence $\mathbf{v}$ is orthogonal to every row of A. So $\mathbf{v}$ is in $(\text{Row } A)^\perp$ by Exercise 57. Thus Null A is contained in $(\text{Row } A)^\perp$, and the result follows.

(b) By (a), $(\text{Row } A^T)^\perp = \text{Null } A^T$. But the rows of A^T are the columns of A, and hence $(\text{Col } A)^\perp = \text{Null } A^T$.

65. Suppose $\mathbf{v}$ is in both Row A and Null A. Because Null $A = (\text{Row } A)^\perp$ by Exercise 61(a), $\mathbf{v}$ is orthogonal to itself, that is, $\mathbf{v} \cdot \mathbf{v} = 0$. So $\mathbf{v} = \mathbf{0}$.

69. By Theorem 6.7, there are unique vectors $\mathbf{w}$ in W and $\mathbf{z}$ in $W^\perp$ such that $\mathbf{u} = \mathbf{w} + \mathbf{z}$. It follows that $\mathbf{u}$ is in $W^\perp$ if and only if $\mathbf{u} = \mathbf{z}$, and hence if and only if $\mathbf{w} = \mathbf{0}$. By Theorem 6.8, $P_W \mathbf{u} = \mathbf{w}$, and hence $P_W \mathbf{u} = \mathbf{0}$ if and only if $\mathbf{u}$ is in $W^\perp$.

ALTERNATE PROOF: By Theorem 6.8, $P_W = C(C^T C)^{-1} C^T$, where C is a matrix whose columns form a basis for W. Now suppose that $\mathbf{u}$ is in $W^\perp$. Then $\mathbf{u}$ is orthogonal to each column of C, and hence $C^T \mathbf{u} = \mathbf{0}$. Therefore
$$\begin{aligned}P_W \mathbf{u} &= C(C^T C)^{-1} C^T \mathbf{u} \\ &= C(C^T C)^{-1}(C^T \mathbf{u}) \\ &= C(C^T C)^{-1} \mathbf{0} = \mathbf{0}.\end{aligned}$$
Conversely, suppose that $P_W \mathbf{u} = \mathbf{0}$. Then
$$\begin{aligned}C(C^T C)^{-1} C^T \mathbf{u} &= \mathbf{0} \\ C^T C(C^T C)^{-1} C^T \mathbf{u} &= C^T \mathbf{0} = \mathbf{0} \\ C^T \mathbf{u} &= \mathbf{0}.\end{aligned}$$
This last equation asserts that $\mathbf{u}$ is orthogonal to the columns of C, a generating set for W. Therefore $\mathbf{u}$ is in $W^\perp$ by Exercise 57.

73. Let $\mathbf{u}$ be a vector in $\mathcal{R}^n$. By Theorem 6.7, there are unique vectors $\mathbf{w}$ and $\mathbf{z}$ in W and $W^\perp$, respectively, such that $\mathbf{u} = \mathbf{w} + \mathbf{z}$. Then
$$\begin{aligned}(P_W + P_{W^\perp})\mathbf{u} &= P_W \mathbf{u} + P_{W^\perp} \mathbf{u} \\ &= \mathbf{w} + \mathbf{z} = \mathbf{u} = I_n \mathbf{u}.\end{aligned}$$
Thus $P_W + P_{W^\perp} = I_n$.

77. (a) We first show that 1 and 0 are eigenvalues of P_W. Since $k \neq 0$, we can choose a nonzero vector $\mathbf{w}$ in W. Then $P_W \mathbf{w} = \mathbf{w}$, and hence $\mathbf{w}$

is an eigenvector with corresponding eigenvalue 1. Since $k \neq n$, we can choose a nonzero vector $\mathbf{z}$ in $W^\perp$. Then $P_W \mathbf{z} = \mathbf{0}$, and hence $\mathbf{z}$ is an eigenvector with corresponding eigenvalue 0.

Next we show that 1 and 0 are the only eigenvalues of P_W. Suppose that λ is a nonzero eigenvalue of P_W, and let $\mathbf{u}$ be an eigenvector corresponding to λ. Then $P_W \mathbf{u} = \lambda \mathbf{u}$, and hence $P_W(\frac{1}{\lambda}\mathbf{u}) = \mathbf{u}$. Thus $\mathbf{u}$ is an image of U_W, and so $\mathbf{u}$ is in W. Therefore

$$\mathbf{u} = P_W\left(\frac{1}{\lambda}\mathbf{u}\right) = \frac{1}{\lambda} P_W \mathbf{u} = \frac{1}{\lambda}\mathbf{u}.$$

Hence $1 = \frac{1}{\lambda}$, that is, $\lambda = 1$.

(b) Since $P_W \mathbf{u} = \mathbf{u}$ if and only $\mathbf{u}$ is in W, we see that W is the eigenspace of P_W corresponding to eigenvalue 1. Similarly, since $P_W \mathbf{u} = \mathbf{0}$ if and only if $\mathbf{u}$ is in $W^\perp$, we have that $W^\perp$ is the eigenspace of P_W corresponding to eigenvalue 0.

(c) Let T be the matrix transformation induced by P_W. Since $T(\mathbf{u}) = 1 \cdot \mathbf{u}$ for all $\mathbf{u}$ in $\mathcal{B}_1$ and $T(\mathbf{u}) = 0 \cdot \mathbf{v}$ for all $\mathbf{v}$ in $\mathcal{B}_2$, we have $[T]_\mathcal{B} = D$, and hence $P_W = BDB^{-1}$ by Theorem 4.12.

81. By computing its reduced row echelon form, we see that the first two columns of A are its pivot columns. Thus the rank of A is 2 and a basis for Col A is

$$\mathcal{B}_1 = \left\{ \begin{bmatrix} 1 \\ 0 \\ -1 \\ 1 \end{bmatrix}, \begin{bmatrix} 0 \\ 1 \\ -2 \\ 1 \end{bmatrix} \right\}.$$

As in Example 2 on pages 390–391, we can find a basis

$$\mathcal{B}_2 = \left\{ \begin{bmatrix} 1 \\ 2 \\ 1 \\ 0 \end{bmatrix}, \begin{bmatrix} -1 \\ -1 \\ 0 \\ 1 \end{bmatrix} \right\}.$$

for $(\text{Col } A)^\perp$. Let

$$B = \begin{bmatrix} 1 & 0 & 1 & -1 \\ 0 & 1 & 2 & -1 \\ -1 & -2 & 1 & 0 \\ 1 & 1 & 0 & 1 \end{bmatrix}$$

and

$$D = \begin{bmatrix} 1 & 0 & 0 & 0 \\ 0 & 1 & 0 & 0 \\ 0 & 0 & 0 & 0 \\ 0 & 0 & 0 & 0 \end{bmatrix}.$$

Notice that B is the matrix whose columns are the vectors in $\mathcal{B}_1 \cup \mathcal{B}_2$. Then by Exercise 77(c),

$$P_W = BDB^{-1}$$

$$= \frac{1}{3}\begin{bmatrix} 2 & -1 & 0 & 1 \\ -1 & 1 & -1 & 0 \\ 0 & -1 & 2 & -1 \\ 1 & 0 & -1 & 1 \end{bmatrix}.$$

85. (a) There is no unique answer. Using Q in the MATLAB function $[Q\ R] = \text{qr}(A,0)$ (see Table D.3 in Appendix D), where A is the matrix whose columns are the vectors in $\mathcal{S}$, we obtain an orthonormal basis containing the vectors

$$\begin{bmatrix} 0 \\ .2914 \\ -.8742 \\ 0 \\ .3885 \end{bmatrix}, \begin{bmatrix} .7808 \\ -.5828 \\ -.1059 \\ 0 \\ .1989 \end{bmatrix},$$

$$\begin{bmatrix} -.0994 \\ -.3243 \\ -.4677 \\ .1082 \\ -.8090 \end{bmatrix}, \text{ and } \begin{bmatrix} -.1017 \\ -.1360 \\ -.0589 \\ -.9832 \\ -.0304 \end{bmatrix}.$$

(b) Let C be the 5×4 matrix whose columns are the vectors in (a). Then the orthogonal projection of $\mathbf{u}$ on W is

$$\mathbf{w} = P_W \mathbf{u} = C(C^T C)^{-1} C^T \mathbf{u}$$

$$= \begin{bmatrix} -6.3817 \\ 6.8925 \\ 7.2135 \\ 1.3687 \\ 2.3111 \end{bmatrix}.$$

(c) The distance from $\mathbf{u}$ to W is

$$\|\mathbf{u} - \mathbf{w}\| = 4.3033.$$

6.4 LEAST-SQUARES APPROXIMATIONS AND ORTHOGONAL PROJECTION MATRICES

1. Using the data, let

$$C = \begin{bmatrix} 1 & 1 \\ 1 & 3 \\ 1 & 5 \\ 1 & 7 \end{bmatrix} \text{ and } \mathbf{y} = \begin{bmatrix} 14 \\ 17 \\ 19 \\ 20 \end{bmatrix}.$$

Then the equation of the least squares line for the given data is $y = a_0 + a_1 x$, where

$$\begin{bmatrix} a_0 \\ a_1 \end{bmatrix} = (C^T C)^{-1} C^T \mathbf{y}$$

$$= \begin{bmatrix} 4 & 16 \\ 16 & 84 \end{bmatrix}^{-1} \begin{bmatrix} 1 & 1 & 1 & 1 \\ 1 & 3 & 5 & 7 \end{bmatrix} \begin{bmatrix} 14 \\ 17 \\ 19 \\ 20 \end{bmatrix}$$

$$= \frac{1}{20} \begin{bmatrix} 21 & -4 \\ -4 & 1 \end{bmatrix} \begin{bmatrix} 70 \\ 300 \end{bmatrix} = \begin{bmatrix} 13.5 \\ 1.0 \end{bmatrix}.$$

Therefore $y = 13.5 + x$.

5. Let

$$C = \begin{bmatrix} 1 & 1 \\ 1 & 3 \\ 1 & 7 \\ 1 & 8 \\ 1 & 10 \end{bmatrix} \text{ and } \mathbf{y} = \begin{bmatrix} 40 \\ 36 \\ 23 \\ 21 \\ 13 \end{bmatrix}.$$

Then the equation of the least squares line for the given data is $y = a_0 + a_1 x$, where

$$\begin{bmatrix} a_0 \\ a_1 \end{bmatrix} = (C^T C)^{-1} C^T \mathbf{y}$$

$$= \frac{1}{274} \begin{bmatrix} 194 & 136 & 20 & -9 & -67 \\ -24 & -14 & 6 & 11 & 21 \end{bmatrix} \begin{bmatrix} 40 \\ 36 \\ 23 \\ 21 \\ 13 \end{bmatrix}$$

$$= \begin{bmatrix} 44 \\ -3 \end{bmatrix}.$$

That is, the equation of the least squares line is $y = 44 - 3x$.

9. Let

$$C = \begin{bmatrix} 1 & 3.5 \\ 1 & 4.0 \\ 1 & 4.5 \\ 1 & 5.0 \end{bmatrix} \text{ and } \mathbf{y} = \begin{bmatrix} 1.0 \\ 2.2 \\ 2.8 \\ 4.3 \end{bmatrix}.$$

and proceed as in Exercises 1 and 5. The equation of the least squares line is $y = -6.35 + 2.1x$, and so the estimates of k and L are 2.1 and

$$L = -\frac{a}{k} = -\frac{(-6.35)}{2.1} \approx 3.02,$$

respectively.

13. Let
$$C = \begin{bmatrix} 1 & 0 & 0 \\ 1 & 1 & 1 \\ 1 & 2 & 4 \\ 1 & 3 & 9 \end{bmatrix} \text{ and } \mathbf{y} = \begin{bmatrix} 2 \\ 3 \\ 5 \\ 8 \end{bmatrix}.$$

Here the matrix C has a third column whose entries are x_i^2 (the square of the first coordinate of each data point). As in Exercises 1 and 5, we compute

$$\begin{bmatrix} a_0 \\ a_1 \\ a_2 \end{bmatrix} = (C^T C)^{-1} C^T \mathbf{y} = \begin{bmatrix} 2 \\ 0.5 \\ 0.5 \end{bmatrix}.$$

The equation of the least squares line is $y = a_0 + a_1 x + a_2 x^2 = 2 + 0.5x + 0.5x^2$.

17. As in Example 3, the vectors that minimize $\|A\mathbf{z} - \mathbf{b}\|$ are the solutions of $A\mathbf{x} = P_W \mathbf{b}$. Let

$$C = \begin{bmatrix} 1 & 2 \\ 1 & -1 \\ 2 & 1 \end{bmatrix}$$

be the matrix whose columns are the pivot columns of A. Then

$$P_W \mathbf{b} = C(C^T C)^{-1} C^T \mathbf{b} = \begin{bmatrix} 0 \\ 2 \\ 2 \end{bmatrix},$$

and the general solution of $A\mathbf{x} = P_W \mathbf{b}$ is

$$\begin{bmatrix} x_1 \\ x_2 \\ x_3 \end{bmatrix} = \frac{1}{3}\begin{bmatrix} 4 \\ -2 \\ 0 \end{bmatrix} + x_3 \begin{bmatrix} -1 \\ 1 \\ 1 \end{bmatrix}.$$

Thus vectors of the form

$$\frac{1}{3}\begin{bmatrix} 4 \\ -2 \\ 0 \end{bmatrix} + x_3 \begin{bmatrix} -1 \\ 1 \\ 1 \end{bmatrix}$$

are the vectors that minimize $\|A\mathbf{z} - \mathbf{b}\|$.

21. The system $A\mathbf{x} = \mathbf{b}$ is consistent, and its general solution has the form

$$\begin{bmatrix} x_1 \\ x_2 \\ x_3 \end{bmatrix} = \begin{bmatrix} 5 \\ -3 \\ 0 \end{bmatrix} + x_3 \begin{bmatrix} -1 \\ 1 \\ 1 \end{bmatrix}.$$

Thus a vector is a solution of $A\mathbf{x} = \mathbf{b}$ if and only if it has the form $\mathbf{v} = \mathbf{v}_0 + \mathbf{z}$, where

$$\mathbf{v}_0 = \begin{bmatrix} 5 \\ -3 \\ 0 \end{bmatrix}$$

and $\mathbf{z}$ is in Null $A = \text{Span} \left\{ \begin{bmatrix} -1 \\ 1 \\ 1 \end{bmatrix} \right\}$. As described on page 408, the solution of $A\mathbf{x} = \mathbf{b}$ having least norm is given by $\mathbf{v}_0 - P_Z \mathbf{v}_0$, where $Z = \text{Null } A$. Let

$$C = \begin{bmatrix} -1 \\ 1 \\ 1 \end{bmatrix}.$$

Then

$$P_Z \mathbf{v}_0 = C(C^T C)^{-1} C^T \mathbf{v}_0$$
$$= -\frac{8}{3}\begin{bmatrix} -1 \\ 1 \\ 1 \end{bmatrix},$$

and so the solution of $A\mathbf{x} = \mathbf{b}$ having least norm is

$$\mathbf{v}_0 - P_Z \mathbf{v}_0 = \frac{1}{3}\begin{bmatrix} 7 \\ -1 \\ 8 \end{bmatrix}.$$

25. In the solution to Exercise 17, we found that the vectors of the form

$$\frac{1}{3}\begin{bmatrix} 4 \\ -2 \\ 0 \end{bmatrix} + x_3 \begin{bmatrix} -1 \\ 1 \\ 1 \end{bmatrix}$$

are the vectors that minimize $\|A\mathbf{x} - \mathbf{b}\|$. We proceed as in Example 4 to find the vector **z** of this form that has the least norm. Let $Z = \text{Null } A$,

$$C = \begin{bmatrix} -1 \\ 1 \\ 1 \end{bmatrix}, \quad \text{and} \quad \mathbf{v}_0 = \frac{1}{3}\begin{bmatrix} 4 \\ -2 \\ 0 \end{bmatrix}.$$

Then

$$P_Z \mathbf{v}_0 = C(C^T C)^{-1} C^T \mathbf{v}_0$$
$$= -\frac{2}{3}\begin{bmatrix} -1 \\ 1 \\ 1 \end{bmatrix},$$

So the vector of least norm that minimizes $\|A\mathbf{x} - \mathbf{b}\|$ is

$$\mathbf{v}_0 - P_Z \mathbf{v}_0 = \frac{1}{3}\begin{bmatrix} 2 \\ 0 \\ 2 \end{bmatrix}.$$

28. False, the least-squares line is the line that minimizes the sum of the *squares* of the vertical distances from the data points to the line.

29. True

30. False, in Example 2, the method is used to approximate data with a polynomial of degree 2.

31. False, the inconsistent system in Example 3 has infinitely many vectors that minimize this distance.

32. True

33. We have

$$\|\mathbf{y} - (a_0 \mathbf{v}_1 + a_1 \mathbf{v}_2)\|^2$$
$$= \left\| \begin{bmatrix} y_1 \\ y_2 \\ \vdots \\ y_n \end{bmatrix} - a_0 \begin{bmatrix} 1 \\ 1 \\ \vdots \\ 1 \end{bmatrix} - a_1 \begin{bmatrix} x_1 \\ x_2 \\ \vdots \\ x_n \end{bmatrix} \right\|^2$$

$$= \left\| \begin{bmatrix} y_1 - (a_0 + a_1 x_1) \\ y_2 - (a_0 + a_1 x_2) \\ \vdots \\ y_n - (a_0 + a_1 x_n) \end{bmatrix} \right\|^2$$

$$= [y_1 - (a_0 + a_1 x_1)]^2 + \cdots$$
$$+ [y_n - (a_0 + a_1 x_n)]^2$$

$$= E.$$

37. Let $A = QR$ be a QR-factorization of A, and let W denote the column space of A. The solutions of $A\mathbf{x} = P_W \mathbf{b}$ minimize $\|A\mathbf{x} - \mathbf{b}\|$. It follows from Exercise 36 that $A\mathbf{x} = P_W \mathbf{b}$ if and only if $QR\mathbf{x} = QQ^T \mathbf{b}$. If we multiply both sides of this equation by Q^T and simplify, we obtain $R\mathbf{x} = Q^T \mathbf{b}$.

41. Following the hint, let W denote the column space of A. Then we have (with entries rounded to 4 places after the decimal)

$$A = \begin{bmatrix} 0.9962 & 0.0872 \\ 0.9848 & 0.1736 \\ 0.9659 & 0.2588 \\ 0.9397 & 0.3420 \\ 0.9063 & 0.4226 \\ 0.8660 & 0.5000 \end{bmatrix}$$

and

$$P_W \mathbf{y} = A(A^T A)^{-1} A^T \mathbf{y} = \begin{bmatrix} 2.8039 \\ 2.6003 \\ 2.3769 \\ 2.1355 \\ 1.8777 \\ 1.6057 \end{bmatrix}.$$

The matrix equation $A\mathbf{x} = P_W \mathbf{y}$ has $a = 2.9862$ and $b = -1.9607$ as its solution (rounded to 4 places after the dec-

142 Chapter 6 Orthogonality

imal). When rounded to 2 significant figures, we have $a = 3.0$ and $b = -2.0$.

6.5 ORTHOGONAL MATRICES AND OPERATORS

1. Since the given matrix is not square, it cannot be an orthogonal matrix.

5. Since
$$\begin{bmatrix} 0 & 1 & 0 \\ 0 & 0 & 1 \\ 1 & 0 & 0 \end{bmatrix}^T \begin{bmatrix} 0 & 1 & 0 \\ 0 & 0 & 1 \\ 1 & 0 & 0 \end{bmatrix}$$
$$= \begin{bmatrix} 0 & 0 & 1 \\ 1 & 0 & 0 \\ 0 & 1 & 0 \end{bmatrix} \begin{bmatrix} 0 & 1 & 0 \\ 0 & 0 & 1 \\ 1 & 0 & 0 \end{bmatrix}$$
$$= I_3,$$
the matrix is orthogonal by Theorem 6.9(b).

9. Let $A = \dfrac{1}{\sqrt{2}} \begin{bmatrix} 1 & 1 \\ 1 & -1 \end{bmatrix}$. Since
$$\det A = \frac{1}{2}(-1 - 1) = -1,$$
Theorem 6.11 shows that the operator is a reflection. The line of reflection is the same as the eigenspace of A corresponding to eigenvalue 1. This is the solution set of $(A - 1I_2)\mathbf{x} = \mathbf{0}$. Because the reduced row echelon form of $A - 1I_2$ is
$$\begin{bmatrix} 1 & -\sqrt{2} - 1 \\ 0 & 0 \end{bmatrix},$$
the general solution of this system is $x = (\sqrt{2} + 1)y$, that is,
$$y = \frac{1}{\sqrt{2} + 1}x = (\sqrt{2} - 1)x.$$
Thus $y = (\sqrt{2} - 1)x$ is the equation of the line of reflection.

13. Let $A = \dfrac{1}{13}\begin{bmatrix} 5 & 12 \\ 12 & -5 \end{bmatrix}$. Since
$$\det A = \frac{1}{13^2}(-25 - 144) = -1,$$
the operator is a reflection. As in Exercise 9, the line of reflection is the eigenspace of A corresponding to eigenvalue 1, which is the solution set of $(A - 1I_2)\mathbf{x} = \mathbf{0}$. Because the reduced row echelon form of $A - 1I_2$ is
$$\begin{bmatrix} 1 & -\frac{3}{2} \\ 0 & 0 \end{bmatrix},$$
the general solution of this system is $x = \frac{3}{2}y$, that is, $y = \frac{2}{3}x$. Thus $y = \frac{2}{3}x$ is the equation of the line of reflection.

17. True

18. False; for example, if T is a translation by a nonzero vector, then T preserves distances, but T is not linear.

19. False, only orthogonal linear operators preserve dot products.

20. True   21. True   22. True

23. False, for example, let $P = I_n$ and $Q = -I_n$.

24. False, for example, let $P = \begin{bmatrix} 1 & 1 \\ 1 & 2 \end{bmatrix}$.

25. True

26. False, consider $\begin{bmatrix} 1 & 1 \\ 1 & -1 \end{bmatrix}$.

27. False, consider P_W from Example 4 in Section 6.3.

28. True   29. True   30. True

31. False, we need $\det Q = 1$.

32. False, for example, if T is a translation by a nonzero vector, then T is a rigid motion, but T is not linear and hence is not orthogonal.

33. False, for example, if T is a translation by a nonzero vector, then T is a rigid motion but T is not linear.

34. True **35.** True **36.** True

37. Let $\mathbf{v} = \frac{1}{7}\begin{bmatrix} 3 \\ -2 \\ 6 \end{bmatrix}$, and suppose that T is a linear operator on $\mathcal{R}^3$ such that $T(\mathbf{v}) = \mathbf{e}_3$. If A is the standard matrix of T, then A is an orthogonal matrix, and so $A^T A = I_3$. Since $A\mathbf{v} = T(\mathbf{v}) = \mathbf{e}_3$, it follows that

$$\mathbf{v} = (A^T A)\mathbf{v} = A^T(A\mathbf{v}) = A^T\mathbf{e}_3,$$

and so $\mathbf{v}$ must be the third column of A^T. As in Example 3, we construct an orthonormal basis for $\{\mathbf{v}\}^\perp$. The vectors in $\{\mathbf{v}\}^\perp$ satisfy

$$3x_1 - 2x_2 + 6x_3 = 0.$$

A basis for the solution set of this equation is

$$\left\{ \begin{bmatrix} 2 \\ 3 \\ 0 \end{bmatrix}, \begin{bmatrix} -2 \\ 0 \\ 1 \end{bmatrix} \right\}.$$

Applying the Gram-Schmidt process to this set, we obtain

$$\left\{ \begin{bmatrix} 2 \\ 3 \\ 0 \end{bmatrix}, \begin{bmatrix} -18 \\ 12 \\ 13 \end{bmatrix} \right\}.$$

which is an orthogonal basis for $\{\mathbf{v}\}^\perp$. Thus

$$\left\{ \frac{1}{\sqrt{13}}\begin{bmatrix} 2 \\ 3 \\ 0 \end{bmatrix}, \frac{1}{7\sqrt{13}}\begin{bmatrix} -18 \\ 12 \\ 13 \end{bmatrix} \right\}$$

is an orthonormal basis for $\{\mathbf{v}\}^\perp$. Hence we may take

$$A^T = \frac{1}{91}\begin{bmatrix} 14\sqrt{13} & -18\sqrt{13} & 39 \\ 21\sqrt{13} & 12\sqrt{13} & -26 \\ 0 & 13\sqrt{13} & 78 \end{bmatrix},$$

so that

$$A = \frac{1}{91}\begin{bmatrix} 14\sqrt{13} & 21\sqrt{13} & 0 \\ -18\sqrt{13} & 12\sqrt{13} & 13\sqrt{13} \\ 39 & -26 & 78 \end{bmatrix}.$$

Thus one possibility for the desired operator is $T = T_A$.

41. We extend $\{\mathbf{v}_1, \mathbf{v}_2\}$ and $\{\mathbf{w}_1, \mathbf{w}_2\}$ to orthonormal bases for $\mathcal{R}^3$ by including

$$\mathbf{v}_3 = \frac{1}{3}\begin{bmatrix} 2 \\ -2 \\ 1 \end{bmatrix} \quad \text{and} \quad \mathbf{w}_3 = \frac{1}{7}\begin{bmatrix} 3 \\ -6 \\ 2 \end{bmatrix},$$

respectively. Let

$$B = \begin{bmatrix} \mathbf{v}_1 & \mathbf{v}_2 & \mathbf{v}_3 \end{bmatrix} = \frac{1}{3}\begin{bmatrix} 1 & 2 & 2 \\ 2 & 1 & -2 \\ 2 & -2 & 1 \end{bmatrix}$$

and

$$C = \begin{bmatrix} \mathbf{w}_1 & \mathbf{w}_2 & \mathbf{w}_3 \end{bmatrix} = \frac{1}{7}\begin{bmatrix} 2 & 6 & 3 \\ 3 & 2 & -6 \\ 6 & -3 & 2 \end{bmatrix},$$

which are orthogonal matrices. Take

$$A = CB^T = \frac{1}{21}\begin{bmatrix} 20 & 4 & -5 \\ -5 & 20 & -4 \\ 4 & 5 & 20 \end{bmatrix}$$

and $T = T_A$ to obtain an orthogonal operator that meets the given requirements.

45. Let $\mathcal{B} = \{\mathbf{v}, \mathbf{w}\}$. Observe that

$$T(\mathbf{v}) = (\mathbf{v}\cdot\mathbf{v}\cos\theta + \mathbf{v}\cdot\mathbf{w}\sin\theta)\mathbf{v}$$
$$+ (-\mathbf{v}\cdot\mathbf{v}\sin\theta + \mathbf{v}\cdot\mathbf{w}\cos\theta)\mathbf{w}$$
$$= \cos\theta\mathbf{v} - \sin\theta\mathbf{w}.$$

Similarly, $T(\mathbf{w}) = \sin\theta\mathbf{v} + \cos\theta\mathbf{w}$, and hence

$$[T]_\mathcal{B} = \begin{bmatrix} \cos\theta & \sin\theta \\ -\sin\theta & \cos\theta \end{bmatrix}.$$

Since $[T]_\mathcal{B}$ is an orthogonal matrix, T is an orthogonal operator.

49. Suppose that λ is an eigenvalue for Q, and let $\mathbf{v}$ be a corresponding eigenvector. Then

$$\|\mathbf{v}\| = \|Q\mathbf{v}\| = \|\lambda\mathbf{v}\| = |\lambda|\|\mathbf{v}\|.$$

Since $\|\mathbf{v}\| \neq 0$, it follows that $|\lambda| = 1$. Therefore $\lambda = \pm 1$.

53. (a)
$$Q_W^T = (2P_W - I_2)^T = 2P_W^T - I_2^T$$
$$= 2P_W - I_2 = Q_W$$

(b)
$$Q_W^2 = (2P_W - I_2)^2$$
$$= 4P_W^2 - 4P_W I_2 + I_2$$
$$= 4P_W - 4P_W + I_2 = I_2$$

(c) By (a) and (b), we have
$$Q_W^T Q_W = Q_W Q_W = Q_W^2 = I_2,$$
and hence Q_W is an orthogonal matrix.

(d)
$$Q_W \mathbf{w} = (2P_W - I_2)\mathbf{w}$$
$$= 2P_W \mathbf{w} - I_2 \mathbf{w}$$
$$= 2\mathbf{w} - \mathbf{w} = \mathbf{w}$$

(e)
$$Q_W \mathbf{v} = (2P_W - I_2)\mathbf{v}$$
$$= 2P_W \mathbf{v} - I_2 \mathbf{v}$$
$$= \mathbf{0} - \mathbf{v} = -\mathbf{v}$$

(f) Select nonzero vectors $\mathbf{w}$ in W and $\mathbf{v}$ in $W^\perp$. Then $\{\mathbf{w}, \mathbf{v}\}$ is a basis for $\mathcal{R}^2$ since it is an orthogonal set of nonzero vectors. Set $P = [\mathbf{w}\ \mathbf{v}]$, and let T be the matrix transformation induced by Q_W. Then Q_W is the standard matrix of T, and T is an orthogonal operator because Q_W is an orthogonal matrix. Also,

$$Q_W = PDP^{-1},$$

where $D = \begin{bmatrix} 1 & 0 \\ 0 & -1 \end{bmatrix}$. Thus

$$\det Q_W = \det(PDP^{-1})$$
$$= (\det P)(\det D)(\det P^{-1})$$
$$= (\det P)(-1)(\det P)^{-1}$$
$$= -1.$$

It follows that T is a reflection. Furthermore, since $T(\mathbf{w}) = \mathbf{w}$, T is the reflection of $\mathcal{R}^2$ about W.

57.
$$\|T(\mathbf{u})\| = \|T(\mathbf{u}) - \mathbf{0}\|$$
$$= \|T(\mathbf{u}) - T(\mathbf{0})\|$$
$$= \|\mathbf{u} - \mathbf{0}\| = \|\mathbf{u}\|$$

61. Since

$$F\left(\begin{bmatrix}1\\0\end{bmatrix}\right) + F\left(\begin{bmatrix}0\\1\end{bmatrix}\right)$$
$$= Q\begin{bmatrix}1\\0\end{bmatrix} + \mathbf{b} + Q\begin{bmatrix}0\\1\end{bmatrix} + \mathbf{b}$$
$$= \mathbf{q}_1 + \mathbf{q}_2 + 2\mathbf{b}$$

and
$$F\left(\begin{bmatrix}1\\1\end{bmatrix}\right) = Q\left(\begin{bmatrix}1\\0\end{bmatrix}+\begin{bmatrix}0\\1\end{bmatrix}\right)+\mathbf{b}$$
$$= \mathbf{q}_1+\mathbf{q}_2+\mathbf{b},$$

it follows that
$$\mathbf{b} = F\left(\begin{bmatrix}1\\0\end{bmatrix}\right)+F\left(\begin{bmatrix}0\\1\end{bmatrix}\right)-F\left(\begin{bmatrix}1\\1\end{bmatrix}\right)$$
$$= \begin{bmatrix}2\\4\end{bmatrix}+\begin{bmatrix}1\\3\end{bmatrix}-\begin{bmatrix}2\\3\end{bmatrix}=\begin{bmatrix}1\\4\end{bmatrix}.$$

Thus
$$\mathbf{q}_1 = Q\begin{bmatrix}1\\0\end{bmatrix} = F\left(\begin{bmatrix}1\\0\end{bmatrix}\right)-\mathbf{b}$$
$$= \begin{bmatrix}2\\4\end{bmatrix}-\begin{bmatrix}1\\4\end{bmatrix}=\begin{bmatrix}1\\0\end{bmatrix}$$

and
$$\mathbf{q}_2 = Q\begin{bmatrix}0\\1\end{bmatrix} = F\left(\begin{bmatrix}0\\1\end{bmatrix}\right)-\mathbf{b}$$
$$= \begin{bmatrix}1\\3\end{bmatrix}-\begin{bmatrix}1\\4\end{bmatrix}=\begin{bmatrix}0\\-1\end{bmatrix}.$$

Therefore $Q = [\mathbf{q}_1\ \mathbf{q}_2] = \begin{bmatrix}1 & 0\\0 & -1\end{bmatrix}$.

65. Let $\mathbf{u}$ and $\mathbf{v}$ be in $\mathcal{R}^n$. Then
$$\|T(\mathbf{u})-T(\mathbf{v})\|^2$$
$$= [T(\mathbf{u})-T(\mathbf{v})]\cdot[T(\mathbf{u})-T(\mathbf{v})]$$
$$= T(\mathbf{u})\cdot T(\mathbf{u})-2T(\mathbf{u})\cdot T(\mathbf{v})$$
$$\qquad + T(\mathbf{v})\cdot T(\mathbf{v})$$
$$= \mathbf{u}\cdot\mathbf{u}-2\mathbf{u}\cdot\mathbf{v}+\mathbf{v}\cdot\mathbf{v}\|\mathbf{u}-\mathbf{v}\|^2.$$

Hence $\|T(\mathbf{u})-T(\mathbf{v})\| = \|\mathbf{u}-\mathbf{v}\|$. It follows that T is a rigid motion. Furthermore,
$$\|T(\mathbf{0})\|^2 = T(\mathbf{0})\cdot T(\mathbf{0}) = \mathbf{0}\cdot\mathbf{0} = 0,$$

and hence $T(\mathbf{0}) = \mathbf{0}$. Therefore T is an orthogonal operator by Theorem 6.13.

69. Let Q be the standard matrix of U, $\mathbf{w} = \begin{bmatrix}1\\m\end{bmatrix}$, and $\mathbf{z} = \begin{bmatrix}-m\\1\end{bmatrix}$. Since $\mathbf{w}$ lies along $\mathcal{L}$ and $\mathbf{z}$ is perpendicular to $\mathcal{L}$, we have $U(\mathbf{w}) = \mathbf{w}$ and $U(\mathbf{z}) = -\mathbf{z}$. Let $\mathcal{B} = \{\mathbf{w},\mathbf{z}\}$ and $P = [\mathbf{w}\ \mathbf{z}]$. Then $\mathcal{B}$ is a basis for $\mathcal{R}^2$, and hence
$$[U]_\mathcal{B} = \begin{bmatrix}1 & 0\\0 & -1\end{bmatrix}\text{ and }Q = P[U]_\mathcal{B}P^{-1}.$$

Thus
$$Q = P[U]_\mathcal{B}P^{-1}$$
$$= \begin{bmatrix}1 & m\\-m & 1\end{bmatrix}\begin{bmatrix}1 & 0\\0 & -1\end{bmatrix}\begin{bmatrix}1 & m\\-m & 1\end{bmatrix}^{-1}$$
$$= \begin{bmatrix}1 & m\\-m & 1\end{bmatrix}\begin{bmatrix}1 & 0\\0 & -1\end{bmatrix}\begin{bmatrix}1 & m\\-m & 1\end{bmatrix}^{-1}$$
$$= \frac{1}{1+m^2}\begin{bmatrix}1-m^2 & 2m\\2m & m^2-1\end{bmatrix}.$$

73. By Exercise 69, the respective standard matrices of the reflections of $\mathcal{R}^2$ about the lines with equations $y = 1.23x$ and $y = -0.24x$ are
$$A = \frac{1}{1+a^2}\begin{bmatrix}1-a^2 & 2a\\2a & a^2-1\end{bmatrix}$$

and
$$B = \frac{1}{1+b^2}\begin{bmatrix}1-b^2 & 2b\\2b & b^2-1\end{bmatrix},$$

where $a = 1.23$ and $b = -0.24$. Thus, by Theorem 2.12, the standard matrix of the composition of these reflections is
$$BA = \begin{bmatrix}-.6262 & .7797\\-.7797 & -.6262\end{bmatrix}.$$

From the form of a rotation matrix, we see that the angle of rotation θ must satisfy $\cos\theta = -.6262$ and $\sin\theta = -.7797$.

Because both of these values are negative, we have $180° < \theta < 270°$. Now $\cos^{-1}(.6262) \approx 51°$, and hence $\theta \approx 180° + 51° = 231°$.

6.6 SYMMETRIC MATRICES

1. (a) The associated quadratic form of the given equation
$$2x^2 - 14xy + 50y^2 - 255 = 0$$
is $2x^2 - 14xy + 50y^2$. As described on pages 428–431, we must find an orthonormal basis for $\mathcal{R}^2$ consisting of eigenvectors of
$$A = \begin{bmatrix} 2 & -7 \\ -7 & 50 \end{bmatrix}.$$

   (b) The eigenvalues of A are $\lambda_1 = 1$ and $\lambda_2 = 51$, and
$$\left\{\begin{bmatrix} 7 \\ 1 \end{bmatrix}\right\} \quad \text{and} \quad \left\{\begin{bmatrix} -1 \\ 7 \end{bmatrix}\right\}$$
are bases for the corresponding eigenspaces. The vectors in these bases are orthogonal, as guaranteed by Theorem 6.14. So
$$\left\{\frac{1}{\sqrt{50}}\begin{bmatrix} 7 \\ 1 \end{bmatrix}, \frac{1}{\sqrt{50}}\begin{bmatrix} -1 \\ 7 \end{bmatrix}\right\}$$
is the desired orthonormal basis for $\mathcal{R}^2$ consisting of eigenvectors of A. From among these basis vectors and their negatives, we choose the vector having two positive components, which is $\frac{1}{\sqrt{50}}\begin{bmatrix} 7 \\ 1 \end{bmatrix}$, and make a rotation matrix with this as its first column. This rotation matrix is
$$P = \frac{1}{\sqrt{50}}\begin{bmatrix} 7 & -1 \\ 1 & 7 \end{bmatrix},$$
and it is the rotation matrix corresponding to the angle
$$\cos^{-1}\frac{7}{\sqrt{50}} \approx 8.13°.$$

   (c) The equations relating x, y and x', y' are given by
$$\begin{bmatrix} x \\ y \end{bmatrix} = P\begin{bmatrix} x' \\ y' \end{bmatrix},$$
that is,
$$x = \frac{7}{\sqrt{50}}x' - \frac{1}{\sqrt{50}}y'$$
$$y = \frac{1}{\sqrt{50}}x' + \frac{7}{\sqrt{50}}y'.$$

   (d) The columns of P are eigenvectors corresponding to the eigenvalues $\lambda_1 = 1$ and $\lambda_2 = 51$, respectively, and so the transformed quadratic form is
$$\lambda_1(x')^2 + \lambda_2(y')^2 = (x')^2 + 51(y')^2.$$
Thus the transformed equation is
$$(x')^2 + 51(y')^2 - 255 = 0$$
$$(x')^2 + 51(y')^2 = 255$$
$$\frac{(x')^2}{255} + \frac{(y')^2}{5} = 1.$$

   (e) This is the equation of an ellipse.

5. (a) As in Exercise 1, take
$$A = \begin{bmatrix} 5 & 2 \\ 2 & 5 \end{bmatrix}.$$

   (b) The eigenvalues of A are 7 and 3, and
$$\left\{\begin{bmatrix} \frac{1}{\sqrt{2}} \\ \frac{1}{\sqrt{2}} \end{bmatrix}, \begin{bmatrix} -\frac{1}{\sqrt{2}} \\ \frac{1}{\sqrt{2}} \end{bmatrix}\right\}$$

is an orthonormal basis for $\mathcal{R}^2$ consisting of corresponding eigenvectors. Choose

$$P = \begin{bmatrix} \frac{1}{\sqrt{2}} & -\frac{1}{\sqrt{2}} \\ \frac{1}{\sqrt{2}} & \frac{1}{\sqrt{2}} \end{bmatrix}.$$

Then P is the rotation matrix corresponding to the angle

$$\cos^{-1}\frac{1}{\sqrt{2}} = 45°.$$

(c) The equations relating x, y and x', y' are given by

$$\begin{bmatrix} x \\ y \end{bmatrix} = P \begin{bmatrix} x' \\ y' \end{bmatrix},$$

that is,

$$x = \tfrac{1}{\sqrt{2}}x' - \tfrac{1}{\sqrt{2}}y'$$
$$y = \tfrac{1}{\sqrt{2}}x' + \tfrac{1}{\sqrt{2}}y'.$$

(d) The columns of P are eigenvectors corresponding to the eigenvalues 7 and 3, respectively, and so the transformed equation is

$$7(x')^2 + 3(y')^2 = 9$$
$$\frac{7(x')^2}{9} + \frac{(y')^2}{3} = 1.$$

(e) The conic section is an ellipse.

9. (a) As in Exercise 1, take

$$A = \begin{bmatrix} 2 & -6 \\ -6 & -7 \end{bmatrix}.$$

(b) The eigenvalues of A are -10 and 5, and

$$\left\{ \begin{bmatrix} \tfrac{1}{\sqrt{5}} \\ \tfrac{2}{\sqrt{5}} \end{bmatrix}, \begin{bmatrix} -\tfrac{2}{\sqrt{5}} \\ \tfrac{1}{\sqrt{5}} \end{bmatrix} \right\}$$

is an orthonormal basis for $\mathcal{R}^2$ consisting of corresponding eigenvectors. Choose

$$P = \begin{bmatrix} \frac{1}{\sqrt{5}} & -\frac{2}{\sqrt{5}} \\ \frac{2}{\sqrt{5}} & \frac{1}{\sqrt{5}} \end{bmatrix}.$$

Then P is the rotation matrix corresponding to the angle

$$\cos^{-1}\frac{1}{\sqrt{5}} \approx 63.43°.$$

(c) The equations relating x, y and x', y' are given by

$$x = \tfrac{1}{\sqrt{5}}x' - \tfrac{2}{\sqrt{5}}y'$$
$$y = \tfrac{2}{\sqrt{5}}x' + \tfrac{1}{\sqrt{5}}y'.$$

(d) The columns of P are eigenvectors corresponding to the eigenvalues -10 and 5, respectively, and so the transformed equation is

$$-10(x')^2 + 5(y')^2 = 200$$
$$-\frac{(x')^2}{20} + \frac{(y')^2}{40} = 1.$$

(e) The conic section is a hyperbola.

13. The characteristic polynomial of A is

$$\det(A - tI_2) = \det\begin{bmatrix} t-3 & 1 \\ 1 & t-3 \end{bmatrix}$$
$$= (t-3)^2 - 1$$
$$= (t-2)(t-4).$$

Thus A has the eigenvalues $\lambda_1 = 2$ and $\lambda_2 = 4$, and $\begin{bmatrix} 1 \\ -1 \end{bmatrix}$ and $\begin{bmatrix} 1 \\ 1 \end{bmatrix}$ are eigenvectors that correspond to these eigenvalues. Normalizing these vectors, we obtain unit vectors

$$\mathbf{u}_1 = \begin{bmatrix} \tfrac{1}{\sqrt{2}} \\ -\tfrac{1}{\sqrt{2}} \end{bmatrix} \quad \text{and} \quad \mathbf{u}_2 = \begin{bmatrix} \tfrac{1}{\sqrt{2}} \\ \tfrac{1}{\sqrt{2}} \end{bmatrix}$$

that form an orthonormal basis $\{\mathbf{u}_1, \mathbf{u}_2\}$ for $\mathcal{R}^2$. Using these eigenvectors and corresponding eigenvalues, we obtain the spectral decomposition

$$A = \lambda_1 \mathbf{u}_1 \mathbf{u}_1^T + \lambda_2 \mathbf{u}_2 \mathbf{u}_2^T$$

$$= 2 \begin{bmatrix} \frac{1}{\sqrt{2}} \\ -\frac{1}{\sqrt{2}} \end{bmatrix} \begin{bmatrix} \frac{1}{\sqrt{2}} & -\frac{1}{\sqrt{2}} \end{bmatrix}$$

$$+ 4 \begin{bmatrix} \frac{1}{\sqrt{2}} \\ \frac{1}{\sqrt{2}} \end{bmatrix} \begin{bmatrix} \frac{1}{\sqrt{2}} & \frac{1}{\sqrt{2}} \end{bmatrix}$$

$$= 2 \begin{bmatrix} 0.5 & -0.5 \\ -0.5 & 0.5 \end{bmatrix} + 4 \begin{bmatrix} 0.5 & 0.5 \\ 0.5 & 0.5 \end{bmatrix}.$$

17. The characteristic polynomial of A is

$$\det(A - tI_3)$$

$$= \det \begin{bmatrix} 3-t & 2 & 2 \\ 2 & 2-7 & 0 \\ 2 & 0 & 4-t \end{bmatrix}$$

$$= -18t + 9t^2 - t^3$$

$$= -t(t-3)(t-6),$$

and so A has the eigenvalues $\lambda_1 = 3$, $\lambda_2 = 6$, and $\lambda_3 = 0$. For each λ_i, select a nonzero solution of $(A - \lambda_i I_3)\mathbf{x} = \mathbf{0}$ to obtain an eigenvector corresponding to each eigenvalue. Since these eigenvalues are distinct, the eigenvectors are orthogonal, and hence normalizing these eigenvectors produces an orthonormal basis for $\mathcal{R}^3$ consisting of eigenvectors of A. Thus if

$$\mathbf{u}_1 = \frac{1}{3}\begin{bmatrix} -1 \\ -2 \\ 2 \end{bmatrix}, \quad \mathbf{u}_2 = \frac{1}{3}\begin{bmatrix} 2 \\ 1 \\ 2 \end{bmatrix},$$

and

$$\mathbf{u}_3 = \frac{1}{3}\begin{bmatrix} -2 \\ 2 \\ 1 \end{bmatrix},$$

then $\{\mathbf{u}_1, \mathbf{u}_2, \mathbf{u}_3\}$ is an orthonormal basis for $\mathcal{R}^3$ consisting of eigenvectors of A. Using these eigenvectors and corresponding eigenvalues, we obtain the spectral decomposition

$$A = \lambda_1 \mathbf{u}_1 \mathbf{u}_1^T + \lambda_2 \mathbf{u}_2 \mathbf{u}_2^T + \lambda_3 \mathbf{u}_3 \mathbf{u}_3^T$$

$$= 3\left(\frac{1}{3}\right)\begin{bmatrix} -1 \\ -2 \\ 2 \end{bmatrix}\frac{1}{3}[-1 \; -2 \; 2]$$

$$+ 6\left(\frac{1}{3}\right)\begin{bmatrix} 2 \\ 1 \\ 2 \end{bmatrix}\frac{1}{3}[2 \; 1 \; 2]$$

$$+ 0\left(\frac{1}{3}\right)\begin{bmatrix} -2 \\ 2 \\ 1 \end{bmatrix}\frac{1}{3}[-2 \; 2 \; 1]$$

$$= 3\begin{bmatrix} \frac{1}{9} & \frac{2}{9} & -\frac{2}{9} \\ \frac{2}{9} & \frac{4}{9} & -\frac{4}{9} \\ -\frac{2}{9} & -\frac{4}{9} & \frac{4}{9} \end{bmatrix} + 6\begin{bmatrix} \frac{4}{9} & \frac{2}{9} & \frac{4}{9} \\ \frac{2}{9} & \frac{1}{9} & \frac{2}{9} \\ \frac{4}{9} & \frac{2}{9} & \frac{4}{9} \end{bmatrix}$$

$$+ 0\begin{bmatrix} \frac{4}{9} & -\frac{4}{9} & -\frac{2}{9} \\ -\frac{4}{9} & \frac{4}{9} & \frac{2}{9} \\ -\frac{2}{9} & \frac{2}{9} & \frac{1}{9} \end{bmatrix}.$$

21. True

22. False, any nonzero vector in $\mathcal{R}^2$ is an eigenvector of the symmetric matrix I_2, but not every 2×2 matrix with nonzero columns is an orthogonal matrix.

23. True

24. False, let $A = \begin{bmatrix} 1 & 4 \\ 1 & 1 \end{bmatrix}$. Then $\begin{bmatrix} 2 \\ 1 \end{bmatrix}$ and $\begin{bmatrix} 2 \\ -1 \end{bmatrix}$ are eigenvectors of A that correspond to the eigenvalues 3 and -1, respectively. But these two eigenvectors are not orthogonal.

6.6 Symmetric Matrices

25. False, if **v** is an eigenvector, then so is 2**v**. But these two eigenvectors are not orthogonal.

26. True   **27.** True

28. False, $\begin{bmatrix} 1 & 0 \\ 0 & -1 \end{bmatrix}$ is not the sum of orthogonal projection matrices.

29. True   **30.** True   **31.** True

32. False, if θ is an acceptable angle of rotation, then so is $\theta \pm \frac{\pi}{2}$.

33. True

34. False, see Exercise 41.

35. False, the matrix must be symmetric.

36. False, the correct matrix is $\begin{bmatrix} a & b \\ b & c \end{bmatrix}$.

37. True   **38.** True

39. False, we also require that $\det P = 1$.

40. False, we need the coefficient of xy to be $2b$.

41. Two different spectral decompositions of $2I_2$ are $2\begin{bmatrix} 1 & 0 \\ 0 & 0 \end{bmatrix} + 2\begin{bmatrix} 0 & 0 \\ 0 & 1 \end{bmatrix}$ and $2\begin{bmatrix} .5 & .5 \\ .5 & .5 \end{bmatrix} + 2\begin{bmatrix} .5 & -.5 \\ -.5 & .5 \end{bmatrix}$.

45. For $i = j$, we have

$$P_i P_i = \mathbf{u}_i \mathbf{u}_i^T \mathbf{u}_i \mathbf{u}_i^T = \mathbf{u}_i (\mathbf{u}_i^T \mathbf{u}_i) \mathbf{u}_i^T$$
$$= \mathbf{u}_i (1) \mathbf{u}_i^T = \mathbf{u}_i \mathbf{u}_i^T = P_i;$$

and for $i \neq j$, we have

$$P_i P_j = \mathbf{u}_i \mathbf{u}_i^T \mathbf{u}_j \mathbf{u}_j^T = \mathbf{u}_i (\mathbf{u}_i^T \mathbf{u}_j) \mathbf{u}_j^T$$
$$= \mathbf{u}_i [0] \mathbf{u}_j^T = O.$$

49. Suppose that $Q_j = P_r + P_{r+1} + \cdots + P_s$. Then

$$Q_j^T = (P_r + P_{r+1} + \cdots + P_s)^T$$
$$= P_r^T + P_{r+1}^T + \cdots + P_s^T$$
$$= P_r + P_{r+1} + \cdots + P_s = Q_j.$$

53. Let $A = \mu_1 Q_1 + \mu_2 Q_2 + \cdots + \mu_k Q_k$ be the spectral decomposition, as in Exercise 47. Then A^s is the sum of all products of s terms (with possible duplication) from the sum above. Any such term containing factors Q_i and Q_j with $i \neq j$ equals O. Otherwise, each factor of the term is of the form $\mu_i Q_i$, and hence the nonzero terms are of the form $\mu_i^s Q_i^s = \mu_i^s Q_i$. Therefore

$$A^s = \mu_1^s Q_1 + \mu_2^s Q_2 + \cdots + \mu_k^s Q_k.$$

57. Let $A = \mu_1 Q_1 + \mu_2 Q_2 + \cdots + \mu_k Q_k$ be the spectral decomposition, as in Exercise 47. By Exercise 55, we have

$$f_j(A)$$
$$= f_j(\mu_1)Q_1 + \cdots$$
$$\quad + f_j(\mu_j)Q_j + \cdots + f_k(\mu_j)Q_k$$
$$= 0Q_1 + \cdots + 1Q_j + \cdots + 0Q_k = Q_j.$$

61. Suppose that A is positive definite. Then A is symmetric, and hence A^{-1} is also symmetric. Furthermore, the eigenvalues of A^{-1} are the reciprocals of the eigenvalues of A. Therefore, since the eigenvalues of A are positive, the eigenvalues of A^{-1} are also positive. It follows that A^{-1} is positive definite by Exercise 59.

65. We prove that the sum of two positive semidefinite $n \times n$ matrices A and B is also positive semidefinite. Note first that $A + B$ is symmetric because both A

and B are symmetric. Also, $\mathbf{v}^T A \mathbf{v} \geq 0$ and $\mathbf{v}^T B \mathbf{v} \geq 0$ because A and B are positive semidefinite. Therefore

$$\mathbf{v}^T(A+B)\mathbf{v} = \mathbf{v}^T A \mathbf{v} + \mathbf{v}^T B \mathbf{v} \geq 0,$$

and hence $A+B$ is positive semidefinite.

69. We prove that if A is a positive semidefinite matrix, then there exists a positive semidefinite matrix B such that $B^2 = A$. Since the eigenvalues of A are nonnegative, each eigenvalue has a nonnegative square root. As in Exercise 47, write

$$A = \mu_1 Q_1 + \mu_2 Q_2 + \cdots + \mu_k Q_k,$$

where the μ_i's are the distinct eigenvalues of A. Define

$$B = \sqrt{\mu_1} Q_1 + \sqrt{\mu_2} Q_2 + \cdots + \sqrt{\mu_k} Q_k.$$

Then $B^2 = A$ by Exercise 53. Moreover, B is symmetric by Theorem 1.2 and Exercise 49. Since the eigenvalues of B are nonnegative, B is a positive semidefinite matrix.

73. Suppose that A is invertible. Let $\mathbf{v}$ be any nonzero vector in $\mathcal{R}^n$. Since nullity $A = 0$, it follows that $A\mathbf{v} \neq \mathbf{0}$. Thus

$$\mathbf{v}^T A^T A \mathbf{v} = (A\mathbf{v})^T(A\mathbf{v})$$
$$= (A\mathbf{v}) \cdot (A\mathbf{v}) > 0,$$

and hence $A^T A$ is positive definite. Similarly, AA^T is positive definite.

6.7 SINGULAR VALUE DECOMPOSITION

1. We wish to write A as $U\Sigma V^T$, where the columns of U and V form orthonormal bases for $\mathcal{R}^2$ satisfying equations (9) and (10) on page 439. We begin by computing

$$A^T A = \begin{bmatrix} 1 & 1 \\ 0 & 0 \end{bmatrix} \begin{bmatrix} 1 & 0 \\ 1 & 0 \end{bmatrix} = \begin{bmatrix} 2 & 0 \\ 0 & 0 \end{bmatrix}.$$

The eigenvalues of $A^T A$ are its diagonal entries, 2 and 0. So the singular value of A is $\sigma_1 = \sqrt{2}$, and the matrix Σ is

$$\Sigma = \begin{bmatrix} \sqrt{2} & 0 \\ 0 & 0 \end{bmatrix}.$$

Because $A^T A$ is a diagonal matrix, $\{\mathbf{e}_1, \mathbf{e}_2\}$ is an orthonormal basis for $\mathcal{R}^2$ consisting of eigenvectors of $A^T A$. So we may take $\mathbf{v}_1 = \mathbf{e}_1$ and $\mathbf{v}_2 = \mathbf{e}_2$ as the columns of V.

Next, we obtain an orthonormal basis for $\mathcal{R}^2$ to serve as the columns of U. From equation (9), we obtain

$$\mathbf{u}_1 = \frac{1}{\sigma_1} A\mathbf{v}_1 = \frac{1}{\sqrt{2}} \begin{bmatrix} 1 \\ 1 \end{bmatrix}.$$

For the second column of U, we can choose any unit vector that is orthogonal to $\mathbf{u}_1$. We can find such a vector by solving $\mathbf{x} \cdot \mathbf{u}_1 = 0$, that is,

$$x_1 + x_2 = 0.$$

For example, we can take $x_1 = -1$ and $x_2 = 1$, and then normalize this vector to obtain $\mathbf{u}_2 = \frac{1}{\sqrt{2}} \begin{bmatrix} -1 \\ 1 \end{bmatrix}$.

Thus $A = U\Sigma V^T$ is a singular value decomposition of A, where

$$U = [\mathbf{u}_1 \ \mathbf{u}_2] = \frac{1}{\sqrt{2}} \begin{bmatrix} 1 & -1 \\ 1 & 1 \end{bmatrix},$$

$$\Sigma = \begin{bmatrix} \sqrt{2} & 0 \\ 0 & 0 \end{bmatrix},$$

and

$$V = [\mathbf{v}_1 \ \mathbf{v}_2] = \begin{bmatrix} 1 & 0 \\ 0 & 1 \end{bmatrix}.$$

5. We begin by finding the eigenvalues of

$$A^T A = \begin{bmatrix} 3 & 2 \\ 2 & 6 \end{bmatrix},$$

which are 7 and 2. (These are ordered from largest to smallest.) Thus the singular values of A are $\sigma_1 = \sqrt{7}$ and $\sigma_2 = \sqrt{2}$. For each of the eigenvalues of $A^T A$, we must find a corresponding eigenvector of $A^T A$. Thus we solve the equations $(A^T A - 7I_2)\mathbf{x} = \mathbf{0}$ and $(A^T A - 2I_2)\mathbf{x} = \mathbf{0}$ to obtain the vectors

$$\begin{bmatrix} 1 \\ 2 \end{bmatrix} \text{ and } \begin{bmatrix} 2 \\ -1 \end{bmatrix}.$$

Normalizing these vectors gives us the columns of V:

$$\mathbf{v}_1 = \frac{1}{\sqrt{5}} \begin{bmatrix} 1 \\ 2 \end{bmatrix} \text{ and } \mathbf{v}_2 = \frac{1}{\sqrt{5}} \begin{bmatrix} 2 \\ -1 \end{bmatrix}.$$

The first two columns of U can be obtained from the vectors $\mathbf{v}_1$ and $\mathbf{v}_2$ using equation (9) on page 439:

$$\mathbf{u}_1 = \frac{1}{\sigma_1} A \mathbf{v}_1 = \frac{1}{\sqrt{7}} \begin{bmatrix} 1 & 1 \\ 1 & -1 \\ 1 & 2 \end{bmatrix} \frac{1}{\sqrt{5}} \begin{bmatrix} 1 \\ 2 \end{bmatrix}$$

$$= \frac{1}{\sqrt{35}} \begin{bmatrix} 3 \\ -1 \\ 5 \end{bmatrix}$$

and

$$\mathbf{u}_2 = \frac{1}{\sigma_1} A \mathbf{v}_2 = \frac{1}{\sqrt{2}} \begin{bmatrix} 1 & 1 \\ 1 & -1 \\ 1 & 2 \end{bmatrix} \frac{1}{\sqrt{5}} \begin{bmatrix} 2 \\ -1 \end{bmatrix}$$

$$= \frac{1}{\sqrt{10}} \begin{bmatrix} 1 \\ 3 \\ 0 \end{bmatrix}.$$

For the third column of U, we can choose any unit vector that is orthogonal to both $\mathbf{u}_1$ and $\mathbf{u}_2$. We can find such a vector by solving the system of linear equations

$$3x_1 - x_2 + 5x_3 = 0$$
$$x_1 + 3x_2 = 0.$$

For example, one possibility is $\begin{bmatrix} -3 \\ 1 \\ 2 \end{bmatrix}$.

Then normalize this vector to obtain the third column of U:

$$\mathbf{u}_3 = \frac{1}{\sqrt{14}} \begin{bmatrix} -3 \\ 1 \\ 2 \end{bmatrix}.$$

Thus if we let

$$U = [\mathbf{u}_1 \ \mathbf{u}_2 \ \mathbf{u}_3] = \begin{bmatrix} \frac{3}{\sqrt{35}} & \frac{1}{\sqrt{10}} & \frac{-3}{\sqrt{14}} \\ \frac{-1}{\sqrt{35}} & \frac{3}{\sqrt{10}} & \frac{1}{\sqrt{14}} \\ \frac{5}{\sqrt{35}} & 0 & \frac{2}{\sqrt{14}} \end{bmatrix},$$

$$\Sigma = \begin{bmatrix} \sigma_1 & 0 \\ 0 & \sigma_2 \\ 0 & 0 \end{bmatrix} = \begin{bmatrix} \sqrt{7} & 0 \\ 0 & \sqrt{2} \\ 0 & 0 \end{bmatrix},$$

and

$$V = [\mathbf{v}_1 \ \mathbf{v}_2] = \begin{bmatrix} \frac{1}{\sqrt{5}} & \frac{2}{\sqrt{5}} \\ \frac{2}{\sqrt{5}} & \frac{-1}{\sqrt{5}} \end{bmatrix},$$

we obtain the singular value decomposition $A = U\Sigma V^T$.

9. We begin by finding the eigenvalues of

$$A^T A = \begin{bmatrix} 6 & 0 & 0 \\ 0 & 2 & 2 \\ 0 & 2 & 5 \end{bmatrix},$$

which are 6, 6, and 1. (These are ordered from largest to smallest.) Thus the singular values of A are $\sigma_1 = \sqrt{6}$, $\sigma_2 = \sqrt{6}$, and $\sigma_3 = 1$. As in Exercise 5, we can obtain an orthonormal basis for $\mathcal{R}^3$ consisting of eigenvectors of $A^T A$, for example,

$$\left\{ \begin{bmatrix} 1 \\ 0 \\ 0 \end{bmatrix}, \frac{1}{\sqrt{5}} \begin{bmatrix} 0 \\ 1 \\ 2 \end{bmatrix}, \frac{1}{\sqrt{5}} \begin{bmatrix} 0 \\ 2 \\ -1 \end{bmatrix} \right\}.$$

Let $\mathbf{v}_1$, $\mathbf{v}_2$, and $\mathbf{v}_3$ denote the vectors in this basis. These vectors are the columns of V.

To obtain the columns of U, we compute

$$\mathbf{u}_1 = \frac{1}{\sigma_1} A \mathbf{v}_1 = \frac{1}{\sqrt{6}} \begin{bmatrix} 1 \\ 2 \\ 1 \end{bmatrix},$$

$$\mathbf{u}_2 = \frac{1}{\sigma_1} A \mathbf{v}_2 = \frac{1}{\sqrt{30}} \begin{bmatrix} 5 \\ -2 \\ -1 \end{bmatrix},$$

and

$$\mathbf{u}_3 = \frac{1}{\sigma_1} A \mathbf{v}_3 = \frac{1}{\sqrt{5}} \begin{bmatrix} 0 \\ 1 \\ -2 \end{bmatrix}.$$

Thus if we take

$$U = \begin{bmatrix} \frac{1}{\sqrt{6}} & \frac{5}{\sqrt{30}} & 0 \\ \frac{2}{\sqrt{6}} & \frac{-2}{\sqrt{30}} & \frac{1}{\sqrt{5}} \\ \frac{1}{\sqrt{6}} & \frac{-1}{\sqrt{30}} & \frac{-2}{\sqrt{5}} \end{bmatrix},$$

$$\Sigma = \begin{bmatrix} \sqrt{6} & 0 & 0 \\ 0 & \sqrt{6} & 0 \\ 0 & 0 & 1 \end{bmatrix},$$

and

$$V = \begin{bmatrix} 1 & 0 & 0 \\ 0 & \frac{1}{\sqrt{5}} & \frac{2}{\sqrt{5}} \\ 0 & \frac{2}{\sqrt{5}} & \frac{-1}{\sqrt{5}} \end{bmatrix},$$

we obtain the singular value decomposition $A = U \Sigma V^T$.

13. From the given characteristic polynomial, we see that the eigenvalues of $A^T A$ are 7, 2, and 0. Thus the singular values of A are $\sigma_1 = \sqrt{7}$ and $\sigma_2 = \sqrt{2}$. It can be shown that

$$\mathbf{v}_1 = \begin{bmatrix} \frac{3}{\sqrt{35}} \\ \frac{-5}{\sqrt{35}} \\ \frac{-1}{\sqrt{35}} \end{bmatrix}, \quad \mathbf{v}_2 = \begin{bmatrix} \frac{1}{\sqrt{10}} \\ 0 \\ \frac{3}{\sqrt{10}} \end{bmatrix},$$

and

$$\mathbf{v}_3 = \begin{bmatrix} \frac{3}{\sqrt{14}} \\ \frac{2}{\sqrt{14}} \\ \frac{-1}{\sqrt{14}} \end{bmatrix}$$

are eigenvectors of $A^T A$ corresponding to the eigenvalues 7, 2, and 0, respectively. Take these vectors to be the columns of V. To obtain the columns of U, let

$$\mathbf{u}_1 = \frac{1}{\sigma_1} A \mathbf{v}_1 = \frac{1}{\sqrt{5}} \begin{bmatrix} 1 \\ -2 \end{bmatrix}$$

and

$$\mathbf{u}_2 = \frac{1}{\sigma_2} A \mathbf{v}_2 = \frac{1}{\sqrt{5}} \begin{bmatrix} 2 \\ 1 \end{bmatrix}.$$

These vectors form an orthonormal basis for $\mathcal{R}^2$. Thus $A = U \Sigma V^T$ is a singular value decomposition of A, where

$$U = \begin{bmatrix} \frac{1}{\sqrt{5}} & \frac{2}{\sqrt{5}} \\ \frac{-2}{\sqrt{5}} & \frac{1}{\sqrt{5}} \end{bmatrix},$$

$$\Sigma = \begin{bmatrix} \sqrt{7} & 0 & 0 \\ 0 & \sqrt{2} & 0 \end{bmatrix},$$

and
$$V = \begin{bmatrix} \frac{3}{\sqrt{35}} & \frac{1}{\sqrt{10}} & \frac{3}{\sqrt{14}} \\ \frac{-5}{\sqrt{35}} & 0 & \frac{2}{\sqrt{14}} \\ \frac{-1}{\sqrt{35}} & \frac{3}{\sqrt{10}} & \frac{-1}{\sqrt{14}} \end{bmatrix}.$$

17. From the given characteristic polynomial, we see that the eigenvalues of $A^T A$ are 21, 18, 0, and 0. Thus the singular values of A are $\sigma_1 = \sqrt{21}$ and $\sigma_2 = \sqrt{18}$. It can be shown that

$$\mathbf{v}_1 = \frac{1}{\sqrt{7}} \begin{bmatrix} 1 \\ 2 \\ 1 \\ 1 \end{bmatrix}, \quad \mathbf{v}_2 = \frac{1}{\sqrt{3}} \begin{bmatrix} 1 \\ -1 \\ 0 \\ 1 \end{bmatrix},$$

$$\mathbf{v}_3 = \frac{1}{\sqrt{11}} \begin{bmatrix} 1 \\ 1 \\ -3 \\ 0 \end{bmatrix}, \text{ and } \mathbf{v}_4 = \frac{1}{\sqrt{2}} \begin{bmatrix} 1 \\ 0 \\ 0 \\ -1 \end{bmatrix}$$

are eigenvectors of $A^T A$ corresponding to the eigenvalues 21, 18, 0, and 0, respectively. Take these vectors to be the columns of V. To obtain the columns of U, let

$$\mathbf{u}_1 = \frac{1}{\sigma_1} A \mathbf{v}_1 = \frac{1}{\sqrt{3}} \begin{bmatrix} 1 \\ 1 \\ -1 \end{bmatrix}$$

and

$$\mathbf{u}_2 = \frac{1}{\sigma_2} A \mathbf{v}_2 = \frac{1}{\sqrt{6}} \begin{bmatrix} 2 \\ -1 \\ 1 \end{bmatrix}.$$

Since $\mathbf{u}_1$ and $\mathbf{u}_2$ are orthonormal, the set of these vectors can be extended to an orthonormal basis $\{\mathbf{u}_1, \mathbf{u}_2, \mathbf{u}_3\}$ for $\mathcal{R}^3$. So we must choose $\mathbf{u}_3$ to be a unit vector that is orthogonal to both $\mathbf{u}_1$ and $\mathbf{u}_2$.

Proceeding as in Exercise 5, we obtain

$$\mathbf{u}_3 = \frac{1}{\sqrt{2}} \begin{bmatrix} 0 \\ 1 \\ 1 \end{bmatrix}.$$

Thus $A = U\Sigma V^T$ is a singular value decomposition of A, where

$$U = \begin{bmatrix} \frac{1}{\sqrt{3}} & \frac{2}{\sqrt{6}} & 0 \\ \frac{1}{\sqrt{3}} & \frac{-1}{\sqrt{6}} & \frac{1}{\sqrt{2}} \\ \frac{-1}{\sqrt{3}} & \frac{1}{\sqrt{6}} & \frac{1}{\sqrt{2}} \end{bmatrix},$$

$$\Sigma = \begin{bmatrix} \sqrt{21} & 0 & 0 & 0 \\ 0 & \sqrt{18} & 0 & 0 \\ 0 & 0 & 0 & 0 \end{bmatrix},$$

and

$$V = \begin{bmatrix} \frac{1}{\sqrt{7}} & \frac{1}{\sqrt{3}} & \frac{1}{\sqrt{11}} & \frac{1}{\sqrt{2}} \\ \frac{2}{\sqrt{7}} & \frac{-1}{\sqrt{3}} & \frac{1}{\sqrt{11}} & 0 \\ \frac{1}{\sqrt{7}} & 0 & \frac{-3}{\sqrt{11}} & 0 \\ \frac{1}{\sqrt{7}} & \frac{1}{\sqrt{3}} & 0 & \frac{-1}{\sqrt{2}} \end{bmatrix}.$$

21. To find the unique solution of $A\mathbf{x} = \mathbf{b}$ with least norm, we find a singular value decomposition of A, where

$$A = \begin{bmatrix} 1 & 1 \\ 2 & 2 \end{bmatrix} \text{ and } \mathbf{b} = \begin{bmatrix} 2 \\ 4 \end{bmatrix}.$$

The eigenvalues of $A^T A$ are 10 and 0, and

$$\mathbf{v}_1 = \frac{1}{\sqrt{2}} \begin{bmatrix} 1 \\ 1 \end{bmatrix} \text{ and } \mathbf{v}_2 = \frac{1}{\sqrt{2}} \begin{bmatrix} 1 \\ -1 \end{bmatrix}.$$

are eigenvectors of $A^T A$ corresponding to the eigenvalues 10 and 0, respectively. These vectors are the columns of V, $\sigma_1 = \sqrt{10}$ is the singular value of A, and

$$\Sigma = \begin{bmatrix} \sqrt{10} & 0 \\ 0 & 0 \end{bmatrix}.$$

Let

$$\mathbf{u}_1 = \frac{1}{\sigma_1} A \mathbf{v}_1 = \frac{1}{\sqrt{5}} \begin{bmatrix} 1 \\ 2 \end{bmatrix},$$

and choose a unit vector $\mathbf{u}_2$ orthogonal to $\mathbf{u}_1$ such as

$$\mathbf{u}_2 = \frac{1}{\sqrt{5}} \begin{bmatrix} 2 \\ -1 \end{bmatrix}.$$

Then, for

$$U = \frac{1}{\sqrt{5}} \begin{bmatrix} 1 & 2 \\ 2 & -1 \end{bmatrix},$$

$A = U \Sigma V^T$ is a singular value decomposition of A.

So, by the boxed result on pages 448–449,

$$\mathbf{z} = V \Sigma^\dagger U^T \mathbf{b}$$

$$= V \begin{bmatrix} \frac{1}{\sqrt{10}} & 0 \\ 0 & 0 \end{bmatrix} \frac{1}{\sqrt{5}} \begin{bmatrix} 1 & 2 \\ 2 & -1 \end{bmatrix}^T \begin{bmatrix} 2 \\ 4 \end{bmatrix}$$

$$= \begin{bmatrix} 1 \\ 1 \end{bmatrix}$$

is the unique solution of $A\mathbf{x} = \mathbf{b}$ with least norm.

25. We proceed as in Exercise 21 with

$$A = \begin{bmatrix} 1 & -2 & 1 \\ -1 & 1 & 2 \end{bmatrix} \text{ and } \mathbf{b} = \begin{bmatrix} 3 \\ -1 \end{bmatrix}.$$

The eigenvalues of $A^T A$ are 7, 5, and 0, and

$$\mathbf{v}_1 = \frac{1}{\sqrt{14}} \begin{bmatrix} -2 \\ 3 \\ 1 \end{bmatrix}, \mathbf{v}_2 = \frac{1}{\sqrt{10}} \begin{bmatrix} 0 \\ -1 \\ 3 \end{bmatrix},$$

and

$$\mathbf{v}_3 = \frac{1}{\sqrt{35}} \begin{bmatrix} 5 \\ 3 \\ 1 \end{bmatrix}$$

are eigenvectors of $A^T A$ corresponding to the eigenvalues 7, 5, and 0, respectively. These vectors are the columns of V, $\sigma_1 = \sqrt{7}$ and $\sigma_2 = \sqrt{5}$ are the singular values of A, and

$$\Sigma = \begin{bmatrix} \sqrt{7} & 0 & 0 \\ 0 & \sqrt{5} & 0 \end{bmatrix}.$$

Let

$$\mathbf{u}_1 = \frac{1}{\sigma_1} A \mathbf{v}_1 = \frac{1}{\sqrt{2}} \begin{bmatrix} -1 \\ 1 \end{bmatrix}$$

and

$$\mathbf{u}_2 = \frac{1}{\sigma_2} A \mathbf{v}_2 = \frac{1}{\sqrt{2}} \begin{bmatrix} 1 \\ 1 \end{bmatrix},$$

so that $A = U \Sigma V^T$ is a singular value decomposition of A, where

$$U = \frac{1}{\sqrt{2}} \begin{bmatrix} -1 & 1 \\ 1 & 1 \end{bmatrix}.$$

So, by the boxed result on pages 448–449,

$$\mathbf{z} = V \Sigma^\dagger U^T \mathbf{b}$$

$$= V \begin{bmatrix} \frac{1}{\sqrt{7}} & 0 \\ 0 & \frac{1}{\sqrt{5}} \\ 0 & 0 \end{bmatrix} \frac{1}{\sqrt{2}} \begin{bmatrix} -1 & 1 \\ 1 & 1 \end{bmatrix}^T \begin{bmatrix} 3 \\ -1 \end{bmatrix}$$

$$= \frac{1}{35} \begin{bmatrix} 20 \\ -37 \\ 11 \end{bmatrix}$$

is the unique solution of $A\mathbf{x} = \mathbf{b}$ with least norm.

29. Let $\mathbf{z}$ be the unique solution of $A\mathbf{x} = \mathbf{b}$ with least norm, where

$$A = \begin{bmatrix} 1 & 2 \\ 2 & 4 \end{bmatrix} \text{ and } \mathbf{b} = \begin{bmatrix} -1 \\ 1 \end{bmatrix}.$$

As in Exercise 5, it can be shown that $A = U\Sigma V^T$ is a singular value decomposition of A, where

$$V = \frac{1}{\sqrt{5}}\begin{bmatrix} 1 & -2 \\ 2 & 1 \end{bmatrix}, \quad \Sigma = \begin{bmatrix} 5 & 0 \\ 0 & 0 \end{bmatrix},$$

and $U = V$. So, by the boxed result on pages 448–449, we have

$$\mathbf{z} = V\Sigma^\dagger U^T \mathbf{b}$$

$$= V\begin{bmatrix} 0.2 & 0 \\ 0 & 0 \end{bmatrix} U^T \mathbf{b} = \begin{bmatrix} 0.04 \\ 0.08 \end{bmatrix}.$$

33. Let $\mathbf{z}$ be the unique solution of $A\mathbf{x} = \mathbf{b}$ with least norm, where

$$A = \begin{bmatrix} 1 & 1 & -1 \\ 1 & 1 & 1 \\ 0 & 0 & 1 \end{bmatrix} \text{ and } \mathbf{b} = \begin{bmatrix} -4 \\ 6 \\ 3 \end{bmatrix}.$$

As in Exercise 5, it can be shown that $A = U\Sigma V^T$ is a singular value decomposition of A, where

$$V = \frac{1}{\sqrt{2}}\begin{bmatrix} 1 & 0 & -1 \\ 1 & 0 & 1 \\ 0 & \sqrt{2} & 0 \end{bmatrix},$$

$$\Sigma = \begin{bmatrix} 2 & 0 & 0 \\ 0 & \sqrt{3} & 0 \\ 0 & 0 & 0 \end{bmatrix},$$

and

$$U = \frac{1}{6}\begin{bmatrix} 3\sqrt{2} & -2\sqrt{3} & \sqrt{6} \\ 3\sqrt{2} & 2\sqrt{3} & -\sqrt{6} \\ 0 & 2\sqrt{3} & 2\sqrt{6} \end{bmatrix}.$$

So, by the boxed result on pages 448–449, we have

$$\mathbf{z} = V\Sigma^\dagger U^T \mathbf{b}$$

$$= V\begin{bmatrix} 0.5 & 0 & 0 \\ 0 & \sqrt{3} & 0 \\ 0 & 0 & 0 \end{bmatrix} U^T \mathbf{b} = \frac{5}{6}\begin{bmatrix} 3 \\ 3 \\ 2 \end{bmatrix}.$$

37. Let $A = \begin{bmatrix} 1 \\ 2 \\ 2 \end{bmatrix}$. By Exercise 3, $A = U\Sigma V^T$ is a singular value decomposition of A, where

$$U = \begin{bmatrix} \frac{1}{3} & \frac{2}{\sqrt{5}} & \frac{2}{3\sqrt{5}} \\ \frac{2}{3} & \frac{-1}{\sqrt{5}} & \frac{4}{3\sqrt{5}} \\ \frac{2}{3} & 0 & \frac{-5}{3\sqrt{5}} \end{bmatrix}, \quad \Sigma = \begin{bmatrix} 3 \\ 0 \\ 0 \end{bmatrix},$$

and

$$V = [1].$$

Thus the pseudoinverse of A is

$$A^\dagger = V\Sigma^\dagger U^T = V\begin{bmatrix} \frac{1}{3} & 0 & 0 \end{bmatrix} U^T$$

$$= \frac{1}{9}[1 \ 2 \ 2].$$

41. A singular value decomposition of the given matrix A was obtained in Exercise 5. Using the matrices U, Σ, and V in the solution to that exercise, we find that the pseudoinverse of A is

$$A^\dagger = V\Sigma^\dagger U^T = \frac{1}{14}\begin{bmatrix} 4 & 8 & 2 \\ 1 & -5 & 4 \end{bmatrix}.$$

45. A singular value decomposition of the given matrix A was obtained in Exercise 13. Using the matrices U, Σ, and V in the solution to that exercise, we find that the pseudoinverse of A is

$$A^\dagger = V\Sigma^\dagger U^T = \frac{1}{14}\begin{bmatrix} 4 & -1 \\ -2 & 4 \\ 8 & 5 \end{bmatrix}.$$

49. Let $A = \begin{bmatrix} 1 & 0 \\ 0 & -1 \\ 1 & 1 \end{bmatrix}$. Using the method of Exercise 5, we see that $U\Sigma V^T$ is a singular value decomposition of A, where

$$U = \frac{1}{6}\begin{bmatrix} \sqrt{6} & -3\sqrt{2} & -2\sqrt{3} \\ -\sqrt{6} & -3\sqrt{2} & 2\sqrt{3} \\ 2\sqrt{6} & 0 & 2\sqrt{3} \end{bmatrix},$$

$$\Sigma = \begin{bmatrix} \sqrt{3} & 0 \\ 0 & 1 \\ 0 & 0 \end{bmatrix},$$

and

$$V = \frac{1}{\sqrt{2}}\begin{bmatrix} 1 & -1 \\ 1 & 1 \end{bmatrix}.$$

Since rank $A = 2$, let

$$D = \begin{bmatrix} 1 & 0 & 0 \\ 0 & 1 & 0 \\ 0 & 0 & 0 \end{bmatrix}.$$

Then using these matrices U and D, we find by equation (13) that

$$P_W = UDU^T = \frac{1}{3}\begin{bmatrix} 2 & 1 & 1 \\ 1 & 2 & -1 \\ 1 & -1 & 2 \end{bmatrix}.$$

53. In Exercise 5, we found a singular value decomposition of

$$A = \begin{bmatrix} 1 & 1 \\ 1 & -1 \\ 1 & 2 \end{bmatrix}.$$

Since rank $A = 2$, let

$$D = \begin{bmatrix} 1 & 0 & 0 \\ 0 & 1 & 0 \\ 0 & 0 & 0 \end{bmatrix}.$$

Using the matrix U in Exercise 5 and this matrix D, we find by equation (13) that

$$P_W = UDU^T = \frac{1}{14}\begin{bmatrix} 5 & 3 & 6 \\ 3 & 13 & -2 \\ 6 & -2 & 10 \end{bmatrix}.$$

55. False, σ^2 is an eigenvalue of A^TA.

56. True

57. False, see Example 7.

58. True

59. False, every matrix has a pseudoinverse.

60. False, $\mathcal{B}_1$ is an orthonormal basis of A^TA.

61. True   **62.** True

63. False, $\mathcal{B}_2$ is an orthonormal basis of AA^T.

64. True   **65.** True

66. False, if $A = U\Sigma V^T$ is a singular value decomposition of A, then $A = (-U)\Sigma(-V)^T$ is also a singular value decomposition of A.

67. False, only the nonzero diagonal entries are singular values.

68. True

69. False, $V\Sigma^T U^T$ is a singular value decomposition of A^T.

70. True   **71.** True

72. False, $\mathbf{u}$ is the unique vector of least norm that minimizes $\|A\mathbf{u} - \mathbf{b}\|$.

73. True

74. False, $A^\dagger = V\Sigma^\dagger U^T$.

75. True

77. (a) Let $\mathcal{B} = \{\mathbf{v}_1, \mathbf{v}_2, \ldots, \mathbf{v}_n\}$ be an orthonormal basis for $\mathcal{R}^n$ satisfying equation (9), and let $\mathbf{v}$ be in $\mathcal{R}^n$. Then

$$\mathbf{v} = a_1\mathbf{v}_1 + a_2\mathbf{v}_2 + \cdots + a_n\mathbf{v}_n$$

for some scalars $a_1, a_2, \ldots, a_n$. Thus

$$\begin{aligned}
\|A\mathbf{v}\|^2 &= (A\mathbf{v}) \cdot (A\mathbf{v}) \\
&= (A\mathbf{v})^T (A\mathbf{v}) \\
&= \mathbf{v}^T A^T A \mathbf{v} \\
&= \mathbf{v} \cdot (A^T A \mathbf{v}) \\
&= (a_1 \mathbf{v}_1 + \cdots + a_n \mathbf{v}_n) \cdot \\
&\quad (a_1 \sigma_1^2 \mathbf{v}_1 + \cdots + a_m \sigma_m^2 \mathbf{v}_m) \\
&= a_1^2 \sigma_1^2 + \cdots + a_m^2 \sigma_m^2 \\
&\leq a_1^2 \sigma_1^2 + \cdots + a_m^2 \sigma_1^2 \\
&\leq (a_1^2 + \cdots + a_n^2) \sigma_1^2 \\
&= \|\mathbf{v}\|^2 \sigma_1^2,
\end{aligned}$$

and hence $\|A\mathbf{v}\| \leq \sigma_1 \|\mathbf{v}\|$. The proof that $\sigma_m \|\mathbf{v}\| \leq \|A\mathbf{v}\|$ is similar.

(b) Let $\mathbf{v} = \mathbf{v}_m$ and $\mathbf{w} = \mathbf{v}_1$, where $\mathbf{v}_1$ and $\mathbf{v}_m$ are as in (a).

81. Since $\Sigma = I_m \Sigma I_n^T$ is a singular value decomposition of Σ, it follows that the pseudoinverse of Σ is $I_n \Sigma^\dagger I_m^T = \Sigma^\dagger$.

85. Suppose that A is a positive semidefinite matrix. Since A is symmetric, there exists an orthogonal matrix V and a diagonal matrix D such that $A = VDV^T$. Furthermore, the diagonal entries of D are the eigenvalues of A, and these are nonnegative by Exercise 60 of Section 6.6. Also, V and D can be chosen so that the diagonal entries are listed in decreasing order of absolute value. Since D has the form given in equation (11), we see that $A = VDV^T$ is a singular value decomposition of A.

Now suppose that A is not positive semidefinite. Then A has a negative eigenvalue. In any factorization of the form $A = V\Sigma V^T$, where V is an orthogonal matrix and Σ is of the form given in equation (11), Σ is a diagonal matrix whose diagonal entries are the eigenvalues of A. Since A has a negative eigenvalue, at least one of the diagonal entries of Σ is negative. This entry cannot be a singular value of A, and it follows that $A = V\Sigma V^T$ is not a singular value decomposition of A.

89. If Σ is an $m \times n$ matrix of the form in equation (11) and $\Sigma^\dagger$ is an $n \times m$ matrix of the form in equation (14), their product is the $m \times m$ diagonal matrix whose first k diagonal entries are $\sigma_i \cdot \dfrac{1}{\sigma_i} = 1$ and whose last $n - k$ diagonal entries are zero.

6.8 PRINCIPAL COMPONENT ANALYSIS

1. By definition, the mean of a set of m observations equals their sum divided by m; so $\bar{x} = \frac{1}{3}[2 - 3 + 4] = 1$.

5.
$$\begin{aligned}
\text{cov}(\mathbf{x}, \mathbf{y}) &= \frac{1}{2}[(2-1)(4-3) \\
&\quad + (-3-1)(2-3) \\
&\quad + (4-1)(3-3)] \\
&= \frac{1}{2}[1 + 4 + 0] = \frac{5}{2}
\end{aligned}$$

9. True

10. False, to obtain the variance, the sum should be divided by $m - 1$.

11. False, the covariance may be any real number.

12. False, the correlation may be any real number between -1 and 1, inclusively.

158 Chapter 6 Orthogonality

13. False, their correlation is either -1 or 1.

14. True

15. False, the covariance may be any real number.

16. True **17.** True **18.** True

19. True **20.** True

21.
$$\text{cov}(\mathbf{x}, \mathbf{y}) = \frac{1}{m-1}(\mathbf{x} - \overline{\mathbf{x}}) \cdot (\mathbf{y} - \overline{\mathbf{y}})$$
$$= \frac{1}{m-1}(\mathbf{y} - \overline{\mathbf{y}}) \cdot (\mathbf{x} - \overline{\mathbf{x}})$$
$$= \text{cov}(\mathbf{y}, \mathbf{x})$$

25. Suppose that $\text{cov}(\mathbf{x}, \mathbf{x}) = 0$. Then $(\mathbf{x} - \overline{\mathbf{x}}) \cdot (\mathbf{x} - \overline{\mathbf{x}}) = 0$. So $\mathbf{x} - \overline{\mathbf{x}} = \mathbf{0}$ or $\mathbf{x} = \overline{\mathbf{x}}$. It follows that $x_i = \overline{x}$ for all i. Now suppose that all the components of $\mathbf{x}$ are equal. Then the mean of $\mathbf{x}$ equals this common value, and so $\mathbf{x} = \overline{\mathbf{x}}$. Thus $\text{cov}(\mathbf{x}, \mathbf{x}) = 0$.

29. By Exercise 28(a), the variance of $c\mathbf{x}$ is
$$\frac{1}{m-1}(c\mathbf{x} - \overline{c\mathbf{x}}) \cdot (c\mathbf{x} - \overline{c\mathbf{x}})$$
$$= \frac{1}{m-1}(c\mathbf{x} - c\overline{\mathbf{x}}) \cdot (c\mathbf{x} - c\overline{\mathbf{x}})$$
$$= \frac{1}{m-1}[c(\mathbf{x} - \overline{\mathbf{x}})] \cdot [c(\mathbf{x} - \overline{\mathbf{x}})]$$
$$= c^2 \frac{1}{m-1}(\mathbf{x} - \overline{\mathbf{x}}) \cdot (\mathbf{x} - \overline{\mathbf{x}})$$
$$= c^2 s_{\mathbf{x}}^2.$$

33. We must show that $\frac{1}{s_{\mathbf{x}}}(\mathbf{x} - \overline{\mathbf{x}})$ has a mean of 0 and a standard deviation of 1. Its mean is
$$\frac{1}{m}\sum_{i=1}^{m}\frac{1}{s_{\mathbf{x}}}(x_i - \overline{x})$$
$$= \frac{1}{ms_{\mathbf{x}}}\left[\sum_{i=1}^{m}x_i - \sum_{i=1}^{m}\overline{x}\right]$$
$$= \frac{1}{ms_{\mathbf{x}}}[m\overline{x} - m\overline{x}] = 0.$$

Using Exercise 29, we see that the variance of $\frac{1}{s_{\mathbf{x}}}(\mathbf{x} - \overline{\mathbf{x}})$ is $\frac{1}{s_{\mathbf{x}}^2}$ times the variance of $(\mathbf{x} - \overline{\mathbf{x}})$. By Exercises 27 and 24, the variance of $\mathbf{x} - \overline{\mathbf{x}}$ is
$$\text{cov}(\mathbf{x} - \overline{\mathbf{x}}, \mathbf{x} - \overline{\mathbf{x}}) = \text{cov}(\mathbf{x}, \mathbf{x}) = s_{\mathbf{x}}^2.$$
So the variance of $\frac{1}{s_{\mathbf{x}}}(\mathbf{x} - \overline{\mathbf{x}})$ equals 1.

37. (a) By Exercises 27 and 22, we have
$$r = \frac{\text{cov}(\mathbf{x}, \mathbf{y})}{s_{\mathbf{x}} s_{\mathbf{y}}}$$
$$= \frac{\text{cov}(\mathbf{x} - \overline{\mathbf{x}}, \mathbf{y} - \overline{\mathbf{y}})}{s_{\mathbf{x}} s_{\mathbf{y}}}$$
$$= \text{cov}\left(\frac{\mathbf{x} - \overline{\mathbf{x}}}{s_{\mathbf{x}}}, \frac{\mathbf{y} - \overline{\mathbf{y}}}{s_{\mathbf{y}}}\right)$$
$$= \text{cov}(\mathbf{x}^*, \mathbf{y}^*).$$

(b) By Exercise 30 and (a) above, we have
$$0 \leq s_{\mathbf{x}^* \pm \mathbf{y}^*}^2$$
$$= s_{\mathbf{x}^*}^2 + s_{\mathbf{y}^*}^2 \pm 2\text{cov}(\mathbf{x}^*, \mathbf{y}^*)$$
$$= 2 \pm 2r.$$
So $\pm r \leq 1$, that is, $|r| \leq 1$.

6.9 ROTATIONS OF $\mathcal{R}^3$ AND COMPUTER GRAPHICS

3. We have

$$M = P_{90°} R_{45°}$$

$$= \begin{bmatrix} 1 & 0 & 0 \\ 0 & 0 & -1 \\ 0 & 1 & 0 \end{bmatrix} \begin{bmatrix} \frac{1}{\sqrt{2}} & -\frac{1}{\sqrt{2}} & 0 \\ \frac{1}{\sqrt{2}} & \frac{1}{\sqrt{2}} & 0 \\ 0 & 0 & 1 \end{bmatrix}$$

$$= \frac{1}{\sqrt{2}} \begin{bmatrix} 1 & -1 & 0 \\ 0 & 0 & -\sqrt{2} \\ 1 & 1 & 0 \end{bmatrix}.$$

7. Let $\mathbf{v}_3 = \frac{1}{\sqrt{2}} \begin{bmatrix} 1 \\ 0 \\ 1 \end{bmatrix}$. We must select nonzero vectors $\mathbf{w}_1$ and $\mathbf{w}_2$ so that $\mathbf{w}_1$, $\mathbf{w}_2$, and $\mathbf{v}_3$ form an orthogonal set and $\mathbf{w}_2$ lies in the direction of the counterclockwise rotation of $\mathbf{w}_1$ by 90° with respect to the orientation defined by $\mathbf{v}_3$. First choose $\mathbf{w}_1$ to be any nonzero vector orthogonal to $\mathbf{v}_3$, say $\mathbf{w}_1 = \begin{bmatrix} 0 \\ 1 \\ 0 \end{bmatrix}$. Then choose $\mathbf{w}_2$ to be a nonzero vector orthogonal to $\mathbf{w}_1$ and $\mathbf{v}_3$. Two possibilities are $\begin{bmatrix} 1 \\ 0 \\ -1 \end{bmatrix}$ and $\begin{bmatrix} -1 \\ 0 \\ 1 \end{bmatrix}$. Since

$$\det \begin{bmatrix} 0 & 1 & 1 \\ 1 & 0 & 0 \\ 0 & -1 & 1 \end{bmatrix} < 0$$

and

$$\det \begin{bmatrix} 0 & -1 & 1 \\ 1 & 0 & 0 \\ 0 & 1 & 1 \end{bmatrix} > 0,$$

we choose $\mathbf{w}_2 = \begin{bmatrix} -1 \\ 0 \\ 1 \end{bmatrix}$ so that the determinant of the matrix $[\mathbf{w}_1 \ \mathbf{w}_2 \ \mathbf{w}_3]$ is positive. (Once we replace $\mathbf{w}_2$ by a unit vector in the same direction, we can apply Theorem 6.20.) Now let $\mathbf{v}_1 = \mathbf{w}_1$,

$$\mathbf{v}_2 = \frac{1}{\|\mathbf{w}_2\|} \mathbf{w}_2 = \frac{1}{\sqrt{2}} \begin{bmatrix} -1 \\ 0 \\ 1 \end{bmatrix},$$

and

$$V = [\mathbf{v}_1 \ \mathbf{v}_2 \ \mathbf{v}_3] = \begin{bmatrix} 0 & -\frac{1}{\sqrt{2}} & \frac{1}{\sqrt{2}} \\ 1 & 0 & 0 \\ 0 & \frac{1}{\sqrt{2}} & \frac{1}{\sqrt{2}} \end{bmatrix}.$$

Since $R_{180°} = \begin{bmatrix} -1 & 0 & 0 \\ 0 & -1 & 0 \\ 0 & 0 & 1 \end{bmatrix}$, we have

$$P = V R_{180°} V^T = \begin{bmatrix} 0 & 0 & 1 \\ 0 & -1 & 0 \\ 1 & 0 & 0 \end{bmatrix}.$$

11. Let $\mathbf{v}_3 = \frac{1}{\sqrt{2}} \begin{bmatrix} 1 \\ -1 \\ 0 \end{bmatrix}$. As in Exercise 7, we select nonzero vectors $\mathbf{w}_1$ and $\mathbf{w}_2$ that are orthogonal to $\mathbf{v}_3$ and to each other so that

$$\det [\mathbf{w}_1 \ \mathbf{w}_2 \ \mathbf{v}_3] > 0.$$

We choose

$$\mathbf{w}_1 = \begin{bmatrix} 1 \\ 1 \\ 0 \end{bmatrix} \quad \text{and} \quad \mathbf{w}_2 = \begin{bmatrix} 0 \\ 0 \\ 1 \end{bmatrix}.$$

Next, set

$$\mathbf{v}_1 = \frac{1}{\|\mathbf{w}_1\|} \mathbf{w}_1 = \frac{1}{\sqrt{2}} \begin{bmatrix} 1 \\ 1 \\ 0 \end{bmatrix},$$

$\mathbf{v}_2 = \mathbf{w}_2$, and

$$V = [\mathbf{v}_1 \ \mathbf{v}_2 \ \mathbf{v}_3] = \begin{bmatrix} \frac{1}{\sqrt{2}} & 0 & \frac{1}{\sqrt{2}} \\ \frac{1}{\sqrt{2}} & 0 & -\frac{1}{\sqrt{2}} \\ 0 & 1 & 0 \end{bmatrix}.$$

Since $R_{30°} = \begin{bmatrix} \frac{\sqrt{3}}{2} & -\frac{1}{2} & 0 \\ \frac{1}{2} & \frac{\sqrt{3}}{2} & 0 \\ 0 & 0 & 1 \end{bmatrix}$, we have

$$P = V R_{30°} V^T$$

$$= \frac{1}{4} \begin{bmatrix} \sqrt{3}+2 & \sqrt{3}-2 & -\sqrt{2} \\ \sqrt{3}-2 & \sqrt{3}+2 & -\sqrt{2} \\ \sqrt{2} & \sqrt{2} & 2\sqrt{3} \end{bmatrix}.$$

15. The rotation matrix described in Exercise 1 is

$$M = \begin{bmatrix} 0 & 1 & 0 \\ 0 & 0 & -1 \\ -1 & 0 & 0 \end{bmatrix}.$$

(a) The axis of rotation is the span of an eigenvector of M corresponding to the eigenvalue 1, and hence we seek a nonzero solution of $(M - I_3)\mathbf{x} = \mathbf{0}$. One such vector is $\begin{bmatrix} -1 \\ -1 \\ 1 \end{bmatrix}$. So the axis of rotation is

$$\text{Span}\left\{ \begin{bmatrix} -1 \\ -1 \\ 1 \end{bmatrix} \right\}.$$

(b) Choose any nonzero vector that is orthogonal to the vector in (a), for example,

$$\mathbf{w} = \begin{bmatrix} 1 \\ 0 \\ 1 \end{bmatrix},$$

and let α be the angle between $\mathbf{w}$ and $M\mathbf{w}$. Notice that $\|M\mathbf{w}\| = \|\mathbf{w}\|$ because M is an orthogonal matrix. Therefore by Exercise 98 in Section 6.1,

$$\cos \alpha = \frac{M\mathbf{w} \cdot \mathbf{w}}{\|M\mathbf{w}\|\|\mathbf{w}\|} = \frac{M\mathbf{w} \cdot \mathbf{w}}{\|\mathbf{w}\|\|\mathbf{w}\|}$$

$$= \frac{\begin{bmatrix} 0 \\ -1 \\ -1 \end{bmatrix} \cdot \begin{bmatrix} 1 \\ 0 \\ 1 \end{bmatrix}}{\left\| \begin{bmatrix} 1 \\ 0 \\ 1 \end{bmatrix} \right\|^2} = -\frac{1}{2}.$$

19. The rotation matrix described in Exercise 5 is

$$M = \frac{1}{4} \begin{bmatrix} 2\sqrt{3} & 0 & 2 \\ 1 & 2\sqrt{3} & -\sqrt{3} \\ -\sqrt{3} & 2 & 3 \end{bmatrix}.$$

(a) The axis of rotation is the span of an eigenvector of M corresponding to the eigenvalue 1, and hence we seek a nonzero solution of $(M - I_3)\mathbf{x} = \mathbf{0}$. One such vector is $\begin{bmatrix} \sqrt{3}+2 \\ \sqrt{3}+2 \\ 1 \end{bmatrix}$. So the axis of rotation is

$$\text{Span}\left\{ \begin{bmatrix} \sqrt{3}+2 \\ \sqrt{3}+2 \\ 1 \end{bmatrix} \right\}.$$

(b) Choose any nonzero vector that is orthogonal to the vector in (a), for example,

$$\mathbf{w} = \begin{bmatrix} -1 \\ 1 \\ 0 \end{bmatrix},$$

and let α be the angle between $\mathbf{w}$ and $M\mathbf{w}$. Because $\|M\mathbf{w}\| = \|\mathbf{w}\|$, we have, by Exercise 98 in Section 6.1,

$$\cos \alpha = \frac{M\mathbf{w} \cdot \mathbf{w}}{\|M\mathbf{w}\|\|\mathbf{w}\|} = \frac{M\mathbf{w} \cdot \mathbf{w}}{\|\mathbf{w}\|\|\mathbf{w}\|}$$

$$= \frac{\frac{1}{4}\begin{bmatrix} -2\sqrt{3} \\ 2\sqrt{3}-1 \\ 2+\sqrt{3} \end{bmatrix} \cdot \begin{bmatrix} -1 \\ 1 \\ 0 \end{bmatrix}}{\left\|\begin{bmatrix} -1 \\ 1 \\ 0 \end{bmatrix}\right\|^2}$$

$$= \frac{4\sqrt{3}-1}{8}.$$

23. Let A be the standard matrix of T_W. Choose a nonzero vector orthogonal to $\begin{bmatrix} 1 \\ 2 \\ 3 \end{bmatrix}$ and $\begin{bmatrix} 1 \\ 0 \\ -1 \end{bmatrix}$, for example, $\begin{bmatrix} 1 \\ -2 \\ 1 \end{bmatrix}$, and let

$$\mathcal{B} = \left\{ \begin{bmatrix} 1 \\ 2 \\ 3 \end{bmatrix}, \begin{bmatrix} 1 \\ 0 \\ -1 \end{bmatrix}, \begin{bmatrix} 1 \\ -2 \\ 1 \end{bmatrix} \right\}.$$

Then $\mathcal{B}$ is a basis for $\mathcal{R}^3$, and

$$T_W\left(\begin{bmatrix} 1 \\ 2 \\ 3 \end{bmatrix}\right) = \begin{bmatrix} 1 \\ 2 \\ 3 \end{bmatrix},$$

$$T_W\left(\begin{bmatrix} 1 \\ 0 \\ -1 \end{bmatrix}\right) = \begin{bmatrix} 1 \\ 0 \\ -1 \end{bmatrix},$$

and

$$T_W\left(\begin{bmatrix} 1 \\ -2 \\ 1 \end{bmatrix}\right) = -\begin{bmatrix} 1 \\ -2 \\ 1 \end{bmatrix}.$$

Let B be the matrix whose columns are the vectors in $\mathcal{B}$. Then

$$[T_W]_\mathcal{B} = \begin{bmatrix} 1 & 0 & 0 \\ 0 & 1 & 0 \\ 0 & 0 & -1 \end{bmatrix},$$

and therefore, by Theorem 4.12,

$$A = B[T_W]_\mathcal{B} B^{-1} = \frac{1}{3}\begin{bmatrix} 2 & 2 & -1 \\ 2 & -1 & 2 \\ -1 & 2 & 2 \end{bmatrix}.$$

27. Let A be the standard matrix of the reflection of $\mathcal{R}^3$ about the plane W with equation $x + 2y - 2z = 0$. A basis for W is

$$\left\{ \begin{bmatrix} -2 \\ 1 \\ 0 \end{bmatrix}, \begin{bmatrix} 2 \\ 0 \\ 1 \end{bmatrix} \right\}.$$

The vector $\begin{bmatrix} 1 \\ 2 \\ -2 \end{bmatrix}$, whose components are the coefficients of the variables in the equation of W, is orthogonal to W. Thus

$$\mathcal{B} = \left\{ \begin{bmatrix} -2 \\ 1 \\ 0 \end{bmatrix}, \begin{bmatrix} 2 \\ 0 \\ 1 \end{bmatrix}, \begin{bmatrix} 1 \\ 2 \\ -2 \end{bmatrix} \right\}.$$

is a basis for $\mathcal{R}^3$, and

$$T_W\left(\begin{bmatrix} -2 \\ 1 \\ 0 \end{bmatrix}\right) = \begin{bmatrix} -2 \\ 1 \\ 0 \end{bmatrix},$$

$$T_W\left(\begin{bmatrix} 2 \\ 0 \\ 1 \end{bmatrix}\right) = \begin{bmatrix} 2 \\ 0 \\ 1 \end{bmatrix},$$

and

$$T_W\left(\begin{bmatrix} 1 \\ 2 \\ -2 \end{bmatrix}\right) = -\begin{bmatrix} 1 \\ 2 \\ -2 \end{bmatrix}.$$

Let B be the matrix whose columns are the vectors in $\mathcal{B}$. Then

$$[T]_\mathcal{B} = \begin{bmatrix} 1 & 0 & 0 \\ 0 & 1 & 0 \\ 0 & 0 & -1 \end{bmatrix},$$

and therefore, by Theorem 4.12,

$$A = B[T]_B B^{-1} = \frac{1}{9}\begin{bmatrix} 7 & -4 & -4 \\ -4 & 1 & 8 \\ 4 & 8 & 1 \end{bmatrix}.$$

31. First, obtain a basis $\{\mathbf{w}_1, \mathbf{w}_2\}$ for W by selecting two linearly independent vectors that are orthogonal to $\mathbf{v}$ such as

$$\mathbf{w}_1 = \begin{bmatrix} 1 \\ 0 \\ 1 \end{bmatrix} \quad \text{and} \quad \mathbf{w}_2 = \begin{bmatrix} 2 \\ -1 \\ 0 \end{bmatrix}.$$

Although we could proceed as in Exercise 23, we will use an alternate approach.

Let A be the standard matrix of T_W. Then

$$A \begin{bmatrix} 1 & 2 & 1 \\ 0 & -1 & 2 \\ 1 & 0 & -1 \end{bmatrix}$$

$$= \begin{bmatrix} A \begin{bmatrix} 1 \\ 0 \\ 1 \end{bmatrix} & A \begin{bmatrix} 2 \\ -1 \\ 0 \end{bmatrix} & A \begin{bmatrix} 1 \\ 2 \\ -1 \end{bmatrix} \end{bmatrix}$$

$$= \begin{bmatrix} 1 & 2 & -1 \\ 0 & -1 & -2 \\ 1 & 0 & 1 \end{bmatrix},$$

and therefore

$$A = \begin{bmatrix} 1 & 2 & -1 \\ 0 & -1 & -2 \\ 1 & 0 & 1 \end{bmatrix} \begin{bmatrix} 1 & 2 & 1 \\ 0 & -1 & 2 \\ 1 & 0 & -1 \end{bmatrix}^{-1}$$

$$= \frac{1}{3}\begin{bmatrix} 2 & -2 & 1 \\ -2 & -1 & 2 \\ 1 & 2 & 2 \end{bmatrix}.$$

35. As in Exercise 31, we begin by finding two vectors that form a basis for W:

$$\mathbf{w}_1 = \begin{bmatrix} -4 \\ 3 \\ 0 \end{bmatrix} \quad \text{and} \quad \mathbf{w}_2 = \begin{bmatrix} -5 \\ 0 \\ 3 \end{bmatrix}.$$

Let A be the standard matrix of T_W. Then, as in Exercise 31, we have

$$A = [\mathbf{w}_1 \ \mathbf{w}_2 \ -\mathbf{v}][\mathbf{w}_1 \ \mathbf{w}_2 \ \mathbf{v}]^{-1}$$

$$= \frac{1}{25}\begin{bmatrix} 16 & -12 & -15 \\ -12 & 9 & -20 \\ -15 & -20 & 0 \end{bmatrix}.$$

39. The given matrix does not have 1 as an eigenvalue. Therefore it is neither a rotation matrix nor the standard matrix of a reflection operator, both of which have 1 as an eigenvalue.

41. (a) Since

$$\det \begin{bmatrix} 1 & 0 & 0 \\ 0 & -1 & 0 \\ 0 & 0 & -1 \end{bmatrix} = 1,$$

the matrix is a rotation matrix by Theorem 6.20.

(b) Observe that $\begin{bmatrix} 1 \\ 0 \\ 0 \end{bmatrix}$ is an eigenvector of the matrix corresponding to eigenvalue 1, and therefore this vector forms a basis for the axis of rotation.

45. Let M denote the given matrix.

(a) Since $\det M = -1$, M is not a rotation matrix by Theorem 6.20. We can establish that M is the standard matrix of a reflection by showing that M has a 2-dimensional eigenspace corresponding to eigenvalue 1. For, in this case, it must have a third eigenvector corresponding to eigenvalue -1 because its determinant equals the product of its eigenvalues. The reduced row echelon form of $M - I_3$

is

$$\begin{bmatrix} 1 & 0 & -(1+\sqrt{2}) \\ 0 & 0 & 0 \\ 0 & 0 & 0 \end{bmatrix},$$

and hence the eigenspace corresponding to eigenvalue 1 is 2-dimensional. Therefore M is the standard matrix of a reflection.

(b) The matrix equation $(M - I_2)\mathbf{x} = \mathbf{0}$ is the system

$$x_1 - (1 + \sqrt{2})x_3 = 0,$$

and hence the vector form of its general solution is

$$\begin{bmatrix} x_1 \\ x_2 \\ x_3 \end{bmatrix} = x_2 \begin{bmatrix} 0 \\ 1 \\ 0 \end{bmatrix} + x_3 \begin{bmatrix} 1+\sqrt{2} \\ 0 \\ 1 \end{bmatrix}.$$

It follows that

$$\left\{ \begin{bmatrix} 0 \\ 1 \\ 0 \end{bmatrix}, \begin{bmatrix} 1+\sqrt{2} \\ 0 \\ 1 \end{bmatrix} \right\}$$

is a basis for the 2-dimensional subspace about which $\mathcal{R}^3$ is reflected.

47. False, consider $P = \begin{bmatrix} -1 & 0 & 0 \\ 0 & 1 & 0 \\ 0 & 0 & 1 \end{bmatrix}$.

48. False, let P be the matrix in the solution to Exercise 47.

49. False, let $Q = \begin{bmatrix} -1 & 0 & 0 \\ 0 & 0 & -1 \\ 0 & 1 & 0 \end{bmatrix}$.

50. False, consider I_3.

51. True

52. False, consider the matrix Q in the solution to Exercise 49.

53. True   54. True   55. True

56. False, for example, if $\phi = \theta = 90°$, then

$$Q_\phi R_\theta = \begin{bmatrix} 0 & 0 & 1 \\ 1 & 0 & 0 \\ 0 & 1 & 0 \end{bmatrix},$$

but

$$R_\theta Q_\phi = \begin{bmatrix} 0 & -1 & 0 \\ 0 & 0 & 1 \\ -1 & 0 & 0 \end{bmatrix}.$$

57. True   58. True

59. False, the matrix is

$$Q_\theta = \begin{bmatrix} \cos\theta & 0 & \sin\theta \\ 0 & 1 & 0 \\ -\sin\theta & 0 & \cos\theta \end{bmatrix}.$$

60. True   61. True   62. True

63. False, the rotation, as viewed from $\mathbf{v}_3$, is counterclockwise.

64. False, the determinant is equal to -1.

65. False, the eigenvector corresponds to the eigenvalue 1.

66. False, any nonzero solution of the matrix equation $(P_\theta - R_\phi^T)\mathbf{x} = \mathbf{0}$ forms a basis for the axis of rotation.

67. True

69. We have

$$\det P_\theta = \det \begin{bmatrix} 1 & 0 & 0 \\ 0 & \cos\theta & -\sin\theta \\ 0 & \sin\theta & \cos\theta \end{bmatrix}$$

$$= \det \begin{bmatrix} \cos\theta & -\sin\theta \\ \sin\theta & \cos\theta \end{bmatrix}$$

$$= \cos^2\theta + \sin^2\theta = 1.$$

The other determinants are computed in a similar manner.

73. (a) Clearly, $T_W(\mathbf{w}) = \mathbf{w}$ for all $\mathbf{w}$ in W, and hence 1 is an eigenvalue of T_W. Let Z denote the eigenspace corresponding to eigenvalue 1. Then W is contained in Z. Observe that $\dim Z < 3$, for otherwise T_W would be the identity transformation, which it is not. Since $\dim W = 2$, it follows that $W = Z$.

    (b) Since $T_W(\mathbf{z}) = -\mathbf{z}$ for all $\mathbf{z}$ in $W^\perp$, -1 is an eigenvalue of T_W and $W^\perp$ is contained in the corresponding eigenspace. Because the eigenspace corresponding to 1 has dimension 2, the eigenspace corresponding to -1 has dimension 1. But $\dim W^\perp = 1$, and hence these two subspaces are equal.

77. By Exercise 72, B and C are orthogonal matrices, and hence BC is an orthogonal matrix by Theorem 6.10. In addition, $\det B = \det C = -1$ by Exercise 75. So

    $$\det BC = (\det B)(\det C)$$
    $$= (-1)(-1) = 1,$$

    and hence BC is a rotation matrix by Theorem 6.20.

81. Let $Q = CB^{-1}$. Then

    $$[Q\mathbf{v}_1 \ Q\mathbf{v}_2 \ Q\mathbf{v}_3] = QB = CB^{-1}B$$
    $$= C = [\mathbf{v}_1 \ \mathbf{v}_2 \ -\mathbf{v}_3],$$

    and hence $Q\mathbf{v}_1 = \mathbf{v}_1$, $Q\mathbf{v}_2 = \mathbf{v}_2$, and $Q\mathbf{v}_3 = -\mathbf{v}_3$. Since $\{\mathbf{v}_1, \mathbf{v}_2\}$ and $\{\mathbf{v}_3\}$ are bases for W and $W^\perp$, respectively, we have $Q\mathbf{v} = \mathbf{v}$ for every vector $\mathbf{v}$ in W, and $Q\mathbf{v} = -\mathbf{v}$ for every vector $\mathbf{v}$ in $W^\perp$. Therefore $Q = CB^{-1}$ is the standard matrix of the reflection of $\mathcal{R}^3$ about W.

CHAPTER 6 REVIEW

1. True
2. True
3. False, the vectors must belong to $\mathcal{R}^n$ for some n.
4. True
5. True
6. True
7. True
8. True
9. False, if W is a 1-dimensional subspace of $\mathcal{R}^3$, then $\dim W^\perp = 2$.
10. False, I_n is an invertible orthogonal projection matrix.
11. True
12. False, let W be the x-axis in $\mathcal{R}^2$, and let $\mathbf{v} = \begin{bmatrix} 1 \\ 2 \end{bmatrix}$. Then $\mathbf{w} = \begin{bmatrix} 1 \\ 0 \end{bmatrix}$, which is not orthogonal to $\mathbf{v}$.
13. False, the least-squares line minimizes the sum of the squares of the *vertical distances* from the data points to the line.
14. False, in addition, each column must have length equal to 1.
15. False, consider $\begin{bmatrix} 1 & 1 \\ 1 & 2 \end{bmatrix}$, which has determinant 1 but is not an orthogonal matrix.
16. True
17. True
18. False, only symmetric matrices have spectral decompositions.
19. True
21. (a) $\|\mathbf{u}\| = \sqrt{3^2 + (-6)^2} = \sqrt{45}$,
    $\|\mathbf{v}\| = \sqrt{4^2 + 2^2} = \sqrt{20}$
    (b) $d = \|\mathbf{u} - \mathbf{v}\| = \left\| \begin{bmatrix} -1 \\ -8 \end{bmatrix} \right\| = \sqrt{65}$
    (c) $\mathbf{u} \cdot \mathbf{v} = 0$

(d) The vectors are orthogonal because their dot product equals 0.

25. Choose a vector $\begin{bmatrix} 1 \\ -2 \end{bmatrix}$ on $\mathcal{L}$ and proceed as in Example 3 on page 366 to obtain $\mathbf{w} = \frac{1}{5}\begin{bmatrix} -1 \\ 2 \end{bmatrix}$. The distance is

$$d = \left\| \begin{bmatrix} 3 \\ 2 \end{bmatrix} - \frac{(-1)}{5}\begin{bmatrix} 1 \\ -2 \end{bmatrix} \right\| = 1.6\sqrt{5}.$$

27.
$$(2\mathbf{u} + 3\mathbf{v}) \cdot \mathbf{w} = 2(\mathbf{u} \cdot \mathbf{w}) + 3(\mathbf{v} \cdot \mathbf{w})$$
$$= 2(5) + 3(-3) = 1$$

31. The dot product of no pair of vectors in S is 0. Thus S is not orthogonal.

Let $\mathbf{u}_1$, $\mathbf{u}_2$, and $\mathbf{u}_3$ denote the vectors in S, listed in the same order. Define $\mathbf{v}_1 = \mathbf{u}_1$,

$$\mathbf{v}_2 = \mathbf{u}_2 - \frac{\mathbf{u}_2 \cdot \mathbf{v}_1}{\|\mathbf{v}_1\|^2}\mathbf{v}_1 = \begin{bmatrix} 0 \\ 0 \\ 1 \\ 1 \end{bmatrix} - \frac{-1}{3}\begin{bmatrix} 1 \\ 1 \\ -1 \\ 0 \end{bmatrix}$$

$$= \frac{1}{3}\begin{bmatrix} 1 \\ 1 \\ 2 \\ 3 \end{bmatrix},$$

and

$$\mathbf{v}_3 = \mathbf{u}_3 - \frac{\mathbf{u}_3 \cdot \mathbf{v}_1}{\|\mathbf{v}_1\|^2}\mathbf{v}_1 - \frac{\mathbf{u}_3 \cdot \mathbf{v}_2}{\|\mathbf{v}_2\|^2}\mathbf{v}_2$$

$$= \begin{bmatrix} 1 \\ 2 \\ 0 \\ 1 \end{bmatrix} - \frac{3}{3}\begin{bmatrix} 1 \\ 1 \\ -1 \\ 0 \end{bmatrix} - \left(\frac{2}{\frac{15}{9}}\right)\left(\frac{1}{3}\right)\begin{bmatrix} 1 \\ 1 \\ 2 \\ 3 \end{bmatrix}$$

$$= \frac{1}{5}\begin{bmatrix} -2 \\ 3 \\ 1 \\ -1 \end{bmatrix}.$$

Therefore an orthogonal basis for S is

$$\left\{ \begin{bmatrix} 1 \\ 1 \\ -1 \\ 0 \end{bmatrix}, \frac{1}{3}\begin{bmatrix} 1 \\ 1 \\ 2 \\ 3 \end{bmatrix}, \frac{1}{5}\begin{bmatrix} -2 \\ 3 \\ 1 \\ -1 \end{bmatrix} \right\}.$$

35. We have

$$\mathbf{w} = (\mathbf{v} \cdot \mathbf{v}_1)\mathbf{v}_1 + (\mathbf{v} \cdot \mathbf{v}_2)\mathbf{v}_2$$

$$= \frac{5}{\sqrt{5}}\frac{1}{\sqrt{5}}\begin{bmatrix} 1 \\ 2 \\ 0 \end{bmatrix} + \frac{-9}{\sqrt{14}}\frac{1}{\sqrt{14}}\begin{bmatrix} -2 \\ 1 \\ 3 \end{bmatrix}$$

$$= \frac{1}{14}\begin{bmatrix} 32 \\ 19 \\ -27 \end{bmatrix}$$

and

$$\mathbf{z} = \mathbf{v} - \mathbf{w}$$

$$= \begin{bmatrix} 1 \\ 2 \\ -3 \end{bmatrix} - \frac{1}{14}\begin{bmatrix} 32 \\ 19 \\ -27 \end{bmatrix} = \frac{1}{14}\begin{bmatrix} -18 \\ 9 \\ -15 \end{bmatrix}.$$

The distance from $\mathbf{v}$ to W is

$$\|\mathbf{z}\| = \frac{3}{14}\sqrt{70}.$$

39. A vector is in W if and only if it is orthogonal to both of the vectors in the given set, that is, if and only if it is a solution of the system

$$\begin{aligned} x_1 - x_2 &= 0 \\ x_1 \quad\quad + x_3 &= 0. \end{aligned}$$

Thus a basis for W is

$$\left\{ \begin{bmatrix} -1 \\ -1 \\ 1 \\ 0 \end{bmatrix}, \begin{bmatrix} 0 \\ 0 \\ 0 \\ 1 \end{bmatrix} \right\}.$$

Let C be the matrix whose columns are the vectors in this basis. Then

$$P_W = C(C^TC)^{-1}C^T$$
$$= \frac{1}{3}\begin{bmatrix} 1 & 1 & -1 & 0 \\ 1 & 1 & -1 & 0 \\ -1 & -1 & 1 & 0 \\ 0 & 0 & 0 & 3 \end{bmatrix},$$

and the vector $\mathbf{w}$ in W closest to $\mathbf{v}$ is

$$\mathbf{w} = P_W\mathbf{v} = \begin{bmatrix} 0 \\ 0 \\ 0 \\ 2 \end{bmatrix}.$$

43. Let A denote the given matrix. Then

$$A^TA = \begin{bmatrix} 0.58 & 0.00 \\ 0.00 & 0.58 \end{bmatrix} \neq \begin{bmatrix} 1 & 0 \\ 0 & 1 \end{bmatrix}.$$

Therefore the matrix is not orthogonal by Theorem 6.9.

47. Since

$$\det \frac{1}{2}\begin{bmatrix} 1 & \sqrt{3} \\ -\sqrt{3} & 1 \end{bmatrix} = \frac{1}{4}(1+3) = 1,$$

the matrix is a rotation matrix. Comparing the first column of this matrix with the first column of the rotation matrix A_θ, we see that

$$A_\theta = \begin{bmatrix} \cos\theta & -\sin\theta \\ \sin\theta & \cos\theta \end{bmatrix} = \begin{bmatrix} \frac{1}{2} & \frac{\sqrt{3}}{2} \\ -\frac{\sqrt{3}}{2} & \frac{1}{2} \end{bmatrix},$$

and hence $\cos\theta = \frac{1}{2}$ and $\sin\theta = -\frac{\sqrt{3}}{2}$. Therefore $\theta = -60°$.

51. The standard matrix of T is

$$Q = \begin{bmatrix} 0 & -1 & 0 \\ 0 & 0 & 1 \\ 1 & 0 & 0 \end{bmatrix}.$$

Since $Q^TQ = I_3$, we see that Q is an orthogonal matrix. Thus T is an orthogonal operator.

55. Let $a_{11} = 1$, $a_{22} = 1$, and $a_{12} = a_{21} = \frac{1}{2}(6) = 3$, so that

$$A = \begin{bmatrix} 1 & 3 \\ 3 & 1 \end{bmatrix}.$$

The eigenvalues of A are 4 and -2, and

$$\left\{\frac{1}{\sqrt{2}}\begin{bmatrix} 1 \\ 1 \end{bmatrix}, \frac{1}{\sqrt{2}}\begin{bmatrix} -1 \\ 1 \end{bmatrix}\right\}$$

is an orthonormal basis of $\mathcal{R}^2$ consisting of corresponding eigenvectors. The first of these basis vectors has both positive components, and so we construct the rotation matrix with it as its first column

$$P = \frac{1}{\sqrt{2}}\begin{bmatrix} 1 & -1 \\ 1 & 1 \end{bmatrix}.$$

This matrix is the rotation matrix corresponding to an angle of

$$\cos^{-1}\left(\frac{1}{\sqrt{2}}\right) = 45°.$$

Thus if we rotate the x- and y-axes by $45°$, the original equation becomes $4(x')^2 - 2(y')^2 = 16$. (The coefficients of $(x')^2$ and $(y')^2$ are the eigenvalues corresponding to the first and second columns of P.) The new equation can be written as

$$\frac{(x')^2}{4} - \frac{(y')^2}{8} = 1,$$

and so the conic section is a hyperbola.

CHAPTER 6 MATLAB EXERCISES

3. Answers are given correct to 4 places after the decimal point.

   (a) The following vectors form an orthonormal basis for W:

$$\mathbf{v}_1 = \begin{bmatrix} -0.1994 \\ 0.1481 \\ -0.1361 \\ -0.6282 \\ -0.5316 \\ 0.4924 \end{bmatrix},$$

$$\mathbf{v}_2 = \begin{bmatrix} 0.1153 \\ 0.0919 \\ -0.5766 \\ 0.6366 \\ -0.4565 \\ 0.1790 \end{bmatrix},$$

and

$$\mathbf{v}_3 = \begin{bmatrix} 0.3639 \\ -0.5693 \\ 0.5469 \\ 0.1493 \\ -0.4271 \\ 0.1992 \end{bmatrix}.$$

(b) For each of the given vectors $\mathbf{u}$, the orthogonal projection of $\mathbf{u}$ on W is given by the boxed formula on page 376:

$$(\mathbf{u} \cdot \mathbf{v}_1)\mathbf{v}_1 + (\mathbf{u} \cdot \mathbf{v}_2)\mathbf{v}_2 + (\mathbf{u} \cdot \mathbf{v}_3)\mathbf{v}_3.$$

These computations produce the following vectors:

(i) $\begin{bmatrix} 1.3980 \\ -1.5378 \\ 1.4692 \\ 2.7504 \\ 1.4490 \\ -1.6574 \end{bmatrix}$ (ii) $\begin{bmatrix} 1 \\ -2 \\ 2 \\ -1 \\ -3 \\ 2 \end{bmatrix}$

(iii) $\begin{bmatrix} 0 \\ 0 \\ 0 \\ 0 \\ 0 \\ 0 \end{bmatrix}$

(c) The vectors are the same.

(d) If M is a matrix whose columns form an orthonormal basis for a subspace W of $\mathcal{R}^n$, then by Theorems 6.8 and 6.9(b),

$$P_W = M(M^T M)^{-1} M^T$$
$$= M(I_n)^{-1} M^T = M M^T.$$

So MM^T is the orthogonal projection matrix for W.

7. Answers are given correct to 4 places after the decimal point.

(a) $P_W = [B \ C]$, where

$$B = \begin{bmatrix} 0.3913 & 0.0730 & -0.1763 \\ 0.0730 & 0.7180 & -0.1688 \\ -0.1763 & -0.1688 & 0.8170 \\ -0.2716 & -0.1481 & -0.2042 \\ 0.2056 & 0.1328 & 0.1690 \\ -0.2929 & 0.3593 & 0.1405 \end{bmatrix}$$

and $C =$

$$\begin{bmatrix} -0.2716 & 0.2056 & -0.2929 \\ -0.1481 & 0.1328 & 0.3593 \\ -0.2042 & 0.1690 & 0.1405 \\ 0.7594 & 0.1958 & 0.0836 \\ 0.1958 & 0.8398 & -0.0879 \\ 0.0836 & -0.0879 & 0.4744 \end{bmatrix}$$

(b) same as (a)

(c) $P_W \mathbf{v} = \mathbf{v}$ for all $\mathbf{v}$ in $\mathcal{S}$.

(d) Let A be the matrix whose columns are the vectors in $\mathcal{S}$. By Exercise 61 of Section 6.3,

$$W^\perp = (\text{Col } A)^\perp = \text{Null } A^T.$$

As described in Table D.2 of Appendix D, the MATLAB command $\texttt{null}(A^T, 'r')$ yields a matrix whose columns form a basis for $W^\perp = \text{Null } A^T$ having ration-al number entries. The resulting basis is

$$\left\{\begin{bmatrix} -1.75 \\ -0.50 \\ -1.00 \\ -1.25 \\ 1.00 \\ 0.00 \end{bmatrix}, \begin{bmatrix} 0.85 \\ -0.60 \\ -0.10 \\ 0.05 \\ 0.00 \\ 1.00 \end{bmatrix}\right\}.$$

In each case, $P_W \mathbf{v} = \mathbf{0}$.

11. Answers are given correct to 4 places after the decimal point.

   (a) Since A is symmetric, we can use the command $[P\ D] = \text{eig}(A)$ to determine the matrices P and D:

   $$P = \begin{bmatrix} -0.5 & -0.5477 & -0.5 & c & 0.0000 \\ 0.5 & -0.5477 & 0.5 & c & 0.0000 \\ -0.5 & 0.3651 & 0.5 & c & 0.4082 \\ 0.0 & 0.3651 & 0.0 & c & -0.8165 \\ 0.5 & 0.3651 & -0.5 & c & 0.4082 \end{bmatrix},$$

   where $c = -0.4472$, and

   $$D = \begin{bmatrix} -4 & 0 & 0 & 0 & 0 \\ 0 & 0 & 0 & 0 & 0 \\ 0 & 0 & -8 & 0 & 0 \\ 0 & 0 & 0 & 5 & 0 \\ 0 & 0 & 0 & 0 & 12 \end{bmatrix}$$

   (b) The columns of P form an orthonormal basis of $\mathcal{R}^5$ consisting of eigenvectors of A, and the diagonal entries of D are the corresponding eigenvalues (in the same order).

   (c) The spectral decomposition of A is a sum of terms of the form $\lambda_j P_j$, where λ_j is the jth diagonal entry of D (that is, the jth eigenvalue obtained in (b)) and $P_j = \mathbf{u}_j \mathbf{u}_j^T$, where $\mathbf{u}_j$ is the jth column of P (which is the jth eigenvector in the basis obtained in (b)). For each j, we can compute $\mathbf{u}_j$ by entering the MATLAB expression $P(:,j)$, as described in Table D.4 of Appendix D. The resulting spectral decomposition is

   $A =$

   $$-4 \begin{bmatrix} 0.25 & -0.25 & 0.25 & 0 & -0.25 \\ -0.25 & 0.25 & -0.25 & 0 & 0.25 \\ 0.25 & -0.25 & 0.25 & 0 & -0.25 \\ 0.00 & 0.00 & 0.00 & 0 & 0.00 \\ -0.25 & 0.25 & -0.25 & 0 & 0.25 \end{bmatrix}$$

   $$+ 0 \begin{bmatrix} .3 & .3 & -.2000 & -.2000 & -.2000 \\ .3 & .3 & -.2000 & -.2000 & -.2000 \\ -.2 & -.2 & .1333 & .1333 & .1333 \\ -.2 & -.2 & .1333 & .1333 & .1333 \\ -.2 & -.2 & .1333 & .1333 & .1333 \end{bmatrix}$$

   $$- 8 \begin{bmatrix} 0.25 & -0.25 & -0.25 & 0 & 0.25 \\ -0.25 & 0.25 & 0.25 & 0 & -0.25 \\ -0.25 & 0.25 & 0.25 & 0 & -0.25 \\ 0.00 & 0.00 & 0.00 & 0 & 0.00 \\ 0.25 & -0.25 & -0.25 & 0 & 0.25 \end{bmatrix}$$

   $$+ 5 \begin{bmatrix} 0.2 & 0.2 & 0.2 & 0.2 & 0.2 \\ 0.2 & 0.2 & 0.2 & 0.2 & 0.2 \\ 0.2 & 0.2 & 0.2 & 0.2 & 0.2 \\ 0.2 & 0.2 & 0.2 & 0.2 & 0.2 \\ 0.2 & 0.2 & 0.2 & 0.2 & 0.2 \end{bmatrix}$$

   $$+ 12 \begin{bmatrix} 0 & 0 & 0.0000 & 0.0000 & 0.0000 \\ 0 & 0 & 0.0000 & 0.0000 & 0.0000 \\ 0 & 0 & 0.1667 & -0.3333 & 0.1667 \\ 0 & 0 & -0.3333 & 0.6667 & -0.3333 \\ 0 & 0 & 0.1667 & -0.3333 & 0.1667 \end{bmatrix}$$

   (d) As on pages 433–435, we obtain

   $$A_2 = \begin{bmatrix} -2 & 2 & 2 & 0 & -2 \\ 2 & -2 & -2 & 0 & 2 \\ 2 & -2 & 0 & -4 & 4 \\ 0 & 0 & -4 & 8 & -4 \\ -2 & 2 & 4 & -4 & 0 \end{bmatrix}.$$

   (e) $\|E_2\| = 6.4031$ and $\|A\| = 15.7797$

   (f) The percentage of information lost by approximating A by A_2 is

   $$\frac{\|E_2\|}{\|A\|} \approx 40.58\%.$$

13. The answer is given correct to 4 places after the decimal point.
$P_W = [B \ C]$, where

$$B = \begin{bmatrix} 0.3913 & 0.0730 & -0.1763 \\ 0.0730 & 0.7180 & -0.1688 \\ -0.1763 & -0.1688 & 0.8170 \\ -0.2716 & -0.1481 & -0.2042 \\ 0.2056 & 0.1328 & 0.1690 \\ -0.2929 & 0.3593 & 0.1405 \end{bmatrix}$$

and

$$C = \begin{bmatrix} -0.2716 & 0.2056 & -0.2929 \\ -0.1481 & 0.1328 & 0.3593 \\ -0.2042 & 0.1690 & 0.1405 \\ 0.7594 & 0.1958 & 0.0836 \\ 0.1958 & 0.8398 & -0.0879 \\ 0.0836 & -0.0879 & 0.4744 \end{bmatrix}$$

17. (a) Observe that

$$\left\{ \begin{bmatrix} -2 \\ 1 \\ 0 \end{bmatrix}, \begin{bmatrix} 1 \\ 0 \\ 1 \end{bmatrix} \right\}$$

is a basis for W, and the vector $\begin{bmatrix} 1 \\ 2 \\ -1 \end{bmatrix}$ is orthogonal to W. So these three vectors form a basis for $\mathcal{R}^3$ consisting of eigenvectors of T_W with corresponding eigenvectors 1, 1, and -1, respectively. Let

$$P = \begin{bmatrix} -2 & 1 & 1 \\ 1 & 0 & 2 \\ 0 & 1 & -1 \end{bmatrix}$$

and

$$D = \begin{bmatrix} 1 & 0 & 0 \\ 0 & 1 & 0 \\ 0 & 0 & -1 \end{bmatrix}.$$

Then $A_W = PDP^{-1}$.

Using the rational format in MAT-LAB, we obtain the exact matrix

$$A_W = PDP^{-1} = \begin{bmatrix} 2/3 & -2/3 & 1/3 \\ -2/3 & -1/3 & 2/3 \\ 1/3 & 2/3 & 2/3 \end{bmatrix}.$$

(b) Since $Q_{23°}$ is a rotation, its determinant is equal to 1. Observe that $\det A_W = -1$. Since A_W and $Q_{23°}$ are orthogonal matrices, $A_W Q_{23°} A_W$ is an orthogonal matrix. Furthermore,

$$\det(A_W Q_{23°} A_W)$$
$$= (\det A_W)(\det Q_{23°})(\det A_W)$$
$$= (-1)(1)(-1) = 1.$$

So $A_W Q_{23°} A_W$ is a rotation matrix by Theorem 6.20.

(c) $\mathbf{v} = \begin{bmatrix} -2 \\ -1 \\ 2 \end{bmatrix}$ is an eigenvector of the matrix in (b) corresponding to eigenvalue 1, and so it is a vector that lies on the axis of rotation. The angle of rotation is $23°$.

(d) Let R be a rotation matrix and A_W be the standard matrix of the reflection of $\mathcal{R}^3$ about a two-dimensional subspace W. If $\mathbf{v}$ is a nonzero vector that lies on the axis of rotation of R, then a nonzero vector $\mathbf{w}$ lies on the axis of rotation of $A_W R A_W$ if and only if $\mathbf{w} = A_W \mathbf{v}$. Furthermore, the angle of rotation of $A_W R A_W$ is equal to the angle of rotation of R.

Chapter 7

Vector Spaces

7.1 VECTOR SPACES AND THEIR SUBSPACES

1. Consider the matrix equation

$$\begin{bmatrix} 0 & 2 & 0 \\ 1 & 1 & 1 \end{bmatrix} = x_1 \begin{bmatrix} 1 & 2 & 1 \\ 0 & 0 & 0 \end{bmatrix}$$
$$+ x_2 \begin{bmatrix} 0 & 0 & 0 \\ 1 & 1 & 1 \end{bmatrix}$$
$$+ x_3 \begin{bmatrix} 1 & 0 & 1 \\ 1 & 2 & 3 \end{bmatrix}.$$

Comparing the corresponding entries of the right and left sides of this matrix equation yields the system of linear equations

$$\begin{aligned} x_1 \quad & + x_3 = 0 \\ 2x_1 \quad & \quad = 2 \\ x_1 \quad & + x_3 = 0 \\ & x_2 + x_3 = 1 \\ & x_2 + 2x_3 = 1 \\ & x_2 + 3x_3 = 1. \end{aligned}$$

The reduced row echelon form of the augmented matrix of this system is

$$R = \begin{bmatrix} 1 & 0 & 0 & 0 \\ 0 & 1 & 0 & 0 \\ 0 & 0 & 1 & 0 \\ 0 & 0 & 0 & 1 \\ 0 & 0 & 0 & 0 \\ 0 & 0 & 0 & 0 \end{bmatrix},$$

indicating that the system is inconsistent. Thus the matrix equation has no solution, and so the given matrix does not lie in the span of the given set.

5. As in Exercise 1, the given matrix lies in the span of the given set if and only if the matrix equation

$$\begin{bmatrix} 2 & 2 & 2 \\ 1 & 1 & 1 \end{bmatrix} = x_1 \begin{bmatrix} 1 & 2 & 1 \\ 0 & 0 & 0 \end{bmatrix}$$
$$+ x_2 \begin{bmatrix} 0 & 0 & 0 \\ 1 & 1 & 1 \end{bmatrix}$$
$$+ x_3 \begin{bmatrix} 1 & 0 & 1 \\ 1 & 2 & 3 \end{bmatrix}$$

has a solution. Comparing the corresponding entries of the right and left sides of this matrix equation yields the system of linear equations

$$\begin{aligned} x_1 \quad & + x_3 = 2 \\ 2x_1 \quad & \quad = 2 \\ x_1 \quad & + x_3 = 2 \\ & x_2 + x_3 = 1 \\ & x_2 + 2x_3 = 1 \\ & x_2 + 3x_3 = 1. \end{aligned}$$

The reduced row echelon form of the augmented matrix of this system is the marix R in the solution to Exercise 1. Therefore the matrix equation has no solution, and so the given matrix does not lie in the span of the given set.

9. Proceeding as in Exercise 1, we obtain the system of equations

$$\begin{aligned} x_1 \phantom{{}+x_2} + x_3 &= -2 \\ 2x_1 \phantom{{}+x_2+x_3} &= -8 \\ x_1 \phantom{{}+x_2} + x_3 &= -2 \\ x_2 + x_3 &= 5 \\ x_2 + 2x_3 &= 7 \\ x_2 + 3x_3 &= 9. \end{aligned}$$

This system has the solution $x_1 = -4$, $x_2 = 3$, $x_3 = 2$, and so the given matrix is a linear combination of the matrices in the given set:

$$\begin{bmatrix} -2 & -8 & -2 \\ 5 & 7 & 9 \end{bmatrix} = (-4) \begin{bmatrix} 1 & 2 & 1 \\ 0 & 0 & 0 \end{bmatrix}$$
$$+ 3 \begin{bmatrix} 0 & 0 & 0 \\ 1 & 1 & 1 \end{bmatrix}$$
$$+ 2 \begin{bmatrix} 1 & 0 & 1 \\ 1 & 2 & 3 \end{bmatrix}.$$

Thus the given matrix lies in the span of the given set.

13. Suppose that c_1, c_2, and c_3 are scalars such that

$$-2 + x + x^2 + x^3 = c_1(1 - x)$$
$$+ c_2(1 + x^2) + c_3(1 + x - x^3).$$

Equating the coefficients of like powers of x on the left and right sides of this equation produces the following system of linear equations:

$$\begin{aligned} c_1 + c_2 + c_3 &= -2 \\ -c_1 \phantom{{}+c_2} + c_3 &= 1 \\ c_2 \phantom{{}+c_3} &= 1 \\ -c_3 &= 1 \end{aligned}$$

This system has the solution $x_1 = -2$, $x_2 = 1$, $x_3 = -1$. Thus

$$-2 + x + x^2 + x^3 = (-2)(1 - x)$$

$$+ 1(1 + x^2)$$
$$+ (-1)(1 + x - x^3).$$

So the given polynomial lies in the span of the given set.

17. As in Exercise 1, the given matrix is in the span of the given set if and only if the matrix equation

$$\begin{bmatrix} 1 & 2 \\ -3 & 4 \end{bmatrix} = x_1 \begin{bmatrix} 1 & 0 \\ -1 & 0 \end{bmatrix} + x_2 \begin{bmatrix} 0 & 1 \\ 0 & 1 \end{bmatrix}$$
$$+ x_3 \begin{bmatrix} 1 & 1 \\ 0 & 0 \end{bmatrix}$$

has a solution. Comparing the right and left sides of the corresponding entries of the matrix equation yields the system of linear equations

$$\begin{aligned} x_1 \phantom{{}+x_2} + x_3 &= 1 \\ x_2 + x_3 &= 2 \\ -x_1 \phantom{{}+x_2+x_3} &= -3 \\ x_2 \phantom{{}+x_3} &= 4. \end{aligned}$$

This system has the unique solution $x_1 = 3$, $x_2 = 4$, and $x_3 = -2$. Therefore

$$\begin{bmatrix} 1 & 2 \\ -3 & 4 \end{bmatrix} = 3 \begin{bmatrix} 1 & 0 \\ -1 & 0 \end{bmatrix} + 4 \begin{bmatrix} 0 & 1 \\ 0 & 1 \end{bmatrix}$$
$$+ (-2) \begin{bmatrix} 1 & 1 \\ 0 & 0 \end{bmatrix}.$$

Thus the given matrix lies in the span of the given set.

21. As in Exercise 17, we find that

$$\begin{bmatrix} 1 & -2 \\ -3 & 0 \end{bmatrix} = 3 \begin{bmatrix} 1 & 0 \\ -1 & 0 \end{bmatrix} + 0 \begin{bmatrix} 0 & 1 \\ 0 & 1 \end{bmatrix}$$
$$+ (-2) \begin{bmatrix} 1 & 1 \\ 0 & 0 \end{bmatrix}.$$

So the given matrix lies in the span of the given set.

25. Any vector in a set lies in the span of that set. So $1 + x$ lies in the span of the given set.

29. A polynomial is a linear combination of the polynomials in S if and only if has the form

$$a(9 + 4x + 5x^2 - 3x^3)$$
$$+ b(-3 - 5x - 2x^2 + x^3)$$
$$= (9a - 3b) + (4a - 5b)x$$
$$+ (5a - 2b)x^2 + (-3a + b)x^3.$$

for some scalars a and b. Comparing the coefficients of the given polynomial with the form above, we obtain the system of linear equations

$$\begin{aligned} 9a - 3b &= 12 \\ 4a - 5b &= -13 \\ 5a - 2b &= 5 \\ -3a + b &= -4. \end{aligned}$$

Since this system has the solution $a = 3$ and $b = 5$, the given polynomial is a linear combination of the polynomials in S with 3 and 5 as its coefficients:

$$12 - 13x + 5x^2 - 4x^3$$
$$= 3(9 + 4x + 5x^2 - 3x^3)$$
$$+ 5(-3 - 5x - 2x^2 + x^3).$$

31. Let W denote the span of the set

$$\{1 + x,\ 1 - x,\ 1 + x^2,\ 1 - x^2\}.$$

Since

$$1 = .5(1 + x) + .5(1 - x),$$

we see that 1 is in W. Since

$$x = .5(1 + x) + (-.5)(1 - x),$$

we see that x is in W. Since

$$x^2 = .5(1 + x^2) + (-.5)(1 - x^2),$$

we see that x^2 is in W. Since $\{1, x, x^2\}$ is a generating set for $\mathcal{P}_2$ that is contained in W, it follows that $W = \mathcal{P}_2$.

33. True

34. False, by Theorem 7.2, the zero vector of a vector space is unique.

35. False, consider $a = 0$ and $\mathbf{v} \ne \mathbf{0}$.

36. True

37. False, any two polynomials can be added.

38. False. For example, if $p(x) = 1 + x^n$ and $q(x) = 1 - x^n$, then $p(x)$ and $q(x)$ each have degree n, but $p(x) + q(x)$ has degree 0. So the set of polynomials of degree n is not closed under addition.

39. True **40.** True **41.** True

42. True **43.** True

44. False, the empty set contains no zero vector.

45. True **46.** True **47.** True

48. True **49.** True **50.** True

51. True **52.** True **53.** True

54. True

55. Let f, g, and h be in $\mathcal{F}(S)$. For any s in S, we have

$$\begin{aligned}[(f + g) + h](s) &= (f + g)(s) + h(s) \\ &= [f(s) + g(s)] + h(s) \\ &= f(s) + [g(s) + h(s)] \\ &= f(s) + (g + h)(s) \\ &= [f + (g + h)](s),\end{aligned}$$

and hence $(f + g) + h = f + (g + h)$.

59. Let f be in $\mathcal{F}(S)$ and a and b be scalars. Then for any s in S, we have

$$[(a+b)f](s) = (a+b)f(s)$$
$$= af(s) + bf(s)$$
$$= (af)(s) + (bf)(s)$$
$$= (af + bf)(s),$$

and hence $(a+b)f = af + bf$.

63. We show that V is a subspace. Since $OB = BO = O$, the zero matrix is in V. Now suppose that A and C are in V. Then

$$(A+C)B = AB + CB = BA + BC$$
$$= B(A+C),$$

and hence $A+C$ is in V. So V is closed under addition. Also, for any scalar c,

$$(cA)B = c(AB) = c(BA) = B(cA),$$

and hence cA is in V. Therefore V is closed under scalar multiplication.

67. Because V is not closed under addition, it is not a subspace. Consider $m=2$, $p(x) = 1 + x^2$ and $q(x) = -1 + 2x^2$. Both $p(x)$ and $q(x)$ are in V, but

$$p(x) + q(x) = 0 + 3x^2$$

is not in V.

71. We show that V is a subspace. Let $\mathbf{0}$ denote the zero function. Then

$$\mathbf{0}(s_1) + \cdots + \mathbf{0}(s_n) = 0 + \cdots + 0 = 0,$$

and hence the zero function is in V. Suppose that f and g are in V. Then

$$(f+g)(s_1) + \cdots + (f+g)(s_n)$$
$$= f(s_1) + \cdots + f(s_n)$$
$$+ g(s_1) + \cdots + g(s_n)$$

$$= 0 + 0 = 0,$$

and hence $f+g$ is in V. Thus V is closed under addition. Let c be any scalar. Then

$$(cf)(s_1) + \cdots + (cf)(s_n)$$
$$= c[f(s_1) + \cdots + f(s_n)]$$
$$= c(0 + \cdots + 0) = 0,$$

and hence cf is in V. Thus V is closed under scalar multiplication.

75. Let

$$p(x) = a_0 + a_1 x + \cdots + a_n x^n$$

and

$$q(x) = b_0 + b_1 x + \cdots + b_n x^n$$

be polynomials (not necessarily of the same degree). With this notation, we verify two of the axioms of a vector space. The others are proved similarly.

Axiom 1 We have

$$p(x) + q(x)$$
$$= (a_0 + b_0) + \cdots + (a_n + b_n)x^n$$
$$= (b_0 + a_0) + \cdots + (b_n + a_n)x^n$$
$$= q(x) + p(x).$$

Axiom 7 Let c be any scalar. Then

$$c[p(x) + q(x)]$$
$$= c(a_0 + b_0) + \cdots + c(a_n + b_n)x^n$$
$$= (ca_0 + cb_0) + \cdots + (ca_n + cb_n)x^n$$
$$= (ca_0 + \cdots + ca_n x^n)$$
$$\quad + (cb_0 + \cdots + cb_n x^n)$$
$$= cp(x) + cq(x).$$

79. Suppose that $\mathbf{u} + \mathbf{v} = \mathbf{u} + \mathbf{w}$. Then $\mathbf{v} + \mathbf{u} = \mathbf{w} + \mathbf{u}$ by axiom 1, and hence $\mathbf{v} = \mathbf{w}$ by Theorem 7.2(a).

83. It follows from axioms 8 and 7, respectively, that

$$(a+b)(\mathbf{u}+\mathbf{v}) = a(\mathbf{u}+\mathbf{v}) + b(\mathbf{u}+\mathbf{v})$$
$$= a\mathbf{u} + a\mathbf{v} + b\mathbf{u} + b\mathbf{v}.$$

87.

$$(-c)(-\mathbf{v}) = (-c)((-1)\mathbf{v})$$
$$= [(-c)(-1)]\mathbf{v} = c\mathbf{v}$$

91. (a) Since $\mathbf{0}(t) = \mathbf{0}(-t) = 0$, the zero function is even. Suppose that f and g are even functions. Then

$$(f+g)(t) = f(t) + g(t)$$
$$= f(-t) + g(-t)$$
$$= (f+g)(-t),$$

and hence $f+g$ is even. Furthermore, for any scalar a,

$$(af)(t) = a[f(t)] = a[f(-t)]$$
$$= (af)(-t),$$

and hence af is even. Thus the subset of even functions is closed under addition and scalar multiplication. Therefore this set is a subspace.

(b) Since $\mathbf{0}(-t) = -\mathbf{0}(t) = 0$, the zero function is odd. Suppose that f and g are odd functions. Then

$$(f+g)(-t) = f(-t) + g(-t)$$
$$= -f(t) - g(t)$$
$$= (-f-g)(t)$$
$$= [-(f+g)](t),$$

and hence $f+g$ is odd. Furthermore, for any scalar a,

$$(af)(-t) = a(f(-t)) = a(-f(t))$$

$$= -(af)(t),$$

and hence af is odd. Thus the subset of odd functions is closed under addition and scalar multiplication. Therefore this set is a subspace.

95. Suppose that W is a subspace of V. Then (i) is satisfied. Let $\mathbf{w}_1$ and $\mathbf{w}_2$ be in W, and let a be a scalar. Then $a\mathbf{w}_1$ is in W because W is closed under scalar multiplication, and hence $a\mathbf{w}_1 + \mathbf{w}_2$ is in W because W is closed under addition. Therefore (ii) is satisfied.

Conversely, suppose conditions (i) and (ii) are satisfied. By (i), the zero vector lies in V. Let $\mathbf{w}_1$ and $\mathbf{w}_2$ be in W. Then $\mathbf{w}_1 + \mathbf{w}_2 = 1 \cdot \mathbf{w}_1 + \mathbf{w}_2$, which is in W by (ii). Hence W is closed under addition. Furthermore, for any scalar a, $a\mathbf{w}_1 = a\mathbf{w}_1 + \mathbf{0}$, which is in W by (ii). Hence W is closed under scalar multiplication. Therefore W is a subspace of V.

7.2 LINEAR TRANSFORMATIONS

1. Yes, T is one-to-one. First, observe that the matrix $C = \begin{bmatrix} 1 & 2 \\ 3 & 4 \end{bmatrix}$ is invertible. Suppose that $T(A) = AC = O$. Then $A = OC^{-1} = O$. Therefore T is one-to-one by Theorem 7.5.

5. No, T is not one-to-one. Since

$$T(1) = x \cdot 0 = 0,$$

we see that T is not one-to-one by Theorem 7.5.

9. Yes, T is onto. Let $C = \begin{bmatrix} 1 & 2 \\ 3 & 4 \end{bmatrix}$, and note that C is invertible. Then for any matrix A in $\mathcal{M}_{2\times 2}$,

$$T(AC^{-1}) = AC^{-1}C = A,$$

and hence A is in the range of T.

13. No. The constant polynomial 1 is not the product of x and any polynomial.

17. Let $C = \begin{bmatrix} 1 & 2 \\ 3 & 4 \end{bmatrix}$. For any A and B in $\mathcal{M}_{2\times 2}$, and for any scalar s,

$$T(A+B) = (A+B)C = AC + BC$$
$$= T(A) + T(B)$$

and

$$T(sA) = (sA)C = s(AC) = sT(A).$$

Therefore T is linear.

21. For any $f(x)$ and $g(x)$ in $\mathcal{P}_2$,

$$T(f(x) + g(x)) = x[f(x) + g(x)]'$$
$$= x[f'(x) + g'(x)]$$
$$= xf'(x) + xg'(x)$$
$$= T(f(x)) + T(g(x)).$$

Similarly, $T(cf(x)) = cT(f(x))$ for any scalar c, and therefore T is linear.

25.
$$UT\left(\begin{bmatrix} a & b \\ c & d \end{bmatrix}\right) = U\left(\begin{bmatrix} a & b \\ c & d \end{bmatrix}\begin{bmatrix} 1 & 2 \\ 3 & 4 \end{bmatrix}\right)$$
$$= U\left(\begin{bmatrix} a+3b & 2a+4b \\ c+3d & 2c+4d \end{bmatrix}\right)$$
$$= \text{trace}\left(\begin{bmatrix} a+3b & 2a+4b \\ c+3d & 2c+4d \end{bmatrix}\right)$$
$$= a + 3b + 2c + 4d.$$

29.
$$TU\left(\begin{bmatrix} s \\ t \\ u \end{bmatrix}\right) = T\left(\begin{bmatrix} s & t \\ t & u \end{bmatrix}\right) = \begin{bmatrix} s & t \\ t & u \end{bmatrix}^T$$
$$= \begin{bmatrix} s & t \\ t & u \end{bmatrix}.$$

33. We show that T does not preserve scalar multiplication and hence is not linear. Let $f(x)$ be any nonzero polynomial. Then

$$T(2f(x)) = [2f(x)]^2$$
$$= 4[f(x)]^2 \neq 2T(f(x)).$$

37. We show that T is linear, but not an isomorphism. Let f and g be in $\mathcal{D}(\mathcal{R})$. Then

$$T(f+g) = \int_0^1 (f(t) + g(t))\,dt$$
$$= \int_0^1 f(t)\,dt + \int_0^1 g(t)\,dt$$
$$= T(f) + T(g).$$

Similarly, $T(cf) = cT(f)$ for any scalar c, and hence T is linear. However, T is not one-to-one. Let $f(t) = 2t - 1$. Then

$$T(f) = \int_0^1 (2t-1)\,dt = t^2 - t\Big|_0^1$$
$$= 0 - 0 = 0,$$

and hence T is not one-to-one by Theorem 7.5.

39. True

40. False, it may fail to be onto.

41. True   **42.** True   **43.** True

44. False, all polynomials are in C^∞.

45. True

46. False, the definite integral of a function in $C([a,b])$ is a real number.

47. True

48. False, the zero function is not in the solution set.

49. *Proof of (a):* The zero vector of $\mathcal{F}(\mathcal{N})$ is the function for which the image of every nonnegative integer is zero. Clearly this function is in V. Let f and g be in V. Then f is nonzero at only finitely many nonnegative integers $a_1, a_2, \ldots, a_r$, and g is nonzero at only finitely many nonnegative integers $b_1, b_2, \ldots, b_s$. Then $f + g$ is zero except possibly at the finitely many nonnegative integers $a_1, a_2, \ldots, a_r, b_1, b_2, \ldots, b_s$. So $f+g$ is in V. Finally, for any scalar c, the function cf can be nonzero at only $a_1, a_2, \ldots, a_r$; so cf is in V. It follows that V is a subspace of $\mathcal{F}(\mathcal{N})$.

Proof of (b): Let f and g be in V, and let n be a positive integer such that $f(k) = g(k) = 0$ for $k > n$. Then

$$T(f+g) = (f+g)(0) + (f+g)(1)x + \cdots + (f+g)(n)x^n$$
$$= [f(0) + f(1)x + \cdots + f(n)x^n]$$
$$+ [g(0) + g(1)x + \cdots + g(n)x^n]$$
$$= T(f) + T(g).$$

Similarly, $T(cf) = cT(f)$ for any scalar c, and hence T is linear.

We now show that T is an isomorphism. Suppose that $T(f) = 0$, the zero polynomial. Then $f(k) = 0$ for all k, and hence f is the zero function. So T is one-to-one. Now consider any polynomial $p(x) = a_0 + a_1 x + \cdots + a_n x^n$. Let $f \colon V \to \mathcal{R}$ be defined by

$$f(k) = \begin{cases} a_k & \text{if } k \leq n \\ 0 & \text{if } k > n. \end{cases}$$

Then $T(f) = p(x)$, and hence T is onto. Therefore T is an isomorphism.

53. Let V, W, and Z be vector spaces and $T \colon V \to W$ and $U \colon W \to Z$ be isomorphisms. Since T and U are both one-to-one and onto, UT is one-to-one and onto. Consider any vectors $\mathbf{u}$ and $\mathbf{v}$ in V. Then

$$UT(\mathbf{u} + \mathbf{v}) = U(T(\mathbf{u} + \mathbf{v}))$$
$$= U(T(\mathbf{u}) + T(\mathbf{v}))$$
$$= UT(\mathbf{u}) + UT(\mathbf{v}).$$

Similarly, $UT(c\mathbf{u}) = cUT(\mathbf{u})$ for any scalar c. Therefore UT is linear, and hence it is an isomorphism.

Let $\mathbf{z}$ be in Z. Then

$$UT(T^{-1}U^{-1}(\mathbf{z})) = U(TT^{-1})U^{-1}(\mathbf{z})$$
$$= UU^{-1}(\mathbf{z})$$
$$= \mathbf{z},$$

and hence $T^{-1}U^{-1}(\mathbf{z}) = (UT)^{-1}(\mathbf{z})$. We conclude that $(UT)^{-1} = T^{-1}U^{-1}$.

57. Since $T(\mathbf{0}) = \mathbf{0}$, it follows that $\mathbf{0}$ is in the range of T. Suppose that $\mathbf{w}_1$ and $\mathbf{w}_2$ are in the range of T and c is a scalar. Then there exist vectors $\mathbf{v}_1$ and $\mathbf{v}_2$ in V such that $T(\mathbf{v}_1) = \mathbf{w}_1$ and $T(\mathbf{v}_2) = \mathbf{w}_2$. Thus

$$T(\mathbf{v}_1 + \mathbf{v}_2) = T(\mathbf{v}_1) + T(\mathbf{v}_2)$$
$$= \mathbf{w}_1 + \mathbf{w}_2,$$

and

$$T(c\mathbf{v}_1) = cT(\mathbf{v}_1) = c\mathbf{w}_1.$$

Hence the range of T is closed under vector addition and scalar multiplication. We conclude that the range of T is a subspace of V.

7.3 BASIS AND DIMENSION

1. Consider the matrix equation

$$x_1 \begin{bmatrix} 1 & 2 \\ 3 & 1 \end{bmatrix} + x_2 \begin{bmatrix} 1 & -5 \\ -4 & 0 \end{bmatrix} + x_3 \begin{bmatrix} 3 & -1 \\ 2 & 2 \end{bmatrix} = \begin{bmatrix} 0 & 0 \\ 0 & 0 \end{bmatrix}.$$

Equating corresponding entries, we obtain the system of linear equations

$$\begin{aligned} x_1 + x_2 + 3x_3 &= 0 \\ 2x_1 - 5x_2 - x_3 &= 0 \\ 3x_1 - 4x_2 + 2x_3 &= 0 \\ x_1 \phantom{{}+x_2} + 2x_3 &= 0. \end{aligned}$$

The reduced row echelon form of the augmented matrix of this system is

$$\begin{bmatrix} 1 & 0 & 2 & 0 \\ 0 & 1 & 1 & 0 \\ 0 & 0 & 0 & 0 \\ 0 & 0 & 0 & 0 \end{bmatrix}.$$

Thus the preceding system has nonzero solutions, for example, $x_1 = -2$, $x_2 = -1$, and $x_3 = 1$, and so the given set of matrices is linearly dependent.

5. Consider the matrix equation

$$x_1 \begin{bmatrix} 1 & 0 & 1 \\ -1 & 2 & 1 \end{bmatrix} + x_2 \begin{bmatrix} -1 & 1 & 2 \\ 2 & -1 & 1 \end{bmatrix} + x_3 \begin{bmatrix} -1 & 0 & 1 \\ 1 & -1 & 0 \end{bmatrix} = \begin{bmatrix} 0 & 0 & 0 \\ 0 & 0 & 0 \end{bmatrix}.$$

Equating corresponding entries, we obtain the system of linear equations

$$\begin{aligned} x_1 - x_2 - x_3 &= 0 \\ x_2 \phantom{{}-x_3} &= 0 \\ x_1 + 2x_2 + x_3 &= 0 \\ -x_1 + 2x_2 + x_3 &= 0 \\ 2x_1 - x_2 - x_3 &= 0 \\ x_1 + x_2 \phantom{{}+x_3} &= 0. \end{aligned}$$

The reduced row echelon form of the augmented matrix of this system is

$$\begin{bmatrix} 1 & 0 & 0 & 0 \\ 0 & 1 & 0 & 0 \\ 0 & 0 & 1 & 0 \\ 0 & 0 & 0 & 0 \\ 0 & 0 & 0 & 0 \\ 0 & 0 & 0 & 0 \end{bmatrix}.$$

So the preceding system has no nonzero solutions, and thus the given set of matrices is linearly independent.

9. Consider the polynomial equation

$$a(1+x) + b(1-x) + c(1+x+x^2) + d(1+x-x^2) = 0.$$

Equating corresponding coefficients, we obtain the system of linear equations

$$\begin{aligned} a + b + c + d &= 0 \\ a - b + c + d &= 0 \\ c - d &= 0. \end{aligned}$$

The reduced row echelon form of the augmented matrix of this system is

$$\begin{bmatrix} 1 & 0 & 0 & 2 & 0 \\ 0 & 1 & 0 & 0 & 0 \\ 0 & 0 & 1 & -1 & 0 \end{bmatrix}.$$

Thus the preceding system has nonzero solutions, for example, $a = -2$, $b = 0$, $c = 1$, and $d = 1$, and so the given set of polynomials is linearly dependent.

13. Consider the polynomial equation

$$a(x^3 + 2x^2) + b(-x^2 + 3x + 1) + c(x^3 - x^2 + 2x - 1) = 0.$$

Equating corresponding coefficients, we obtain the system of linear equations

$$\begin{aligned} a \phantom{{}-b} + c &= 0 \\ 2a - b - c &= 0 \\ 3b + 2c &= 0 \\ b - c &= 0. \end{aligned}$$

The reduced row echelon form of the augmented matrix of this system is

$$\begin{bmatrix} 1 & 0 & 0 & 0 \\ 0 & 1 & 0 & 0 \\ 0 & 0 & 1 & 0 \\ 0 & 0 & 0 & 0 \end{bmatrix}.$$

Thus the preceding system has no nonzero solutions, and so the given set of polynomials is linearly independent.

17. Assume that $\{t, t\sin t\}$ is linearly dependent. Since these are both nonzero functions, there exists a nonzero scalar a such that $t\sin t = at$ for all t in $\mathcal{R}$. Setting $t = \frac{\pi}{2}$, we obtain $\frac{\pi}{2}\sin\frac{\pi}{2} = a\frac{\pi}{2}$, from which we see that $a = 1$. Setting $t = \frac{\pi}{4}$, we obtain $\frac{\pi}{4}\sin\frac{\pi}{4} = a\frac{\pi}{4}$, from which we see that $a = 1/\sqrt{2}$. This is a contradiction, and it follows that the set is linearly independent.

21. We show that for any positive integer n, any subset consisting of n functions is linearly independent. This is certainly true for $n = 1$ because any set consisting of a single nonzero function is linearly independent.

   Now suppose that we have established that any subset consisting of k functions is linearly independent, where k is a fixed positive integer. Consider any subset consisting of $k + 1$ functions

   $$\{e^{n_1 t}, e^{n_2 t}, \ldots, e^{n_{k+1} t}\}.$$

   Let $a_1, \ldots, a_{k+1}$ be scalars such that

   $$a_1 e^{n_1 t} + \cdots + a_k e^{n_k t} + a_{k+1} e^{n_{k+1} t} = 0$$

   for all t. We form two equations from the equation above. The first equation is obtained by taking the derivative of both sides with respect to t, and the second equation is obtained by multiplying both sides of the equation by n_{k+1}. The results are

   $$n_1 a_1 e^{n_1 t} + \cdots + n_k a_k e^{n_k t} + n_{k+1} a_{k+1} e^{n_{k+1} t} = 0$$

   and

   $$n_{k+1} a_1 e^{n_1 t} + \cdots + n_{k+1} a_k e^{n_k t} + n_{k+1} a_{k+1} e^{n_{k+1} t} = 0.$$

   Now subtract both sides of the second equation from both sides of the first equation to obtain

   $$(n_1 - n_{k+1}) a_1 e^{n_1 t} + \cdots + (n_k - n_{k+1}) a_k e^{n_k t} = 0.$$

   Since this last equation involves a linear combination of a set of k functions, which is assumed to be linearly independent, each coefficient $(n_i - n_{k+1}) a_i$ is zero. But $n_i \neq n_{k+1}$ for each i, $1 \leq i \leq k$, and hence each $a_i = 0$. Thus the original equation reduces to $a_{k+1} e^{n_{k+1} t} = 0$, from which we conclude that $a_{k+1} = 0$. It follows that any subset consisting of $k + 1$ functions is linearly independent.

   Since a set of 1 function is linearly independent, the preceding paragraph implies that a set of 2 functions is linearly independent. Repeating this reasoning, we see that any set of 3 functions is linearly independent. Continuing this argument $n - 1$ times, we conclude that any set of n functions is linearly independent.

   (This proof can also be written using mathematical induction.)

25. Let

    $$p_1(x) = \frac{(x-1)(x-2)}{(0-1)(1-2)},$$

$$p_2(x) = \frac{(x-0)(x-2)}{(1-0)(1-2)},$$

and

$$p_3(x) = \frac{(x-0)(x-1)}{(2-0)(2-1)}.$$

Then

$$p(x) = 1p_1(x) + 0p_2(x) + 3p_3(x)$$
$$= \frac{1}{2}(x-1)(x-2) + \frac{3}{2}x(x-1)$$
$$= 2x^2 - 3x + 1.$$

29. Let

$$p_1(x) = \frac{(x-0)(x-1)(x-2)}{(-1-0)(-1-1)(-1-2)},$$
$$= \frac{1}{6}(-x^3 + 3x^2 - 2x)$$

$$p_2(x) = \frac{(x+1)(x-1)(x-2)}{(0+1)(0-1)(0-2)},$$
$$= \frac{1}{6}(3x^3 - 6x^2 - 3x + 6)$$

$$p_3(x) = \frac{(x+1)(x-0)(x-2)}{(1+1)(1-0)(1-2)},$$
$$= \frac{1}{6}(-3x^3 + 3x^2 + 6x)$$

and

$$p_4(x) = \frac{(x+1)(x-0)(x-1)}{(2+1)(2-0)(2-1)}$$
$$= \frac{1}{6}(x^3 - x).$$

Then

$$p(x) = 5p_1(x) + 2p_2(x)$$
$$\quad + (-1)p_3(x) + 2p_4(x)$$
$$= x^3 - 4x + 2.$$

31. False. For example, the infinite set $\{1, x, x^2, \ldots\}$ is a linearly independent subset of $\mathcal{P}$.

32. False. Only finite-dimensional vector spaces have finite bases.

33. False. The dimension of $\mathcal{P}_n$ is equal to $n+1$.

34. False. For example, $\mathcal{P}_2$ is a 3-dimensional subspace of the infinite-dimensional vector space $\mathcal{P}$.

35. False. Finite-dimensional vector spaces only have finite bases, and infinite-dimensional vector spaces only have infinite bases.

36. True   37. True   38. True

39. False, the set is linearly independent.

40. True   41. True

42. False, its dimension is mn.

43. True   44. True

45. False, its dimension is mn.

46. True   47. True

48. False. For example $\{1, x, 1+x\}$ is a finite linearly dependent subset of $\mathcal{P}$, but $\mathcal{P}$ is infinite-dimensional.

49. This set is linearly dependent. To show this, we find a nonzero solution of

$$(af + bg + ch)(n)$$
$$= a(n+1) + b + c(2n-1)$$
$$= (a+2c)n + (a+b-c) = 0$$

for all n. Thus we set the coefficients equal to zero to obtain the system

$$a + 2c = 0$$
$$a + b - c = 0,$$

which has the nonzero solution $a = -2$, $b = 3$, and $c = 1$. So $-2f + 3g + h = 0$.

53. A matrix $A = \begin{bmatrix} x_1 & x_2 \\ x_3 & x_4 \end{bmatrix}$ is in W if and only if $x_1 + x_4 = 0$. For any such matrix A,

$$A = \begin{bmatrix} x_1 & x_2 \\ x_3 & -x_1 \end{bmatrix} = x_1 \begin{bmatrix} 1 & 0 \\ 0 & -1 \end{bmatrix}$$
$$+ x_2 \begin{bmatrix} 0 & 1 \\ 0 & 0 \end{bmatrix} + x_3 \begin{bmatrix} 0 & 0 \\ 1 & 0 \end{bmatrix}.$$

It follows that

$$\left\{ \begin{bmatrix} 1 & 0 \\ 0 & -1 \end{bmatrix}, \begin{bmatrix} 0 & 1 \\ 0 & 0 \end{bmatrix}, \begin{bmatrix} 0 & 0 \\ 1 & 0 \end{bmatrix} \right\}$$

is a basis for W.

57. Observe that the polynomials in W have degrees less than 2. Hence $\{1, x\}$ is a basis for W.

61. (a) Since every entry of C_n is $\frac{1}{n}$, the sum of any row, column, and diagonal is equal to 1. Therefore C_n is a magic square with sum 1.
    (b) Let A be an $n \times n$ magic square with sum s and $B = A - sC_n$. Since A and C_n are in V_n, it follows that B is in V_n, and B has sum $s - s \cdot 1 = 0$. So B is in W_n and $A = B + sC_n$. Since $B = A - sC_n$, it is necessarily unique.

65. Consider the system of linear equations that results from equating the n row sums, the n column sums, the sum of the entries of the diagonal, and the sum of the entries of the secondary diagonal to 0. The coefficient matrix of this homogeneous system contains $2n + 2$ rows (because there are $2n + 2$ equations) and n^2 columns (because there are n^2 variables, which are the n^2 entries of an $n \times n$ matrix). Add rows 1 through $n - 1$ to row n, creating a row with all entries equal to 1. Now subtract rows $n + 1$ through $2n$ from this new nth row of 1s to obtain a zero row. The other $2n + 1$ rows are linearly independent, and hence the coefficient matrix has rank $2n + 1$. Therefore the dimension of the solution space is $n^2 - (2n + 1) = n^2 - 2n - 1$.

Define $\Phi \colon \mathcal{R}^{n^2} \to \mathcal{M}_{n \times n}$ by

$$\Phi \left(\begin{bmatrix} x_{11} \\ x_{12} \\ \vdots \\ x_{nn} \end{bmatrix} \right) =$$
$$\begin{bmatrix} x_{11} & x_{12} & \cdots & x_{1n} \\ x_{21} & x_{22} & \cdots & x_{2n} \\ \vdots & \vdots & & \vdots \\ x_{n1} & x_{n2} & \cdots & x_{nn} \end{bmatrix}.$$

This mapping is an isomorphism, and W_n is the image of the solution space of the homogeneous system described in the preceding paragraph. Since an isomorphism preserves dimensions, it follows that $\dim W_n = n^2 - 2n - 1$.

69. Choose a basis $\mathcal{B}$ for a vector space V of dimension n, and let $\Phi_\mathcal{B} \colon V \to \mathcal{R}^n$ be the isomorphism defined on page 513.

(a) Consider any subset $\mathcal{S}$ of V containing more than n vectors, and let $\mathcal{S}'$ be the set of images of these vectors under $\Phi_\mathcal{B}$. Then $\mathcal{S}'$ is a subset of $\mathcal{R}^n$ consisting of more than n vectors, and hence is linearly dependent. As a consequence, $\mathcal{S}$ is linearly dependent. For otherwise, the images of vectors in a linearly independent set under an isomorphism would be linearly dependent, contrary to Theorem 7.8.

(b) Part (b) follows from (a) and Exercise 68.

73. Suppose that $\mathcal{B} = \{\mathbf{v}_1, \mathbf{v}_2, \ldots, \mathbf{v}_n\}$ is a basis for a vector space V, and let $\mathbf{u}$ and $\mathbf{v}$ be in V. Then there exist unique scalars $a_1, a_2, \ldots, a_n$ and $b_1, b_2, \ldots, b_n$ such that

$$\mathbf{u} = a_1\mathbf{v}_1 + a_2\mathbf{v}_2 + \cdots + a_n\mathbf{v}_n$$

and

$$\mathbf{v} = b_1\mathbf{v}_1 + b_2\mathbf{v}_2 + \cdots + b_n\mathbf{v}_n.$$

Thus

$$\Phi_\mathcal{B}(\mathbf{u} + \mathbf{v})$$
$$= \Phi_\mathcal{B}((a_1 + b_1)\mathbf{v}_1 + \cdots + (a_n + b_n)\mathbf{v}_n)$$
$$= \begin{bmatrix} a_1 + b_1 \\ a_2 + b_2 \\ \vdots \\ a_n + b_n \end{bmatrix}$$
$$= \begin{bmatrix} a_1 \\ a_2 \\ \vdots \\ a_n \end{bmatrix} + \begin{bmatrix} b_1 \\ b_2 \\ \vdots \\ b_n \end{bmatrix}$$
$$= \Phi_\mathcal{B}(\mathbf{u}) + \Phi_\mathcal{B}(\mathbf{v}).$$

Similarly $\Phi_\mathcal{B}(c\mathbf{u}) = c\Phi_\mathcal{B}(\mathbf{u})$ for every $\mathbf{u}$ in V, and hence $\Phi_\mathcal{B}$ is linear.

77. Let $T: \mathcal{P}_n \to \mathcal{R}$ be defined by

$$T(f(x)) = \int_a^b f(x)\, dx.$$

Then T is linear because the definite integral of a sum of polynomials is the sum of the definite integrals, and the definite integral of a scalar multiple of a polynomial is the same multiple of its definite integral. Therefore T is in $\mathcal{L}(\mathcal{P}_n, \mathcal{R})$. By Exercise 76, $\{T_0, T_1, \ldots, T_n\}$ is a basis for $\mathcal{L}(\mathcal{P}_n, \mathcal{R})$, and hence there exist unique scalars $c_1, c_1, \ldots, c_n$ such that

$$T = c_0T_0 + c_1T_1 + \cdots + c_nT_n.$$

Thus, for any polynomial $f(x)$ in $\mathcal{P}_n$,

$$\int_a^b f(x)\, dx = T(f(x))$$
$$= (c_0T_0 + c_1T_1 + \cdots + c_nT_n)(f(x))$$
$$= c_0T_0(f(x)) + c_1T_1(f(x)) + \cdots$$
$$\quad + c_nT_n(f(x))$$
$$= c_0 f(0) + c_1 f(1) + \cdots + c_n f(n).$$

81. As in Exercise 1, the given set of matrices is linearly dependent, and

$$M_3 = (-3)M_1 + 2M_2 + 0M_4,$$

where M_j is the jth matrix in the set.

7.4 MATRIX REPRESENTATIONS OF LINEAR OPERATORS

1. Since

$$\begin{bmatrix} 1 & 2 \\ 3 & 4 \end{bmatrix} = 1\begin{bmatrix} 1 & 0 \\ 0 & 0 \end{bmatrix} + 3\begin{bmatrix} 0 & 0 \\ 1 & 0 \end{bmatrix}$$
$$+ 4\begin{bmatrix} 0 & 0 \\ 0 & 1 \end{bmatrix} + 2\begin{bmatrix} 0 & 1 \\ 0 & 0 \end{bmatrix},$$

it follows that

$$[A]_\mathcal{B} = \begin{bmatrix} 1 \\ 3 \\ 4 \\ 2 \end{bmatrix}.$$

5. It is easy to see that

$$\mathbf{u} = (-3)\begin{bmatrix} -1 \\ 1 \\ 0 \\ 0 \end{bmatrix} + 2\begin{bmatrix} 1 \\ 0 \\ 1 \\ 0 \end{bmatrix} + 1\begin{bmatrix} 1 \\ 0 \\ 0 \\ 1 \end{bmatrix}.$$

So $[\mathbf{u}]_\mathcal{B} = \begin{bmatrix} -3 \\ 2 \\ 1 \end{bmatrix}.$

9. Since

$$D(e^t) = 1e^t, \quad D(e^{2t}) = 2e^{2t},$$

and

$$D(e^{3t}) = 3e^{3t},$$

we see that

$$[T]_\mathcal{B} = [D]_\mathcal{B} = \begin{bmatrix} 1 & 0 & 0 \\ 0 & 2 & 0 \\ 0 & 0 & 3 \end{bmatrix}.$$

13. The images of the basis vectors are

$$T(1) = 1' - 1'' = 0 - 0 = 0$$
$$T(x) = x' - x'' = 1 - 0 = 1$$
$$T(x^2) = (x^2)' - (x^2)'' = 2x - 2$$
$$T(x^3) = (x^3)' - (x^3)'' = 3x^2 - 6x.$$

Thus the coordinate vectors relative to $\mathcal{B}$ of these images are

$$[T(1)]_\mathcal{B} = \begin{bmatrix} 0 \\ 0 \\ 0 \\ 0 \end{bmatrix}, \quad [T(x)]_\mathcal{B} = \begin{bmatrix} 1 \\ 0 \\ 0 \\ 0 \end{bmatrix},$$

$$[T(x^2)]_\mathcal{B} = \begin{bmatrix} -2 \\ 2 \\ 0 \\ 0 \end{bmatrix},$$

and

$$[T(x^3)]_\mathcal{B} = \begin{bmatrix} 0 \\ -6 \\ 3 \\ 0 \end{bmatrix}.$$

Hence $[T]_\mathcal{B} = \begin{bmatrix} 0 & 1 & -2 & 0 \\ 0 & 0 & 2 & -6 \\ 0 & 0 & 0 & 3 \\ 0 & 0 & 0 & 0 \end{bmatrix}.$

17. (a) Let $\mathcal{B} = \{1, x, x^2\}$. Then, from Example 3, we have

$$[D]_\mathcal{B} = \begin{bmatrix} 0 & 1 & 0 \\ 0 & 0 & 2 \\ 0 & 0 & 0 \end{bmatrix}.$$

So

$$[p'(x)]_\mathcal{B} = [D(p(x))]_\mathcal{B}$$
$$= [D]_\mathcal{B}[p(x)]_\mathcal{B}$$
$$= \begin{bmatrix} 0 & 1 & 0 \\ 0 & 0 & 2 \\ 0 & 0 & 0 \end{bmatrix}\begin{bmatrix} 6 \\ 0 \\ -4 \end{bmatrix}$$
$$= \begin{bmatrix} 0 \\ -8 \\ 0 \end{bmatrix},$$

and hence $p'(x) = -8x$.

(b) As in (a),

$$[p'(x)]_\mathcal{B} = [D(p(x))]_\mathcal{B}$$
$$= [D]_\mathcal{B}[p(x)]_\mathcal{B}$$
$$= \begin{bmatrix} 0 & 1 & 0 \\ 0 & 0 & 2 \\ 0 & 0 & 0 \end{bmatrix}\begin{bmatrix} 2 \\ 3 \\ 5 \end{bmatrix}$$
$$= \begin{bmatrix} 3 \\ 10 \\ 0 \end{bmatrix}.$$

Thus $p'(x) = 3 + 10x$.

(c) Taking $\mathcal{B} = \{1, x, x^2, x^3\}$, we have
$$[D]_\mathcal{B} = \begin{bmatrix} 0 & 1 & 0 & 0 \\ 0 & 0 & 2 & 0 \\ 0 & 0 & 0 & 3 \\ 0 & 0 & 0 & 0 \end{bmatrix}.$$

So
$$[p'(x)]_\mathcal{B} = [D(p(x))]_\mathcal{B}$$
$$= [D]_\mathcal{B}[p(x)]_\mathcal{B}$$
$$= \begin{bmatrix} 0 & 1 & 0 & 0 \\ 0 & 0 & 2 & 0 \\ 0 & 0 & 0 & 3 \\ 0 & 0 & 0 & 0 \end{bmatrix} \begin{bmatrix} 0 \\ 0 \\ 0 \\ 1 \end{bmatrix}$$
$$= \begin{bmatrix} 0 \\ 0 \\ 3 \\ 0 \end{bmatrix},$$

Thus $p'(x) = 3x^2$.

21. Let $T = D$, the differential operator on $V = \text{Span}\{e^t, t^{2t}, e^{3t}\}$. By Exercise 9,
$$[T]_\mathcal{B} = [D]_\mathcal{B} = \begin{bmatrix} 1 & 0 & 0 \\ 0 & 2 & 0 \\ 0 & 0 & 3 \end{bmatrix},$$
and hence 1, 2, and 3 are the eigenvalues of D with corresponding bases $\{e^t\}$, $\{e^{2t}\}$, $\{e^{3t}\}$.

25. In Exercise 13, we saw that if $\mathcal{B} = \{1, x, x^2, x^3\}$, then
$$[T]_\mathcal{B} = \begin{bmatrix} 0 & 1 & -2 & 0 \\ 0 & 0 & 2 & -6 \\ 0 & 0 & 0 & 3 \\ 0 & 0 & 0 & 0 \end{bmatrix}.$$

Since this is an upper triangular matrix, its diagonal entries are the eigenvalues of T. So 0 is the only eigenvalue of T. As in Chapter 5, we find that a basis for the eigenspace of this matrix corresponding to eigenvalue 0 is $\{\mathbf{e}_1\}$. By the boxed result on page 527, a basis for the corresponding eigenspace of T consists of the polynomial whose coordinate vector relative to $\mathcal{B}$ is $\mathbf{e}_1$, which is the polynomial 1.

28. False, let T be the 90°-rotation operator on $\mathcal{R}^2$.

29. False, the vector space on which the operator is defined must be finite-dimensional.

30. True   31. True   32. True

33. True

34. False. Let D be the linear operator on C^∞ defined in Example 2 of Section 7.2. Then, as is shown in Example 5 of this section, every real number is an eigenvalue of D.

35. False, the eigenspace is the set of *symmetric* matrices.

36. False,
$$[T]_\mathcal{B} = \begin{bmatrix} [T(\mathbf{v}_1)]_\mathcal{B} & \cdots & [T(\mathbf{v}_n)]_\mathcal{B} \end{bmatrix}.$$

37. False. As written, this expression may make no sense. What is true is that $[T(\mathbf{v})]_\mathcal{B} = [T]_\mathcal{B}[\mathbf{v}]_\mathcal{B}$.

38. True   39. True

41. For $\mathcal{B} = \{1, x, x^2\}$, which is a basis for $\mathcal{P}_2$, we see that
$$[D]_\mathcal{B} = \begin{bmatrix} 0 & 1 & 0 \\ 0 & 0 & 2 \\ 0 & 0 & 0 \end{bmatrix}.$$

(a) Since the eigenvalues of D are the diagonal entries of $[D]_\mathcal{B}$, D has only one eigenvalue, which is 0.

(b) Since

$$\{[1]_\mathcal{B}\} = \left\{ \begin{bmatrix} 1 \\ 0 \\ 0 \end{bmatrix} \right\}$$

is a basis for the eigenspace of $[D]_\mathcal{B}$ corresponding to the eigenvalue 0, it follows that $\{1\}$ is a basis for the eigenspace of D corresponding to the eigenvalue 0.

45. (a) For any matrices A and C in $M_{2\times 2}$,

$$\begin{aligned} T(A+C) &= (\text{trace } (A+C))B \\ &= (\text{trace } A + \text{trace } C)B \\ &= (\text{trace } A)B + (\text{trace } C)B \\ &= T(A) + T(C). \end{aligned}$$

Similarly, $T(cA) = cT(A)$ for any scalar c, and therefore T is linear.

(b) Let $\mathcal{B}$ be as in Example 9. Then

$$T\left(\begin{bmatrix} 1 & 0 \\ 0 & 0 \end{bmatrix}\right)$$

$$= \left(\text{trace } \begin{bmatrix} 1 & 0 \\ 0 & 0 \end{bmatrix}\right)\begin{bmatrix} 1 & 2 \\ 3 & 4 \end{bmatrix}$$

$$= 1\begin{bmatrix} 1 & 2 \\ 3 & 4 \end{bmatrix}$$

$$= 1\begin{bmatrix} 1 & 0 \\ 0 & 0 \end{bmatrix} + 2\begin{bmatrix} 0 & 1 \\ 0 & 0 \end{bmatrix}$$

$$+ 3\begin{bmatrix} 0 & 0 \\ 1 & 0 \end{bmatrix} + 4\begin{bmatrix} 0 & 0 \\ 0 & 1 \end{bmatrix},$$

$$T\left(\begin{bmatrix} 0 & 1 \\ 0 & 0 \end{bmatrix}\right)$$

$$= \left(\text{trace } \begin{bmatrix} 0 & 1 \\ 0 & 0 \end{bmatrix}\right)\begin{bmatrix} 1 & 2 \\ 3 & 4 \end{bmatrix}$$

$$= 0\begin{bmatrix} 1 & 2 \\ 3 & 4 \end{bmatrix} = \begin{bmatrix} 0 & 0 \\ 0 & 0 \end{bmatrix},$$

$$T\left(\begin{bmatrix} 0 & 0 \\ 1 & 0 \end{bmatrix}\right)$$

$$= \left(\text{trace } \begin{bmatrix} 0 & 0 \\ 1 & 0 \end{bmatrix}\right)\begin{bmatrix} 1 & 2 \\ 3 & 4 \end{bmatrix}$$

$$= 0\begin{bmatrix} 1 & 2 \\ 3 & 4 \end{bmatrix} = \begin{bmatrix} 0 & 0 \\ 0 & 0 \end{bmatrix},$$

and

$$T\left(\begin{bmatrix} 0 & 0 \\ 0 & 1 \end{bmatrix}\right)$$

$$= \left(\text{trace } \begin{bmatrix} 0 & 0 \\ 0 & 1 \end{bmatrix}\right)\begin{bmatrix} 1 & 2 \\ 3 & 4 \end{bmatrix}$$

$$= 1\begin{bmatrix} 1 & 2 \\ 3 & 4 \end{bmatrix}$$

$$= 1\begin{bmatrix} 1 & 0 \\ 0 & 0 \end{bmatrix} + 2\begin{bmatrix} 0 & 1 \\ 0 & 0 \end{bmatrix}$$

$$+ 3\begin{bmatrix} 0 & 0 \\ 1 & 0 \end{bmatrix} + 4\begin{bmatrix} 0 & 0 \\ 0 & 1 \end{bmatrix}.$$

Hence

$$[T]_\mathcal{B} = \begin{bmatrix} 1 & 0 & 0 & 1 \\ 2 & 0 & 0 & 2 \\ 3 & 0 & 0 & 3 \\ 4 & 0 & 0 & 4 \end{bmatrix}.$$

(c) Suppose that A is a nonzero matrix with trace equal to zero. Then

$$\begin{aligned} T(A) &= (\text{trace } A)B \\ &= 0B = O = 0A, \end{aligned}$$

and hence A is an eigenvector of T with corresponding eigenvalue equal to 0.

(d) Suppose that A is an eigenvector of T with a corresponding nonzero

eigenvalue λ. Then
$$\lambda A = T(A) = (\text{trace } A)B,$$
and so $A = \left(\frac{\text{trace } A}{\lambda}\right)B$ since $\lambda \neq 0$.

49. Let W be the eigenspace of T corresponding to λ. Since $T(\mathbf{0}) = \mathbf{0} = \lambda \mathbf{0}$, the zero vector is in W. For any $\mathbf{u}$ and $\mathbf{v}$ in W,
$$T(\mathbf{u} + \mathbf{v}) = T(\mathbf{u}) + T(\mathbf{v})$$
$$= \lambda \mathbf{u} + \lambda \mathbf{v} = \lambda(\mathbf{u} + \mathbf{v}),$$
and hence $\mathbf{u} + \mathbf{v}$ is in W. Similarly, any scalar multiple of $\mathbf{u}$ is in W, and hence W is closed under addition and scalar multiplication. Therefore W is a subspace of V. For any nonzero vector $\mathbf{u}$ in V, the equation $T(\mathbf{u}) = \lambda \mathbf{u}$ is satisfied if and only if $\mathbf{u}$ is an eigenvector corresponding to λ, and hence if and only if $\mathbf{u}$ is in W.

53. (a) For any polynomials $f(x)$ and $g(x)$ in $\mathcal{P}_2$,
$$T(f(x) + g(x)) = \begin{bmatrix} f(1) + g(1) \\ g(1) + g(2) \end{bmatrix}$$
$$= \begin{bmatrix} f(1) \\ f(2) \end{bmatrix} + \begin{bmatrix} g(1) \\ g(2) \end{bmatrix}$$
$$= T(f(x)) + T(g(x)).$$
Similarly $T(cf(x)) = cT(f(x))$. So T is linear.

(b) Since
$$T(1) = \begin{bmatrix} 1 \\ 1 \end{bmatrix}, \quad T(x) = \begin{bmatrix} 1 \\ 2 \end{bmatrix},$$
and
$$T(x^2) = \begin{bmatrix} 1 \\ 4 \end{bmatrix},$$
it follows that $[T]_\mathcal{B}^\mathcal{C} = \begin{bmatrix} 1 & 1 & 1 \\ 1 & 2 & 4 \end{bmatrix}$.

(c) (i) We have
$$T(f(x)) = \begin{bmatrix} a + b + c \\ a + 2b + 4b \end{bmatrix}$$
$$= [T(f(x))]_\mathcal{C}.$$

(ii) Clearly, $[f(x)]_\mathcal{B} = \begin{bmatrix} a \\ b \\ c \end{bmatrix}$ and
$$[T]_\mathcal{B}^\mathcal{C} [f(x)]_\mathcal{B} = \begin{bmatrix} 1 & 1 & 1 \\ 1 & 2 & 4 \end{bmatrix} \begin{bmatrix} a \\ b \\ c \end{bmatrix}$$
$$= \begin{bmatrix} a + b + c \\ a + 2b + 4b \end{bmatrix}$$
$$= [T(f(x))]_\mathcal{C}.$$

7.5 INNER PRODUCT SPACES

3. We have
$$\langle f, g \rangle = \int_1^2 f(t)g(t)\, dt$$
$$= \int_1^2 t(t^2 + 1)\, dt$$
$$= \left. \frac{1}{4}t^4 + \frac{1}{2}t^2 \right|_1^2$$
$$= (4 + 2) - \left(\frac{1}{4} + \frac{1}{2}\right) = \frac{21}{4}.$$

7. We have
$$\langle f, g \rangle = \int_1^2 f(t)g(t)\, dt = \int_1^2 te^t\, dt$$
$$= \left. te^t - e^t \right|_1^2$$
$$= (2e^2 - e^2) - (e - e) = e^2.$$

11.
$$\langle A, B \rangle = \text{trace}\left(AB^T\right)$$

$$= \text{trace}\left(\begin{bmatrix} 1 & -1 \\ 2 & 3 \end{bmatrix} \begin{bmatrix} 2 & 1 \\ 4 & 0 \end{bmatrix}\right)$$

$$= \text{trace}\left(\begin{bmatrix} -2 & 1 \\ 16 & 2 \end{bmatrix}\right) = 0.$$

15.

$$\langle A, B \rangle = \text{trace}\left(AB^T\right)$$

$$= \text{trace}\left(\begin{bmatrix} 3 & 2 \\ 1 & -1 \end{bmatrix} \begin{bmatrix} -1 & 0 \\ 2 & 4 \end{bmatrix}\right)$$

$$= \text{trace}\left(\begin{bmatrix} 1 & 8 \\ -3 & -4 \end{bmatrix}\right) = -3.$$

19. We have

$$\langle f(x), g(x) \rangle = \int_{-1}^{1} (x^2 - 2)(3x + 5)\, dx$$

$$= \int_{-1}^{1} (3x^3 + 5x^2 - 6x - 10)\, dx$$

$$= x^4 + \frac{5}{3}x^3 - 3x^2 - 10x \Big|_{-1}^{1}$$

$$= -\frac{31}{3} - \frac{19}{3} = -\frac{50}{3}.$$

23. We have

$$\langle f(x), g(x) \rangle = \int_{-1}^{1} (x^2 + 1)(x - 1)\, dx$$

$$= \int_{-1}^{1} (x^3 - x^2 + x - 1)\, dx$$

$$= \frac{1}{4}x^4 - \frac{1}{3}x^3 + \frac{1}{2}x^2 - x \Big|_{-1}^{1}$$

$$= -\frac{7}{12} - \frac{25}{12} = -\frac{8}{3}.$$

25. False, it is a scalar.

26. True

27. False, an inner product has scalar values.

28. False, any positive scalar multiple of an inner product is an inner product.

29. True

30. False, if the set contains the zero vector, it is linearly dependent.

31. True   **32.** True   **33.** True

34. True

35. False, the indefinite integral of functions is not a scalar.

36. True   **37.** True

38. False, the norm of a vector equals $\sqrt{\langle \mathbf{v}, \mathbf{v} \rangle}$.

39. False, the equality must hold for *every* vector **u**.

40. False, $\langle A, B \rangle = \text{trace}(AB^T)$.

41. True   **42.** True

43. False, to obtain the normalized Legendre polynomials, these polynomials must be normalized.

44. False, $\mathcal{B}$ must be an orthonormal basis.

45. Let f, g, and h be in $\mathsf{C}([a, b])$.

Axiom 3 We have

$$\langle f + g, h \rangle = \int_{a}^{b} (f + g)(t)h(t)\, dt$$

$$= \int_{a}^{b} f(t)h(t)\, dt$$

$$+ \int_{a}^{b} g(t)h(t)\, dt$$

$$= \langle f, h \rangle + \langle g, h \rangle.$$

Axiom 4 For any scalar c, we have

$$\langle cf, g \rangle = \int_a^b (cf(t))g(t)\, dt$$
$$= c \int_a^b f(t)g(t)\, dt = c \langle f, g \rangle.$$

49. Let $\mathbf{u}$, $\mathbf{v}$, and $\mathbf{w}$ be in $\mathcal{R}^n$. If $\mathbf{u} \neq \mathbf{0}$, then

$$\langle \mathbf{u}, \mathbf{u} \rangle = A\mathbf{u} \cdot \mathbf{u}$$
$$= (A\mathbf{u})^T \mathbf{u} = \mathbf{u}^T A \mathbf{u} > 0$$

because A is positive definite, establishing axiom 1.

We have

$$\langle \mathbf{u}, \mathbf{v} \rangle = (A\mathbf{u}) \cdot \mathbf{v} = (A\mathbf{u})^T \mathbf{v} = \mathbf{u}^T A \mathbf{v}$$
$$= \mathbf{u} \cdot (A\mathbf{v}) = (A\mathbf{v}) \cdot \mathbf{u} = \langle \mathbf{v}, \mathbf{u} \rangle,$$

establishing axiom 2.

Notice that

$$\langle \mathbf{u} + \mathbf{v}, \mathbf{w} \rangle = (A(\mathbf{u} + \mathbf{v})) \cdot \mathbf{w}$$
$$= (A\mathbf{u} + A\mathbf{v}) \cdot \mathbf{w}$$
$$= (A\mathbf{u}) \cdot \mathbf{w} + (A\mathbf{v}) \cdot \mathbf{w}$$
$$= \langle \mathbf{u}, \mathbf{w} \rangle + \langle \mathbf{u}, \mathbf{w} \rangle,$$

establishing axiom 3.

Finally, for any scalar a,

$$\langle a\mathbf{u}, \mathbf{v} \rangle = (A(a\mathbf{u})) \cdot \mathbf{v}$$
$$= a((A\mathbf{u}) \cdot \mathbf{v}) = a \langle \mathbf{u}, \mathbf{v} \rangle,$$

establishing axiom 4.

53. We show that the rule is not an inner product because axiom 1 is not satisfied. Let $f: [0, 2] \to \mathcal{R}$ be defined by

$$f(t) = \begin{cases} 0 & \text{if } 0 \leq t \leq 1 \\ t - 1 & \text{if } 1 < t \leq 2. \end{cases}$$

Since f is continuous, it is in V. Furthermore, f is not the zero function. However

$$\langle f, f \rangle = \int_0^1 f(t)^2\, dt = \int_0^1 0\, dt = 0.$$

57. This rule defines an inner product, as shown below.

Axiom 1 Let $\mathbf{u}$ be a nonzero vector in V. Then

$$\langle \mathbf{u}, \mathbf{u} \rangle = a \langle \mathbf{u}, \mathbf{u} \rangle_1 + b \langle \mathbf{u}, \mathbf{u} \rangle_2 > 0$$

since $\langle \mathbf{u}, \mathbf{u} \rangle_1 > 0$, $\langle \mathbf{u}, \mathbf{u} \rangle_2 > 0$, and a and b are positive.

Axiom 2 Let $\mathbf{u}$ and $\mathbf{v}$ be in V. Then

$$\langle \mathbf{u}, \mathbf{v} \rangle = a \langle \mathbf{u}, \mathbf{v} \rangle_1 + b \langle \mathbf{u}, \mathbf{v} \rangle_2$$
$$= a \langle \mathbf{v}, \mathbf{u} \rangle_1 + b \langle \mathbf{v}, \mathbf{u} \rangle_2$$
$$= \langle \mathbf{v}, \mathbf{u} \rangle.$$

Axiom 3 Let $\mathbf{u}$, $\mathbf{v}$, and $\mathbf{w}$ be in V. Then

$$\langle \mathbf{u} + \mathbf{v}, \mathbf{w} \rangle$$
$$= a \langle \mathbf{u} + \mathbf{v}, \mathbf{w} \rangle_1 + b \langle \mathbf{u} + \mathbf{v}, \mathbf{w} \rangle_2$$
$$= a \langle \mathbf{u}, \mathbf{w} \rangle_1 + a \langle \mathbf{v}, \mathbf{w} \rangle_1$$
$$\quad + b \langle \mathbf{u}, \mathbf{w} \rangle_2 + b \langle \mathbf{v}, \mathbf{w} \rangle_2$$
$$= a \langle \mathbf{u}, \mathbf{w} \rangle_1 + b \langle \mathbf{u}, \mathbf{w} \rangle_2$$
$$\quad + a \langle \mathbf{v}, \mathbf{w} \rangle_1 + b \langle \mathbf{v}, \mathbf{w} \rangle_2$$
$$= \langle \mathbf{u}, \mathbf{w} \rangle + \langle \mathbf{v}, \mathbf{w} \rangle.$$

Axiom 4 Let $\mathbf{u}$ and $\mathbf{v}$ be in V, and let c be a scalar. Then

$$\langle c\mathbf{u}, \mathbf{v} \rangle = a \langle c\mathbf{u}, \mathbf{v} \rangle_1 + b \langle c\mathbf{u}, \mathbf{v} \rangle_2$$
$$= ac \langle \mathbf{u}, \mathbf{v} \rangle_1 + bc \langle \mathbf{u}, \mathbf{v} \rangle_2$$
$$= c(a \langle \mathbf{u}, \mathbf{v} \rangle_1 + b \langle \mathbf{u}, \mathbf{v} \rangle_2)$$
$$= c \langle \mathbf{u}, \mathbf{v} \rangle.$$

61. Let $u_1 = 1$, $u_2 = e^t$, and $u_3 = e^{-t}$. We apply the Gram-Schmidt process to $\{u_1, u_2, u_3\}$ to obtain an orthogonal basis $\{v_1, v_2, v_3\}$. Let $v_1 = u_1 = 1$,

$$v_2 = u_2 - \frac{\langle u_2, v_1 \rangle}{\|v_1\|^2} v_1$$

$$= e^t - \frac{\int_0^1 e^t 1 \, dt}{\int_0^1 1^2 \, dt} 1$$

$$= e^t - \frac{e-1}{1} 1 = e^t - e + 1,$$

and

$$v_3 = u_3 - \frac{\langle u_3, v_1 \rangle}{\|v_1\|^2} v_1 - \frac{\langle u_3, v_2 \rangle}{\|v_2\|^2} v_2$$

$$= e^{-t} - \frac{\int_0^1 e^{-t} 1 \, dt}{\int_0^1 1^2 \, dt} 1 -$$

$$\frac{\int_0^1 e^{-t}(e^t - e + 1) \, dt}{\int_0^1 (e^t - e + 1)^2 \, dt}(e^t - e + 1)$$

$$= e^{-t} - \frac{e-1}{e}$$

$$- \frac{2(e^2 - 3e + 1)}{e(e-1)(e-3)}(e^t - e + 1)$$

$$= e^{-t} + \frac{e^2 - 2e - 1}{e(e-3)}$$

$$- \frac{2(e^2 - 3e + 1)}{e(e-1)(e-3)} e^t.$$

Thus $\{v_1, v_2, v_3\}$ is an orthogonal basis for the subspace generated by $\{1, e^t, e^{-t}\}$.

65. We have

$$\langle u, 0 \rangle = \langle 0, u \rangle$$
$$= \langle 0 \cdot 0, u \rangle = 0 \langle 0, u \rangle = 0.$$

69. Suppose that $\langle u, v \rangle = 0$ for all u in V. Since w is in V, we have $\langle w, w \rangle = 0$, and hence $w = 0$ by axiom 1.

73. Observe that

$$AB^T = \begin{bmatrix} a_{11} & a_{12} \\ a_{21} & a_{22} \end{bmatrix} \begin{bmatrix} b_{11} & b_{21} \\ b_{12} & b_{22} \end{bmatrix}$$

$$= \begin{bmatrix} a_{11}b_{11} + a_{12}b_{12} & a_{11}b_{21} + a_{12}b_{22} \\ a_{21}b_{11} + a_{22}b_{12} & a_{21}b_{21} + a_{22}b_{22} \end{bmatrix},$$

and hence

$$\langle A, B \rangle = \text{trace}\,(AB^T)$$
$$= a_{11}b_{11} + a_{12}b_{12} + a_{21}b_{21} + a_{22}b_{22}.$$

77. If u or v is the zero vector, then both sides of the equality have the value zero. So suppose that $u \neq 0$ and $v \neq 0$. Then there exists a scalar c such that $v = cu$. Hence $\langle u, v \rangle^2 = \langle u, cu \rangle^2 = c^2 \langle u, u \rangle^2$, and

$$\langle u, u \rangle \langle v, v \rangle = \langle u, u \rangle \langle cu, cu \rangle$$
$$= \langle u, u \rangle c^2 \langle u, u \rangle$$
$$= c^2 \langle u, u \rangle^2.$$

Therefore $\langle u, v \rangle^2 = \langle u, u \rangle \langle v, v \rangle$.

81. (a) This is identical to Exercise 72 in Section 6.6.
(b) Let $\mathcal{B}$ be a basis for $\mathcal{R}^n$ that is orthonormal with respect to the given inner product, let B be the $n \times n$ matrix whose columns are the vectors in $\mathcal{B}$, and let $A = (B^{-1})^T B^{-1}$. (Although $\mathcal{B}$ is orthonormal with respect to the given inner product, it need not be orthonormal with respect to the usual dot product on $\mathcal{R}^n$.) Then A is positive definite by Exercise 72 in Section 6.6. Furthermore, by Theorem 4.11, $[u]_\mathcal{B} = B^{-1} u$ for any

vector $\mathbf{u}$ in $\mathcal{R}^n$. So, for any vectors $\mathbf{u}$ and $\mathbf{v}$ in $\mathcal{R}^n$, we may apply Exercise 71 to obtain

$$\begin{aligned}\langle \mathbf{u}, \mathbf{v}\rangle &= [\mathbf{u}]_\mathcal{B} \cdot [\mathbf{v}]_\mathcal{B} \\ &= (B^{-1}\mathbf{u}) \cdot (B^{-1}\mathbf{v}) \\ &= (B^{-1}\mathbf{u})^T (B^{-1}\mathbf{v}) \\ &= \mathbf{u}^T (B^{-1})^T (B^{-1}\mathbf{v}) \\ &= \mathbf{u}^T A \mathbf{v} = (A\mathbf{u})^T \mathbf{v} \\ &= (A\mathbf{u}) \cdot \mathbf{v}.\end{aligned}$$

85. Let $\mathbf{u}$ and $\mathbf{v}$ be in W. By Exercise 84,

$$\begin{aligned}\mathbf{u} + \mathbf{v} &= \langle \mathbf{u}+\mathbf{v}, \mathbf{w}_1\rangle \mathbf{w}_1 + \cdots \\ &\quad + \langle \mathbf{u}+\mathbf{v}, \mathbf{w}_n\rangle \mathbf{w}_n \\ &= (\langle \mathbf{u}, \mathbf{w}_1\rangle + \langle \mathbf{v}, \mathbf{w}_1\rangle)\mathbf{w}_1 + \cdots \\ &\quad + (\langle \mathbf{u}, \mathbf{w}_n\rangle + \langle \mathbf{v}, \mathbf{w}_n\rangle)\mathbf{w}_n.\end{aligned}$$

CHAPTER 7 REVIEW

1. False, for example, C^∞ is not a subset of $\mathcal{R}^n$ for any n.

2. True

3. False, the dimension is mn.

4. False, it is an $mn \times mn$ matrix.

5. True

6. False, for example, let $\mathbf{u}$ and $\mathbf{w}$ be any vectors in an inner product space that are not orthogonal, and let $\mathbf{v} = \mathbf{0}$.

7. True

11. Yes, V is a vector space. We verify some of the axioms of a vector space.

Axiom 2 Let f, g, and h be in V. Then, for any x in $\mathcal{R}$,

$$\begin{aligned}((f \oplus g) \oplus h)(x) &= (f \oplus g)(x)h(x) \\ &= (f(x)g(x))h(x) \\ &= f(x)(g(x)h(x)) \\ &= f(x)(g \oplus h)(x) \\ &= (f \oplus (g \oplus h))(x).\end{aligned}$$

So $(f \oplus g) \oplus h = f \oplus (g \oplus h)$.

Axiom 7 Let f and g be in V, and let a be a scalar. Then, for any x in $\mathcal{R}$,

$$\begin{aligned}(a \odot (f \oplus g))(x) &= ((f \oplus g)(x))^a \\ &= (f(x)g(x))^a \\ &= f(x)^a g(x)^a \\ &= (a \odot f)(x)(a \odot g)(x) \\ &= ((a \odot f) \oplus (a \odot g))(x).\end{aligned}$$

Thus $a \odot (f \oplus g) = (a \odot f) \oplus (a \odot g)$.

15. No, W is not a subspace. Since $\lambda \neq 0$, it follows that λ is not an eigenvalue of O, and hence O is not in W. Therefore W is not a subspace of V.

19. Consider the matrix equation

$$x_1 \begin{bmatrix} 1 & 2 \\ 1 & -1 \end{bmatrix} + x_2 \begin{bmatrix} 0 & 1 \\ 2 & 0 \end{bmatrix} + x_3 \begin{bmatrix} -1 & 3 \\ 1 & 1 \end{bmatrix} = \begin{bmatrix} 4 & 1 \\ -2 & -4 \end{bmatrix}.$$

Comparing the corresponding entries on both sides of this equation, we obtain the system

$$\begin{aligned}x_1 \quad\quad\quad - x_3 &= 4 \\ 2x_1 + x_2 + 3x_3 &= 1 \\ x_1 + 2x_2 + x_3 &= -2 \\ -x_1 \quad\quad + x_3 &= -4,\end{aligned}$$

190   Chapter 7   Vector Spaces

whose augmented matrix has the reduced row echelon form

$$\begin{bmatrix} 1 & 0 & 0 & 3 \\ 0 & 1 & 0 & -2 \\ 0 & 0 & 1 & -1 \\ 0 & 0 & 0 & 0 \end{bmatrix}.$$

Therefore the system has the solution $x_1 = 3$, $x_2 = -2$, and $x_3 = -1$. These are coefficients of a linear combination that produces the given matrix.

23. A polynomial $f(x) = a + bx + cx^2 + dx^3$ is in W if and only if

$$f(0) + f'(0) + f''(0) = a + b + 2c = 0,$$

that is,

$$a = -b - 2c.$$

So $f(x)$ is in W if and only if

$$f(x) = (-b - 2c) + bx + cx^2 + dx^3$$
$$= b(-1 - x) + c(-2 + x^2) + dx^3.$$

It follows that W is the span of

$$\{-1 + x, -2 + x^2, x^3\}.$$

Since this set is linearly independent, it is a basis for W. Therefore $\dim W = 3$.

27. T is both linear and an isomorphism. Let $f(x)$ and $g(x)$ be in $\mathcal{P}_2$. Then

$$T(f(x) + g(x)) = \begin{bmatrix} (f+g)(0) \\ (f+g)'(0) \\ \int_0^1 (f+g)(t)\,dt \end{bmatrix}$$

$$= \begin{bmatrix} f(0) + g(0) \\ f'(0) + g'(0) \\ \int_0^1 f(t)\,dt + \int_0^1 g(t)\,dt \end{bmatrix}$$

$$= \begin{bmatrix} f(0) \\ f'(0) \\ \int_0^1 f(t)\,dt \end{bmatrix} + \begin{bmatrix} g(0) \\ g'(0) \\ \int_0^1 g(t)\,dt \end{bmatrix}$$

$$= T(f(x)) + T(g(x)).$$

Thus T preserves addition. Furthermore, for any scalar c,

$$T(cf(x)) = \begin{bmatrix} (cf)(0) \\ (cf)'(0) \\ \int_0^1 cf(t)\,dt \end{bmatrix}$$

$$= \begin{bmatrix} cf(0) \\ cf'(0) \\ c\int_0^1 f(t)\,dt \end{bmatrix}$$

$$= c \begin{bmatrix} f(0) \\ f'(0) \\ \int_0^1 f(t)\,dt \end{bmatrix}$$

$$= cT(f(x)),$$

and hence T preserves scalar multiplication. Therefore T is linear.

To show that T is an isomorphism, it suffices to show that T is one-to-one because the domain and the codomain of T are finite-dimensional vector spaces with the same dimension. Suppose $f(x) = a + bx + cx^2$ is a polynomial in $\mathcal{P}_2$ such that $T(f(x)) = 0$, the zero polynomial. Comparing components in this vector equation, we have

$$f(0) = 0, \quad f'(0) = 0,$$

and

$$\int_0^1 f(t)\,dt = 0.$$

Since $f(0) = a + b0 + c0^2 = a$, we have $a = 0$. Similarly, we obtain $b = 0$ from the second equation, and $c = 0$ from the third equation. Therefore f is the zero polynomial, and so the null space of T is the zero subspace. We conclude that T is one-to-one, and hence T is an isomorphism.

31. We have

$$T\left(\begin{bmatrix} 1 & 0 \\ 0 & 0 \end{bmatrix}\right) = 2\begin{bmatrix} 1 & 0 \\ 0 & 0 \end{bmatrix} + \begin{bmatrix} 1 & 0 \\ 0 & 0 \end{bmatrix}^T$$

$$= 3\begin{bmatrix} 1 & 0 \\ 0 & 0 \end{bmatrix},$$

$$T\left(\begin{bmatrix} 0 & 1 \\ 0 & 0 \end{bmatrix}\right) = 2\begin{bmatrix} 0 & 1 \\ 0 & 0 \end{bmatrix} + \begin{bmatrix} 0 & 1 \\ 0 & 0 \end{bmatrix}^T$$

$$= 2\begin{bmatrix} 0 & 1 \\ 0 & 0 \end{bmatrix} + 1\begin{bmatrix} 0 & 0 \\ 1 & 0 \end{bmatrix},$$

$$T\left(\begin{bmatrix} 0 & 0 \\ 1 & 0 \end{bmatrix}\right) = 2\begin{bmatrix} 0 & 0 \\ 1 & 0 \end{bmatrix} + \begin{bmatrix} 0 & 0 \\ 1 & 0 \end{bmatrix}^T$$

$$= 2\begin{bmatrix} 0 & 0 \\ 1 & 0 \end{bmatrix} + \begin{bmatrix} 0 & 1 \\ 0 & 0 \end{bmatrix},$$

$$= 1\begin{bmatrix} 0 & 1 \\ 0 & 0 \end{bmatrix} + 2\begin{bmatrix} 0 & 0 \\ 1 & 0 \end{bmatrix},$$

and

$$T\left(\begin{bmatrix} 0 & 0 \\ 0 & 1 \end{bmatrix}\right) = 2\begin{bmatrix} 0 & 0 \\ 0 & 1 \end{bmatrix} + \begin{bmatrix} 0 & 0 \\ 0 & 1 \end{bmatrix}^T$$

$$= 3\begin{bmatrix} 0 & 0 \\ 0 & 1 \end{bmatrix}.$$

Therefore

$$[T]_\mathcal{B} = \begin{bmatrix} 3 & 0 & 0 & 0 \\ 0 & 2 & 1 & 0 \\ 0 & 1 & 2 & 0 \\ 0 & 0 & 0 & 3 \end{bmatrix}.$$

35. Using the matrix computed in Exercise 31, we have

$$[T^{-1}]_\mathcal{B} = \begin{bmatrix} 3 & 0 & 0 & 0 \\ 0 & 2 & 1 & 0 \\ 0 & 1 & 2 & 0 \\ 0 & 0 & 0 & 3 \end{bmatrix}^{-1}$$

$$= \frac{1}{3}\begin{bmatrix} 1 & 0 & 0 & 0 \\ 0 & 2 & -1 & 0 \\ 0 & -1 & 2 & 0 \\ 0 & 0 & 0 & 1 \end{bmatrix}.$$

Hence for any matrix $\begin{bmatrix} a & b \\ c & d \end{bmatrix}$ in $\mathcal{M}_{2 \times 2}$,

$$\left[T^{-1}\left(\begin{bmatrix} a & b \\ c & d \end{bmatrix}\right)\right]_\mathcal{B}$$

$$= [T^{-1}]_\mathcal{B}\left[\begin{bmatrix} a & b \\ c & d \end{bmatrix}\right]_\mathcal{B}$$

$$= \frac{1}{3}\begin{bmatrix} 1 & 0 & 0 & 0 \\ 0 & 2 & -1 & 0 \\ 0 & -1 & 2 & 0 \\ 0 & 0 & 0 & 1 \end{bmatrix}\begin{bmatrix} a \\ b \\ c \\ d \end{bmatrix}$$

$$= \frac{1}{3}\begin{bmatrix} a \\ 2b - c \\ -b + 2c \\ d \end{bmatrix}.$$

Therefore

$$T^{-1}\left(\begin{bmatrix} a & b \\ c & d \end{bmatrix}\right)$$

$$= \frac{1}{3}\begin{bmatrix} a & 2b - c \\ -b + 2c & d \end{bmatrix}.$$

39. Let $A = [T]_\mathcal{B}$. By Exercise 31,

$$A = \begin{bmatrix} 3 & 0 & 0 & 0 \\ 0 & 2 & 1 & 0 \\ 0 & 1 & 2 & 0 \\ 0 & 0 & 0 & 3 \end{bmatrix}.$$

We begin by finding the eigenvalues of A. The characteristic polynomial of A is $(t-3)^3(t-1)$, and hence the eigenvalues of A are 3 and 1.

Next, we find a basis for the eigenspace of A corresponding to the eigenvalue 3. Since the reduced row echelon form of $A - 3I_4$ is

$$\begin{bmatrix} 0 & 1 & -1 & 0 \\ 0 & 0 & 0 & 0 \\ 0 & 0 & 0 & 0 \\ 0 & 0 & 0 & 0 \end{bmatrix},$$

a basis for the eigenspace of A corresponding to the eigenvalue 3 is

$$\left\{ \begin{bmatrix} 1 \\ 0 \\ 0 \\ 0 \end{bmatrix}, \begin{bmatrix} 0 \\ 1 \\ 1 \\ 0 \end{bmatrix}, \begin{bmatrix} 0 \\ 0 \\ 0 \\ 1 \end{bmatrix} \right\}.$$

The matrices whose coordinate vectors relative to $\mathcal{B}$ are the vectors in the preceding basis form the basis

$$\left\{ \begin{bmatrix} 1 & 0 \\ 0 & 0 \end{bmatrix}, \begin{bmatrix} 0 & 1 \\ 1 & 0 \end{bmatrix}, \begin{bmatrix} 0 & 0 \\ 0 & 1 \end{bmatrix} \right\}$$

for the eigenspace of T corresponding to the eigenvalue 3.

Finally, we find a basis for the eigenspace of A corresponding to the eigenvalue 1. Since the reduced row echelon form of $A - 1I_4$ is

$$\begin{bmatrix} 1 & 0 & 0 & 0 \\ 0 & 1 & 1 & 0 \\ 0 & 0 & 0 & 1 \\ 0 & 0 & 0 & 0 \end{bmatrix},$$

a basis for this eigenspace is

$$\left\{ \begin{bmatrix} 0 \\ 1 \\ -1 \\ 0 \end{bmatrix} \right\}.$$

Since $\begin{bmatrix} 0 & 1 \\ -1 & 0 \end{bmatrix}$ is the matrix whose coordinate vector relative to $\mathcal{B}$ is the preceding basis vector, we see that

$$\left\{ \begin{bmatrix} 0 & 1 \\ -1 & 0 \end{bmatrix} \right\}$$

is a basis for the eigenspace of T corresponding to the eigenvalue 1.

43. For any matrix $\begin{bmatrix} a & b \\ c & d \end{bmatrix}$ in $\mathcal{M}_{2 \times 2}$,

$$\text{trace}\left(\begin{bmatrix} 0 & 1 \\ 1 & 0 \end{bmatrix} \begin{bmatrix} a & b \\ c & d \end{bmatrix} \right)$$

$$= \text{trace}\begin{bmatrix} c & d \\ a & b \end{bmatrix}$$

$$= c + b,$$

and so the matrix is in W if and only if $c = -b$. Thus the matrix is in W if and only if it has the form

$$\begin{bmatrix} a & b \\ -b & d \end{bmatrix}$$

$$= a\begin{bmatrix} 1 & 0 \\ 0 & 0 \end{bmatrix} + b\begin{bmatrix} 0 & 1 \\ -1 & 0 \end{bmatrix} + d\begin{bmatrix} 0 & 0 \\ 0 & 1 \end{bmatrix}.$$

The matrices in this linear combination form an orthogonal set. If we divide each matrix by its length, we obtain

$$M_1 = \begin{bmatrix} 1 & 0 \\ 0 & 0 \end{bmatrix}, \quad M_2 = \frac{1}{\sqrt{2}}\begin{bmatrix} 0 & 1 \\ -1 & 0 \end{bmatrix}$$

and

$$M_3 = \begin{bmatrix} 0 & 0 \\ 0 & 1 \end{bmatrix}.$$

So $\{M_1, M_2, M_3\}$ is an orthonormal basis for W.

Therefore the orthogonal projection of

$A = \begin{bmatrix} 2 & 5 \\ 9 & -3 \end{bmatrix}$ on W is

$\langle M_1, A \rangle M_1 + \langle M_2, A \rangle M_2$
$\quad + \langle M_2, A \rangle M_2$

$= 2 \begin{bmatrix} 1 & 0 \\ 0 & 0 \end{bmatrix} + \left(-\frac{4}{\sqrt{2}}\right) \frac{1}{\sqrt{2}} \begin{bmatrix} 0 & 1 \\ -1 & 0 \end{bmatrix}$

$\quad + (-3) \begin{bmatrix} 0 & 0 \\ 0 & 1 \end{bmatrix}$

$= \begin{bmatrix} 2 & -2 \\ 2 & -3 \end{bmatrix}.$

47. We use the orthonormal basis $\{\mathbf{w}_1, \mathbf{w}_2, \mathbf{w}_3\}$ from Exercise 45 to obtain the desired orthogonal projection. Let $\mathbf{w}$ denote the function $\mathbf{w}(x) = \sqrt{x}$. By equation (2) on page 538, we have

$$\mathbf{w} = \langle \mathbf{w}, \mathbf{w}_1 \rangle \mathbf{w}_1 + \langle \mathbf{w}, \mathbf{w}_2 \rangle \mathbf{w}_2$$
$$\quad + \langle \mathbf{w}, \mathbf{w}_3 \rangle \mathbf{w}_3.$$

Now

$\langle \mathbf{w}, \mathbf{w}_1 \rangle = \int_0^1 1\sqrt{x}\, dx = \frac{2}{3},$

$\langle \mathbf{w}, \mathbf{w}_2 \rangle = \int_0^1 \sqrt{3}(2x-1)\sqrt{x}\, dx$
$\quad = \frac{2\sqrt{3}}{15},$

and

$\langle \mathbf{w}, \mathbf{w}_3 \rangle$
$\quad = \int_0^1 \sqrt{5}(6x^2 - 6x + 1)\sqrt{x}\, dx$
$\quad = \frac{-2\sqrt{5}}{105}.$

Thus

$\mathbf{w} = \frac{2}{3}(1) + \frac{2}{5}(2x - 1)$

$\quad + \frac{(-2)}{21}(6x^2 - 6x + 1)$

$= \frac{6}{35} + \frac{48}{35}x - \frac{4}{7}x^2.$

CHAPTER 7 MATLAB EXERCISES

1. The given set contains polynomials from $\mathcal{P}_4$. Let $\mathcal{B} = \{1, x, x^2, x^3, x^4\}$, which is a basis for $\mathcal{P}_4$. Then

$[1 + 2x + x^2 - x^3 + x^4]_\mathcal{B} = \begin{bmatrix} 1 \\ 2 \\ 1 \\ -1 \\ 1 \end{bmatrix},$

$[2 + x + x^3 + x^4]_\mathcal{B} = \begin{bmatrix} 2 \\ 1 \\ 0 \\ 1 \\ 1 \end{bmatrix},$

$[1 - x + x^2 + 2x^3 + 2x^4]_\mathcal{B} = \begin{bmatrix} 1 \\ -1 \\ 1 \\ 2 \\ 2 \end{bmatrix},$

and

$[1 + 2x + 2x^2 - x^3 - 2x^4]_\mathcal{B} = \begin{bmatrix} 1 \\ 2 \\ 2 \\ -1 \\ -2 \end{bmatrix}.$

Note that $[p(x)]_\mathcal{B} = \Phi_\mathcal{B}(p(x))$. Since $\Phi_\mathcal{B} \colon \mathcal{P}_4 \to \mathcal{R}^5$ is an isomorphism, it follows from Theorem 7.8 that the given set of polynomials is linearly independent if and only if the set of coordinate vectors

of these polynomials is linearly independent. So because the reduced row echelon form of
$$\begin{bmatrix} 1 & 2 & 1 & 1 \\ 2 & 1 & -1 & 2 \\ 1 & 0 & 1 & 2 \\ -1 & 1 & 2 & -1 \\ 1 & 1 & 2 & -2 \end{bmatrix}$$
is $[\mathbf{e}_1 \ \mathbf{e}_2 \ \mathbf{e}_3 \ \mathbf{e}_4]$, the given set is linearly independent.

5. Let $\mathcal{B} = \{E_{11}, E_{12}, E_{13}, E_{21}, E_{22}, E_{23}\}$, where E_{ij} is the 2×3 matrix whose (i,j)-entry is 1 and whose other entries are 0. As explained on page 516, $\mathcal{B}$ is a basis for $\mathcal{M}_{2 \times 3}$. For any 2×3 matrix A, we have $T(A) = BAC$, where
$$B = \begin{bmatrix} 1 & 3 \\ 1 & -1 \end{bmatrix}$$
and
$$C = \begin{bmatrix} 4 & -2 & 0 \\ 3 & -1 & 3 \\ -3 & 3 & 1 \end{bmatrix}.$$
Thus
$$T(E_{11}) = \begin{bmatrix} 4 & -2 & 0 \\ 4 & -2 & 0 \end{bmatrix},$$
$$T(E_{12}) = \begin{bmatrix} 3 & -1 & 3 \\ 3 & -1 & 3 \end{bmatrix},$$
$$T(E_{13}) = \begin{bmatrix} -3 & 3 & 1 \\ -3 & 3 & 1 \end{bmatrix},$$
$$T(E_{21}) = \begin{bmatrix} 12 & -6 & 0 \\ -4 & 2 & 0 \end{bmatrix},$$
$$T(E_{22}) = \begin{bmatrix} 9 & -3 & 9 \\ -3 & 1 & -3 \end{bmatrix},$$
and
$$T(E_{23}) = \begin{bmatrix} -9 & 9 & 3 \\ 3 & -3 & -1 \end{bmatrix},$$

Therefore the respective coordinate vectors relative to $\mathcal{B}$ of these matrices are
$$\begin{bmatrix} 4 \\ -2 \\ 0 \\ 4 \\ -2 \\ 0 \end{bmatrix}, \begin{bmatrix} 3 \\ -1 \\ 3 \\ 3 \\ -1 \\ 3 \end{bmatrix}, \begin{bmatrix} -3 \\ 3 \\ 1 \\ -3 \\ 3 \\ 1 \end{bmatrix}, \begin{bmatrix} 12 \\ -6 \\ 0 \\ -4 \\ 2 \\ 0 \end{bmatrix}, \begin{bmatrix} 9 \\ -3 \\ 9 \\ -3 \\ 1 \\ -3 \end{bmatrix},$$
and
$$\begin{bmatrix} -9 \\ 9 \\ 3 \\ 3 \\ -3 \\ -1 \end{bmatrix}.$$
So $[T]_\mathcal{B}$ is the 6×6 matrix having these coordinate vectors as its columns.

(a) The characteristic polynomial of $[T]_\mathcal{B}$ is
$$(t-8)(t-4)^2(t+4)^2(t+8),$$
and so the eigenvalues of $[T]_\mathcal{B}$ and T are 8, 4, -4, and -8.

(b) By using the MATLAB command $\mathtt{null}(A, \text{'r'})$, we can obtain bases for each of the corresponding eigenspaces. The resulting bases are
$$\left\{ \begin{bmatrix} -3 & 3 & 3 \\ -1 & 1 & 1 \end{bmatrix} \right\},$$
$$\left\{ \begin{bmatrix} 3 & 0 & 3 \\ 0 & 1 & 0 \end{bmatrix}, \begin{bmatrix} 2 & 1 & 2 \\ 1 & 0 & 1 \end{bmatrix} \right\},$$
$$\left\{ \begin{bmatrix} -3 & 2 & -3 \\ 0 & 1 & 0 \end{bmatrix}, \begin{bmatrix} 0 & -1 & 0 \\ 1 & 0 & 1 \end{bmatrix} \right\},$$
and
$$\left\{ \begin{bmatrix} 1 & -1 & -1 \\ -1 & 1 & 1 \end{bmatrix} \right\}.$$
Combining these four eigenspace bases yields a basis for $\mathcal{M}_{2 \times 3}$ consisting of eigenvectors of T.